AF785892

Current Topics in Microbiology and Immunology

248

Editors

R.W. Compans, Atlanta/Georgia
M. Cooper, Birmingham/Alabama
J.M. Hogle, Boston/Massachusetts · Y. Ito, Kyoto
H. Koprowski, Philadelphia/Pennsylvania · F. Melchers, Basel
M. Oldstone, La Jolla/California · S. Olsnes, Oslo
M. Potter, Bethesda/Maryland · H. Saedler, Cologne
P.K. Vogt, La Jolla/California · H. Wagner, Munich

Springer
Berlin
Heidelberg
New York
Barcelona
Hong Kong
London
Milan
Paris
Singapore
Tokyo

Origin and Evolution of the Vertebrate Immune System

Edited by L. Du Pasquier and G.W. Litman

With 81 Figures and 17 Tables

Springer

Professor Dr. LOUIS DU PASQUIER
Basel Institute for Immunlogy
Grenzacherstr. 487
CH-4005 Basel
Switzerland
e-mail: dupasquier@dial.eunet.ch

Professor GARY W. LITMAN, M.D.
University of South Florida
All Children's Hospital
801 6th Street South
St. Petersburg, FL 33701
USA
e-mail: litmang@allkids.org

Cover Illustration: Sometimes during his trip across Persia, Alexander the Great encountered a wonderful talking tree with the heads of all sorts of vertabrates. The editors of this volume hope that this "phylogenetic" tree will also talk to us and bring messages of truth concerning the immune system. (Hand colored copy of a persian miniature, found in Delhi during the 10th International Immunology Meeting, and representing the wonderful talking tree of Alexander)

Additional material to this book can be downloaded from http://extras.springer.com.

ISBN 978-3-642-64078-0 ISBN 978-3-642-59674-2 (eBook)
DOI 10.1007/978-3-642-59674-2

This work is subject to copyright. All rights are reserved, whether the whole or part of the material is concerned, specifically the rights of translation, reprinting, reuse of illustrations, recitation, broadcasting, reproduction on microfilm or in any other way, and storage in data banks. Duplication of this publication or parts thereof is permitted only under the provisions of the German Copyright Law of September 9, 1965, in its current version, and permission for use must always be obtained from Springer-Verlag. Violations are liable for prosecution under the German Copyright Law.

© Springer-Verlag Berlin Heidelberg 2000
Softcover reprint of the hardcover 1st edition 2000

Library of Congress Catalog Card Number 15-12910

The use of general descriptive names, registered names, trademarks, etc. in this publication does not imply, even in the absence of a specific statement, that such names are exempt from the relevant protective laws and regulations and therefore free for general use.

Product liability: The publishers cannot guarantee the accuracy of any information about dosage and application contained in this book. In every individual case the user must check such information by consulting other relevant literature.

Cover Design: *design & production GmbH*, Heidelberg
Typesetting: Scientific Publishing Services (P) Ltd, Madras
Production Editor: Angélique Gcouta
Printed on acid-free paper SPIN: 10682082 27/3020GC 5 4 3 2 1 0

Preface

The comparative approach to immunology can be traced to the era of Pasteur and Metchnikov in which observations regarding foreign recognition in invertebrates was a factor in the development of the principal concepts that created the foundation of what now is the broad field of immunology. With each major experimental and conceptual breakthrough, the classical, albeit essential, question has been asked "are the immune systems of phylogenetically primitive vertebrates and invertebrates similar to that of mammals?" Somewhat surprisingly for the jawed vertebrates, the general answer has been a qualified form of "yes", whereas for agnathans and invertebrate phyla it has been "no" so far. The apparent abruptness in the appearance of the immune system of vertebrates is linked to the introduction of the somatic generation of the diversity of its antigen specific receptors. Therefore the questions regarding the origin and evolution of the specific immune system revolve around this phenomenon.

With respect to the origin of the system (aside from the origin of the rearranging machinery itself, the study of which is still in its infancy) one can ask questions about the cellular and molecular contexts in which the mechanism was introduced. Were MHC class I and II molecules already there for selection or did they come later in evolution as a necessary consequence? How were the lymphocyte lineages committed to their various tasks; how could clonal selection be introduced? What was the role of the genes that were going to become the target for the introduction of the somatic rearrangement? Were they already lymphocyte receptors or were they involved in a totally different function in other tissues of the body? How far can one trace the ancestors of the T-cell receptor and immunoglobulin architectures? What was the size of the gene pool devoted to the immune system at its beginning and what was its level of duplication?

With respect to later evolution, the following questions can be asked: What will happen to the constellation of genes assembled in the primordial immune system? How did the families of genes evolve in the context of each class of vertebrate? What has been respected and what has been found merely accessory? How were large families of diverse genes maintained? Is somatic diversity used in the same way in all species? What has happened to the innate immune system, elements of which have been inherited from some invertebrate phyla?

Without giving all the answers, this book does offer the outcome of several current lines of research aimed at elucidating these questions. With the advent of sophisticated biochemical and molecular genetic approaches, the delineation between the adaptive immune system that is found in all jawed vertebrates and its apparent absence in jawless vertebrates becomes increasingly clearer as experimental strategies and thresholds have moved closer to being capable of delineating possible relationships. "Relationship" is perhaps the key word in the modern approaches presented in this book. Indeed, instead of locking themselves into a systematic and sterile opposition between vertebrates and invertebrates, many authors in this volume have established bridges across phyla. They have paved the way for those who seek to monitor the change in commitment of cell lineage, specific receptor expression, exon shuffling and so on that no doubt occurred during the establishment of the gnathostome immune system.

Today's approach mostly involves description of newly discovered genes. It is less functional, less biochemical, and less cellular than in the past, but at the same time it sets the rules of the game by clearly establishing what has been conserved throughout evolution. This is hopefully a temporary situation. Returning to function, and especially regulation, will certainly be necessary when structural data leave us with a paradoxical situation, such as in the case of antibody diversity expression across vertebrates. While all of them have an ample amount of diversity available, only warm-blooded animals exploit its full potential. Why is this the case? For ultimate understanding, more coordination among scientists of different disciplines will be required.

We hope that these few chapters will stimulate research by showing where we stand and by offering new avenues to explore, some of which are presented here in an evolutionary context for the first time.

This book is by no means comprehensive and can be complemented by reading recent issues of Immunological Reviews (nos. 166, Immune systems of ectothermic vertebrates, and 167, Genomic organisation of the MHC: structure, origin and function).

L. Du Pasquier
G. Litman

List of Contents

Evolution of Receptors

Elasmobranchs

Somatic Diversification

TCR/CD3 Complex

List of Contributors

(Their addresses can be found at the beginning of their respective chapters.)

Bridge to Invertebrates

New Approaches Towards an Understanding of Deuterostome Immunity

J.P. Rast, Z. Pancer, and E.H. Davidson

1 Introduction

The vertebrate immune system is distinguished by an unusual propensity for genetic invention. For example, three forms of programmed somatic DNA recombination (V-D-J recombination, class switching and a highly targeted form of gene conversion) have arisen entirely independently in the course of immunoglobulin (Ig) heavy-chain gene evolution. Similar phenomena are virtually unknown in other metazoan genetic systems. Genes that mediate immunity are further characterized by accelerated sequence divergence rates when compared to nonimmune genes in studies of mouse and human gene orthologs (Hughes 1997; Murphy 1993). Both of these attributes, the tendency towards mechanistic novelty and a high rate of sequence evolution, may emerge from the dynamic nature of host-pathogen interactions and thus be a universal characteristic of immune systems. To investigate this possibility it is necessary to characterize immunity in animal phyla where the vertebrate forms of adaptive immunity are absent. A number of molecular advances have been made in recent years in the study of arthropod immunity (e.g., Hoffmann et al. 1996; Iwanaga and Kawabata 1998). As these data accumulate, in combination with similar work on an invertebrate deuterostome that is described here, a more general understanding of immunity will emerge.

Division of Biology 156-29, California Institute of Technology, Pasadena, CA 91125, USA

We are investigating the immune system of an echinoderm, the purple sea urchin (*Strongylocentrotus purpuratus*). The three phyla, Echinodermata, Hemichordata (which includes the acorn worms), and the Chordata comprise the deuterostomes, a monophyletic subgroup of the bilaterian Metazoa (Fig. 1). Invertebrate members of this group occupy an advantageous phylogenetic position from which to compare the molecular systems that have been well studied in vertebrate immunity. Homologs of key vertebrate immune system genes are more readily isolated and issues of orthology (i.e., establishing true relationships among the members of a multigene family) are more easily solved in comparisons among the deuterostomes than in comparisons between protostome invertebrates and deuterostomes. This last point is very important as the orthology of many arthropod and vertebrate immune system molecules has been called into question (HUGHES 1998a). Furthermore, even when homology is present, the fast pace of immune gene divergence may quickly erode any useful phylogenetic signal, especially with regard to the expansive multigene families that are a major feature of vertebrate immunity.

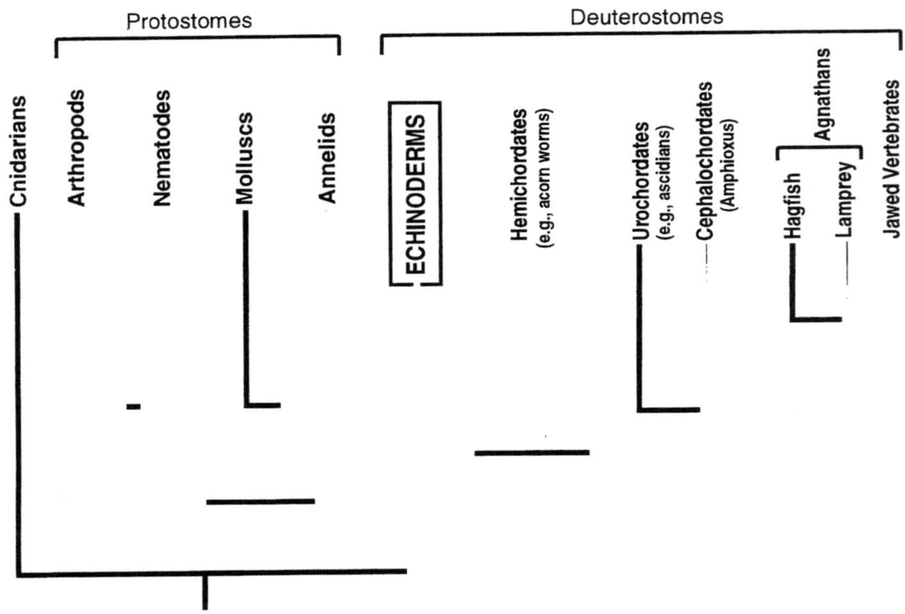

Fig. 1. A phylogenetic tree relating selected representative metazoan phyla and chordate subgroups with the echinoderms (*boxed*). All deuterostome phyla are shown. The vertebrates (lamprey, hagfish, gnathostomes) are illustrated as an unresolved trichotomy as there is some controversy regarding their interrelationships (see FOREY and JANVIER 1993; MALLATT and SULLIVAN 1998; STOCK and WHITT 1992). Protostome monophyly and subgroup relations are according to GIRIBET and RIBERA (1998), AGUINALDO et al. (1997), BALAVOINE (1997) and HALANYCH et al. (1995). The hemichordates are a sister group to the echinoderms (WADA and SATOH 1994; TURBEVILLE et al. 1994; CASTRESANA et al. 1998). The Cnidaria (e.g., jellyfishes) are shown to root the bilaterian groups (protostomes, deuterostomes). Deuterostome monophyly and chordate interrelationships are according to EERNISSE (1997) and PETERSON (1995). Interrelationships among these groups is reviewed by BALAVOINE and ADOUTTE (1998)

We approach this problem with three primary objectives: (a) to expand the knowledge base of metazoan immune mechanisms by applying unbiased strategies to look at molecular aspects of basic immune functions (e.g., bacterial clearance), (b) to use immunity in the sea urchin in a comparative approach to infer the primitive condition of the deuterostome immune system, and (c) to characterize the deuterostome immune system as a general model for the evolution of complex genetic systems.

The phylogenetic distribution of the genes now known to be responsible for the vertebrate forms of adaptive immunity (i.e., the rearranging T cell receptors (TCR) and Ig, and the recombination activating genes, RAG-1 and RAG-2) suggests that the mammalian-type adaptive immune system is limited to the jawed vertebrates (gnathostomes: Davidson 1994; Litman et al. 1999). As far as it is currently understood, this phylogenetic boundary applies only to the rearranging receptors themselves and to some of the specialized systems necessary for their rearrangement. The effector systems through which these targeting systems operate may exist as orthologous systems in species outside of the vertebrates (e.g., the complement system described below; Al-Sharif et al. 1998; Smith et al. 1996, 1998). The Ig/TCR recombination machinery itself (i.e., the RAG genes) may have been acquired by an early vertebrate through a horizontal gene transfer event (Agrawal et al. 1998). In contrast, the Ig-like proteins that today make up the modern rearranging receptors and antigen presentation genes were clearly derived vertically from ancestral genes in the common deuterostome ancestor. Whether nonrearranging homologs of these genes are still present or, if present, retain their ancestral function outside of the jawed vertebrates is an open question. Genes may be retained in the sea urchin that by comparison will tell us something of the primitive function of the vertebrate antigen binding immune receptors, prior to their capacity to rearrange.

The quantity of information concerning vertebrate immunity from cellular, developmental and molecular perspectives is enormous, and offers an excellent opportunity to study the evolutionary behavior of complex molecular networks. As more sea urchin genes that are homologous to genes of the vertebrate immune system are isolated, it is becoming clear that the phylogenetic position of the echinoderms will truly enable the creation and testing of hypotheses of large scale immune system evolution.

2 Sea Urchin Immunity

From a practical standpoint, sea urchins are an excellent invertebrate group in which to explore immunity. Their relatively large size allows the collection of sufficient numbers of immunologically relevant cells. Their larval development is well studied from both classical and molecular standpoints, and the developmental origins of their adult body plan, which is most relevant to ontological questions regarding the sea urchin immune system, is increasingly well defined (Davidson et al. 1998). The sea urchin species that we study, as most other echinoderms, is a relatively long-lived animal even by vertebrate standards, and there is no reason to believe that it is exempt from the necessity of a sophisticated immune system.

Immunity in the sea urchin is effected largely by coelomocytes, which are free roaming, leukocytelike-cells that occupy the coelomic cavity of the adult animal. Comprised of at least four morphologically distinct cell types, they have the capacity to recognize and engulf and digest bacteria and other foreign particles, to release antibacterial compounds, and to form cellular clots in response to injury and infection (reviewed in Smith and Davidson 1992). About two-thirds of these cells are phagocytes, and the rest are vibratile cells, red spherule cells and colorless spherule cells. Previous investigations have demonstrated the capability of coelomocytes to respond transcriptionally to lipopolysaccharides (Smith et al. 1995) and to live bacterial challenge (Pancer et al. 1999). Coelomocytes express effector proteins that are clear homologs of the vertebrate complement system proteins (Al-Sharif et al. 1998; Smith et al. 1998) and they utilize transcription factors that are homologous to those employed in vertebrate lymphoid and myeloid cells (Pancer et al. 1999).

Consistent with the absence of gnathostome-type adaptive immunity, no immunological memory has been demonstrated in the echinoderms. While second allografts are rejected at a significantly increased rate over primary allografts in the sea urchin, this rate is similarly accelerated for third party allografts, indicative of a persistent but nonspecific upregulated mechanism. There is no evidence for specific immune memory (Smith and Davidson 1994, and references therein). Since it is now generally accepted that sea urchins lack a mammalian-type TCR/MHC adaptive immune system, the absence of acute allograft rejection should come as no surprise. Acute allograft rejection in the jawed vertebrates is largely a byproduct of the disruption of a developmentally selected clonal affinity of TCR for self-MHC. Thus acute allograft rejection is in a sense a very particular "artifact" of the jawed vertebrate recognition system and its absence may say little about the potential sophistication of unrelated recognition systems.

The vertebrate rearranging antigen receptor genes represent only one of multiple mechanisms used to target the effectors of the immune system towards pathogens and pathogen infected cells. Other systems that come under the single general heading of "innate immunity" are clearly present in all metazoans. Innate immunity, however, is comprised of many heterogeneous levels of function, from transcriptional regulatory systems to recognition and effector proteins, and cannot be thought of as a single mechanism. Any of these systems may well be expanded in a manner analogous to the vertebrate rearranging receptors (i.e., display some form of multiantigen targeting potential). For example, a tremendous multigene superfamily of both secreted and membrane-bound scavenger receptor-cysteine rich domain containing proteins are expressed exclusively by the coelomocytes of the sea urchin (Pancer et al. 1999) and may play a role in antigen recognition.

Comparative studies of immunity outside of the jawed vertebrates have led to difficulties, many of which are related to the pace at which immune recognition and effector molecules appear to evolve (Hughes 1997; Murphy 1993). Attempts to isolate these genes using cross-hybridization or degenerate PCR methods over large phylogenetic distances are often fruitless, and most of the cell surface receptors and accessory molecules that are intensely studied in mammalian immunology are entirely unknown outside of this vertebrate class. At the same time, phenomenological assays that have been optimized in mammalian studies may fail in divergent species for trivial reasons, and such failures are uninformative. New efforts are therefore needed to obtain and characterize the function of echinoid homologs of known vertebrate immune system genes As well as to identify novel genes involved in sea urchin immunity.

3 New Approaches

We have initiated studies along three lines in order to identify genes involved in sea urchin immunity. These include: (a) the isolation of transcription factors that are homologs of genes known to be relevant to immunity in the vertebrates, (b) directed expressed sequence tag projects to isolate genes that respond to immune challenge, and (c) the use of chromosomal context to isolate sea urchin homologs of key vertebrate immune mediators. These methods depend largely on three technologies: the use of high-density arrayed libraries, subtractive hybridization techniques that enable the generation of complex cDNA probes representing the transcriptional states of differentially treated cells or animals, and the ability to efficiently analyze the genomic structure of large chromosomal regions using stable large-insert bacterial cloning techniques (Amemiya et al. 1996). All of these techniques at some point rely on efficient automated sequencing technologies that over the past few years have lent feasibility to large scale cDNA characterization projects which may now be the most efficient means to isolate highly divergent genes. Two of these methods initially rely on the recognition of similarity with immune proteins previously identified as important in other organisms; however, the differential cDNA screening methods will lead directly to entirely novel immune system proteins.

3.1 Transcription Factors and the Conservation of Regulatory Networks

While many of the vertebrate immune recognition and effector genes rapidly diverge and are therefore inaccessible to standard hybridization- and PCR-based gene isolation techniques, transcription factors generally possess well conserved DNA binding domains that enable the isolation of homologs from highly divergent animals. Unlike effector molecules, however, the expression patterns of these genes per se does not lead directly to an understanding of the immune functions that they regulate. Nearly all transcription factors are expressed in multiple contexts both spatially and temporally. Nonetheless, many transcription factors are restricted to a limited number of adult tissues and combinatorial aspects of these genes and their targets may indicate homology at the regulatory network level. We are interested in characterizing not only the evolution of the genes involved in effecting immunity but also the transcriptional networks to which they belong. For this purpose we have isolated a number of sea urchin transcription factor genes, homologs of which have been shown to be important in the immune systems of vertebrates and in some cases protostome invertebrates.

An NFκB factor homolog, SpNFκB, isolated from coelomocytes, is most closely related to the vertebrate p65 and p105 proteins (Pancer et al. 1999). NFκB/Rel domain proteins are widespread mediators of immune response, characterized in both protostome invertebrates and from the vertebrates (Ghosh et al. 1998). SpNFκB message levels are strongly increased in coelomocytes upon bacterial injection. A similar, though delayed, increase in SpNFκB levels is elicited also by sea water injection. Acute upregulation of SpNFκB expression levels may be useful as a gauge of immune response.

A sea urchin representative of the vertebrate hematopoietic GATA transcription factor family, SpGATAc, has also been isolated (Pancer et al. 1999). This gene is clearly orthologous with both the vertebrate GATA-2 and GATA-3 factors, while its exact relationship with the GATA-1 genes is somewhat obscured by their divergent sequence. In the vertebrate hematopoietic system GATA-2 functions in stem cells and GATA-3 is expressed in developing and mature T-cells. GATA-1 is expressed in developing erythrocytes and megakaryocytes (Weiss and Orkin 1995). The sea urchin SpGATAc gene appears to have diverged from a common ancestor with the vertebrate GATA-2/3 genes prior to their duplication. SpGATAc is expressed transiently during embryogenesis, then is restricted almost entirely to coelomocytes in the adult. Message from this gene becomes almost undetectable by RNA blot hybridization shortly after bacterial challenge, but, unlike many other genes analyzed, its levels are nearly unaffected upon sea water injection (Pancer et al. 1999). Thus the down-regulation of SpGATAc upon bacterial challenge probably represents a separate response pathway than that which drives SpNFκB and SpRunt expression (described below). The occurrence of this class of GATA factors in both the sea urchin coelomocytes and the vertebrate hematopoietic system suggests that some form of homology is manifested within the transcriptional network of these cells relative to the hematopoietic system of vertebrates. A single sea urchin homolog (SpGATAe) of the vertebrate GATA-4,-5,-6 genes, which are typically expressed in endoderm and developing heart in vertebrates, has also been identified but is not expressed in coelomocytes.

Transcription factors containing Runt domains heterodimerize with a non-DNA binding β subunit and regulate transcription in vertebrate lymphoid and myeloid cells (Tenen et al. 1997). A sea urchin homolog, SpRunt-1, first described from embryos (Coffman et al. 1996), is expressed also in coelomocytes (Pancer et al. 1999). As with SpNFκB but in a more pronounced way, SpRunt-1 is upregulated in coelomocytes upon injection of bacteria, and is also upregulated to a lesser extent and with a time delay upon sea water injection. Here again the expression of this gene may be used in the differential screening (described in the next section) as a marker of coelomocyte activation.

As with the expression of the profilin gene described previously (SMITH et al. 1995), these transcription factor expression experiments open the door to a molecular approach to immunity in the sea urchin. The sometimes strong response to sea water injection, albeit with slower kinetics, was initially surprising. However, the influx of sea water into the coelomic cavity in these assays, along with potential contaminating bacterial metabolites, is a clear signal of injury and would be expected to elicit a response. Although these bacterial challenge assays are being refined with respect to the method of inoculation, timing of sampling and the determination of the subset of responding cell types, they are already clear on one point: upon challenge the cells are responding by the activation (or repression) of gene batteries. The effectors of sea urchin immunity, whether homologous to those of vertebrates or completely novel, are sure to be among the genes in these batteries. The strategies that we are taking to isolate these effector genes, along with other transcription factors that may be part of these regulatory networks are described below.

3.2 Differential and Subtractive Screening

A more general approach that we have initiated towards understanding immunity in the sea urchin is based on assessing coelomocyte gene expression patterns in response to immune challenge. These analyses rely on the use of high density arrayed coelomocyte cDNA libraries. These arrays carry 18,432 cDNA clones in duplicate on 22 × 22cm nylon filters (GRESS et al. 1992). Complex cDNA probes are generated either simply by labeling cDNA derived from coelomocyte mRNA isolated from animals that have been differentially treated, or by labeling cDNA following subtractive hybridization between cDNA samples from different treatments. Arrayed libraries offer a crucial advantage over traditional libraries in that complex expression data can be accumulated over multiple experiments. Screening with reciprocally subtracted probes furthermore permits the exclusion of artifacts that result from nondifferentially expressed material, which nonetheless survives the subtraction process.

An RNA blot hybridization analysis of three unrelated messages isolated in a subtractive screen of bacterial challenged and uninjected animal coelomocytes is shown (Fig. 2) to illustrate preliminary trials of this method. These clones encode short proteins that have no obvious homologs in the gene databases. Message from each of the three genes accumulate with different kinetics after immune challenge. Two of the genes (DD104 and DD185) respond in a time delayed fashion also to control sea water injections, while the third clone (DD186) is upregulated almost exclusively in response to bacterial injection. The figure illustrates the potential use of differential screenings to isolate genes that react transcriptionally in specific ways to immune challenge. When applied more broadly to this problem, this method may lead to the identification of specific recognition pathways. Unlike the other strategies discussed here, this differential screening approach assumes no similarity with the vertebrate immune system, and entirely unexpected pathways of immune recognition may be revealed.

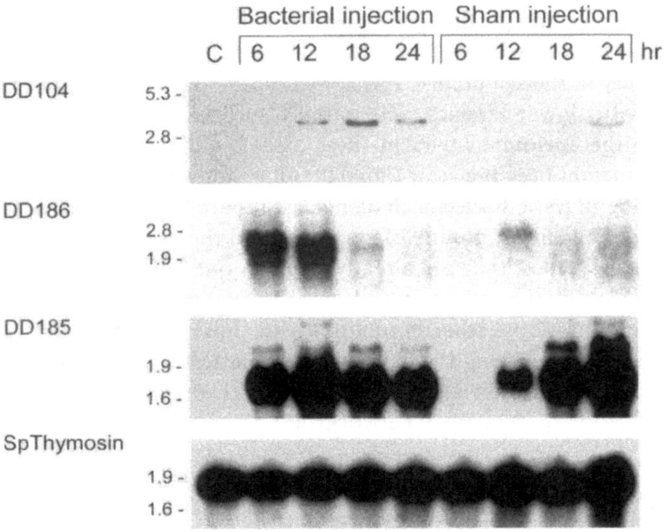

Fig. 2. RNA blot hybridization of bacterial injected and sham injected (sea water injected) sea urchin coelomocytes with probes generated from three different genes isolated by subtractive probe differential screening. *Numbers along top*, hours after treatment. A control lane (*C*) is taken from uninjected animals. SpThymosin, a sea urchin β thymosin homolog that is present at generally constant levels in coelomocytes, was hybridized to the same blot as a loading control (*bottom*)

3.3 Synteny Across The Vertebrate-Invertebrate Boundary

Conservation of synteny (homologous chromosomal gene position) over evolutionary time may serve as a basis for the localization of genes that are too divergent to clone by traditional gene isolation techniques. Genes that are relatively conserved and readily isolated can be used as genomic markers for a given chromosomal context. Three main lines of reasoning strongly support the feasibility of this strategy to find genes even among the most divergent deuterostomes. Studies comparing mouse and human gene positions have led to the estimate that only about 140 chromosomal rearrangement events are necessary to explain the chromosomal differences between these two species (Eppig and Nadeau 1995) which diverged about 100 million years (MY) ago from a common ancestor (Hedges et al. 1996). Consistent with a relatively low rate of chromosomal rearrangement is the description of a number of conserved chromosomal regions between the pufferfish, *Fugu rubripes* and mammals (e.g., Miles et al. 1998) which diverged from a common ancestor approximately 425 MY ago (Carroll 1988). Perhaps most encouraging are the findings that some duplicated chromosomal regions appear to have been preserved within the vertebrate genome itself as blocks of paralogous genes (e.g., see Kandil et al. 1996; Kasahara et al. 1996; Pebusque et al. 1998). As described below, some syntenic regions have probably been preserved for a large proportion of the time separating the echinoderms and the chordates.

The vertebrate genome is proposed to have undergone two rounds of duplication following the divergence of the cephalochordates and probably before the divergence of the modern agnathans and jawed vertebrates (Lundin 1993; Ohno 1970), although the precise scenario is arguable (Pebusque et al. 1998). The evidence for these duplications take the form of paralogous sets of chromosomally contiguous genes. Examples include the four vertebrate HOX clusters and genes that lie adjacent to them (Bailey et al. 1997) and, of particular immunological interest, genes encoded in the vertebrate MHC (Kandil et al. 1996; Kasahara et al. 1996, 1997; Katsanis et al. 1996). When exposed to phylogenetic scrutiny, some candidate MHC paralog genes appear to have actually duplicated prior to the proposed vertebrate genome duplications (Hughes 1998b); nonetheless, for other genes phylogenetic analysis is consistent with an early vertebrate duplication. The reasons for the persistence of these blocks are not obvious, but whether they testify to the rarity of chromosomal rearrangement or to functional constraints imposed by their mechanisms of expression, their utility as markers is clear. Syntenic analysis of genes in echinoderms, which should typically possess a single locus that corresponds to multiple paralogous loci of vertebrates, is likely to be less ambiguous than similar analyses among the vertebrates.

With this type of analysis in mind and as part of the ongoing *S. pupuratus* Genome Project we have constructed *S. purpuratus* P1 artificial chromosome and bacterial artificial chromosome large insert libraries. Beginning with the sea urchin complement homologs of the vertebrate C3/4/5 and factor B genes (BF) (Al-Sharif et al. 1998; Smith et al. 1996, 1998), and a number of other relevant sea urchin homologs that we have isolated, we have begun to analyze the linkage of these genes relative to those encoded in the mammalian MHC class III and class I regions. Preliminary information both from the analysis of large insert clones and from probe hybridization to DNA blots strongly suggest that the syntenic relationships of at least some of these genes have been preserved over the long evolutionary period since the divergence of the echinoderms and vertebrates. These include linkage between the *S. purpuratus* C3/4/5 and Bf complement genes and between *S. purpuratus* Notch and PBX homologs (Fig. 3). With the massive influx of positional data emerging from vertebrate genome projects, other chromosomal regions relevant to immunity will increasingly become available for this type of analysis.

4 The Evolution of Complex Systems

Gene sequences and expression patterns suggest that sea urchins and vertebrates share important molecular features of immunity which are likely to be homologous in the sense that they derive directly by descent from similar features in the common deuterostome ancestor. To gain a deeper understanding of the evolution of this system the organizational levels at which this homology exists need to be more

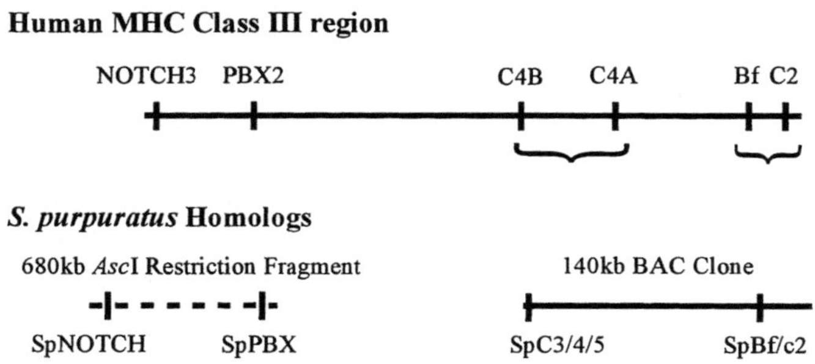

Fig. 3. Linkage between two pairs of *S. purpuratus* homologs of genes encoded in the vertebrate MHC class III region. Cohybridization of *S. purpuratus* Notch and PBX (a homeobox containing transcription factor) probes on DNA blots of pulsed-field gel electrophoresis (PFGE) separated *Asc*I restriction fragments suggest that these genes are linked within 680kb. This interpretation is supported by cohybridization to larger fragments in similar analyses using different restriction enzymes. Genes encoding the *S. purpuratus* C3/4/5 and Bf/C2 complement factor homologs are contained within the 140-kb insert of a BAC clone. Probes corresponding to these genes also cohybridize to restriction fragments on PFGE blots. Distances are not shown to scale. *Brackets*, tandem duplications that probably occurred within the vertebrates. The linkage status between the two *S. purpuratus* fragments is unknown

clearly defined. The cellular level is a tempting candidate unit of homology. Coelomocytes have both functional and molecular similarities to the leukocytes of vertebrates, but what evidence do we have that the immune cells of the sea urchin and vertebrates derives from a similar cell in their common ancestor? The coelomocytes of the echinoderm and blood cells of the vertebrate immune systems emerge from entirely different developmental backgrounds; i.e., there is no obvious similarity between these two groups with respect to "hematopoiesis." Development in these two phyla, in fact, takes place under very different sets of rules (Davidson et al. 1995; Peterson et al. 1997). While these cells share a subset of molecular systems, significant morphological differences suggest that a collection of corresponding molecular differences also are present. Because all cells possess the same genomic complement of genes, entire genetic networks can in be redirected to new cells during the course of evolution. Thus the cells themselves may represent a collection of both homologous and nonhomologous genetic networks with respect to the vertebrate immune cells. While the cellular level of organization may be a valid unit of homology in more closely related groups, given the extent of divergence between the echinoderms and vertebrates, it may be more fruitful to search for homology in networks at noncellular levels.

Subsystems of a complex immune system may evolve independently from one another, and as a first step towards understanding system-wide evolution these points of disjunction must be characterized. Various potential mechanisms of evolutionary system remodeling can be envisioned. Novel effectors of sea urchin immunity may have been independently "plugged" into an underlying homologous immune transcriptional network. This could be accomplished by the acquisition of immune transcription factor binding sites within the *cis*-regulatory regions of the new effector genes themselves or into the *cis*-regulatory regions of their regulators.

Effector gene function may be greatly expanded by gene duplication and divergence, as was the case for the vertebrate rearranging immune receptors. Cell signals may be transduced to a new set of nuclear targets either by changes in the cytosolic pathways that lead to transcription factor activation or by the acquisition of binding sites for these terminal transcription factors by novel genes. Importantly, the developmental input that controls the deployment of immune networks may redirect entire systems to new spatial domains within the developing animal. Delimiting the units of this system which are preserved between echinoderms and vertebrates is a necessary first step in any deep understanding of the evolution of this network.

The genes that are shared between vertebrates and echinoderm immune cells suggest that the vertebrate and echinoderm immune systems possess transcriptional regulatory systems and terminal genes in common. Thus it is probable that homology in the immune system of deuterostomes is carried, at least, at the level of gene networks. Whether this extends to molecular developmental determinants of these cells is open to question. It is plausible that the conserved features of these transcriptional networks exist entirely within the more terminal regulation of the effector genes (i.e., the effector genes themselves and their, more or less, direct regulators of transcription). The molecular pathways of development that address these networks to sea urchin coelomocytes and vertebrate blood cells may be nonhomologous.

Figure 4 illustrates a simplified model of a hypothetical conserved immune system gene battery. Four terminal genes are represented, two of which are primitive and shared among the deuterostomes, while the other two have been co-opted within a particular deuterostome lineage. A transcription factor, which in this context is dedicated to immune function, directly activates these genes. This immune transcription factor is itself under the control of both developmental and inductive inputs. A developmental transcription factor determines what cell may express the immune transcription factor and the inductive transcription factor determines when and where it is expressed by this cell. A "core" portion of the overall system, which includes the dedicated immune system transcription factor, its inductive regulator and a subset of the terminal genes that they run is conserved and can be called homologous among deuterostomes. The developmental input, which controls the cell type in which the immune function regulator is expressed and some terminal genes are not homologous with those of the vertebrates. Further characterization of the genes in the sea urchin system by the methods described here will identify the genes that play these roles and will better define the extent of homology with the vertebrate system.

One clear message emerges from the study of echinoderm immunity. Homology between the immune systems of protostomes and vertebrates may be difficult to pin down and some systems, even though similar in function, may be independently derived (Hughes 1998a). Echinoderms, which entirely lack a vertebrate-type adaptive immune system and exhibit a radically different body plan and mode of development, nonetheless maintain a number of homologous molecular features relative to the vertebrate immune system. Importantly, these sim-

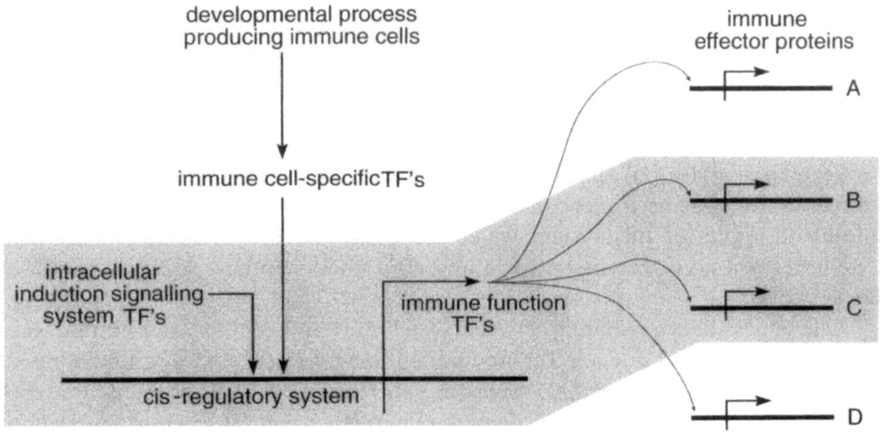

Fig. 4. Diagrammatic presentation of regulatory elements conserved in immune system evolution. *Right*, a battery of genes encoding effector proteins (*A–D*) under the control of a transcription factor (*TF*) produced by a regulatory gene dedicated to immune functions. This gene is intended to symbolize a set of such genes, just as the gene battery symbolizes a set of such gene batteries. Two regulatory inputs are shown, integrated in the *cis*-regulatory system of the gene encoding the transcription factor. One is developmental and consists of transcription factors which are presented only in the cells which during development constitute the immune system progenitors. The second is inductive, and this consists of the terminus of a signal transduction system which responds to immune challenge. Both inputs are required for gene expression. *Shading*, the aspects of this simplified network which can be considered a shared, conserved features of deuterostome immune system regulatory machinery. This consists essentially of the signal system, the *cis*-regulation control system of the dedicated transcription factor gene, and several of the downstream genes (*B, C*); genes *A* and *D* might be added in a given branch of evolution, and the whole might be activated in a developmentally different set of cells

ilarities include both the genes that encode effector proteins and those that are likely to regulate their transcription. Linking these genes together into networks will lead to a more precise understanding of the extent to which the deuterostomes share immune mechanisms and the nature of the evolution of these systems. Eventually this work will help define the evolutionary underpinnings of vertebrate adaptive immunity.

Acknowledgements. We would like to thank M.K. Anderson for comments on the manuscript, K.J. Peterson for advice on metazoan phylogenetics, L.C. Smith for prepublication use of an *S. purpuratus* factor B probe and Paola Oliveri for help with the sea urchin PBX homeobox gene analysis. We also thank Jina Yun and Miki Yun for invaluable technical assistance. J.P.R. is supported by NIH Individual NRSA GM-18478, and Z.P. by NIH Training Grant HD-07257.

References

Agrawal A, Eastman QM, Schatz DG (1998) Transposition mediated by RAG1 and RAG2 and its implications for the evolution of the immune system Nature 394:744–751

Aguinaldo AM, Turbeville JM, Linford LS, Rivera MC, Garey JR, Raff RA, Lake JA (1997) Evidence for a clade of nematodes, arthropods and other moulting animals. Nature 387:489–493

Al-Sharif WZ, Sunyer JO, Lambris JD, Smith LC (1998) Sea urchin coelomocytes specifically express a homologue of the complement component C3. J Immunol 160:2983–2997
Amemiya CT, Ota T, Litman GW (1996) Construction of P1 Artificial Chromosome (PAC) libraries from lower vertebrates. In: Birren, B and Lai, E (ed) Nonmammalian Genomic Analysis: A practical Guide. Academic Press, New York, pp 223–256
Bailey WJ, Kim J, Wagner GP, Ruddle FH (1997) Phylogenetic reconstruction of vertebrate Hox cluster duplications. Mol Biol Evol 14:843–853
Balavoine G (1997) The early emergence of platyhelminths is contradicted by the agreement between 18 s rRNA and Hox genes data. C R Acad Sci III 320:83–94
Balavoine G, Adoutte A (1998) One or three Cambrian radiations? Science 280:397–398
Carroll RL (1988) Vertebrate Paleontology and Evolution. Freeman, New York
Castresana J, Feldmaier-Fuchs G, Yokobori S, Satoh N, Paabo S (1998) The mitochondrial genome of the hemichordate Balanoglossus carnosus and the evolution of deuterostome mitochondria. Genetics 150:1115–1123
Coffman JA, Kirchhamer CV, Harrington MG, Davidson EH (1996) SpRunt-1, a new member of the runt domain family of transcription factors, is a positive regulator of the aboral ectoderm-specific CyIIIA gene in sea urchin embryos. Dev Biol 174:43–54
Davidson EH (1994) Stepwise evolution of major functional systems in vertebrates, including the immune system. In: Hoffmann JA, Janeway CA, Shunji N (ed) Phylogenetic Perspectives in Immunity: the Insect Host Defense. R.G. Landes Company, Austin, pp 133–142
Davidson EH, Cameron RA, Ransick A (1998) Specification of cell fate in the sea urchin embryo: summary and some proposed mechanisms. Development 125:3269–3290
Davidson EH, Peterson KJ, Cameron RA (1995) Origin of bilaterian body plans: evolution of developmental regulatory mechanisms. Science 270:1319–1325
Eernisse DJ (1997) Arthropod and annelid relationships re-examined. In: Fortey, RA and Thomas, RH (ed) Arthropod Relationships, Systematics Association Special Volume Series 55. Chapman and Hall, London, pp 43–56
Eppig JT, Nadeau JH (1995) Comparative maps: the mammalian jigsaw puzzle. Curr Opin Genet Dev 5:709–716
Forey P, Janvier P (1993) Agnathans and the origin of jawed vertebrates. Nature 361:129–134
Ghosh S, May MJ, Kopp EB (1998) NF-kappa B and Rel proteins: evolutionarily conserved mediators of immune responses. Annu Rev Immunol 16:225–260
Giribet G, Ribera C (1998) The position of arthropods in the animal kingdom: A search for a reliable outgroup for internal arthropod phylogeny. Mol Phylogenet Evol 9:481–488
Gress TM, Hoheisel JD, Lennon GG, Zehetner G, Lehrach H (1992) Hybridization fingerprinting of high-density cDNA-library arrays with cDNA pools derived from whole tissues. Mamm Genome 3:609–619
Halanych KM, Bacheller JD, Aguinaldo AM, Liva SM, Hillis DM, Lake JA (1995) Evidence from 18 s ribosomal DNA that the lophophorates are protostome animals. Science 267:1641–1643
Hedges SB, Parker PH, Sibley CG, Kumar S (1996) Continental breakup and the ordinal diversification of birds and mammals. Nature 381:226–229
Hoffmann JA, Reichhart JM, Hetru C (1996) Innate immunity in higher insects. Curr Opin Immunol 8: 8–13
Hughes AL (1997) Rapid evolution of immunoglobulin superfamily C2 domains expressed in immune system cells. Mol Biol Evol 14:1–5
Hughes AL (1998a) Protein phylogenies provide evidence of a radical discontinuity between arthropod and vertebrate immune systems. Immunogenetics 47:283–296
Hughes AL (1998b) Phylogenetic tests of the hypothesis of block duplication of homologous genes on human chromosomes. 6:9, and 1. Mol Biol Evol 15:854–870
Iwanaga S, Kawabata S (1998) Evolution and phylogeny of defense molecules associated with innate immunity in horseshoe crab. Front Biosci 3:973–984
Kandil E, Namikawa C, Nonaka M, Greenberg AS, Flajnik MF, Ishibashi T, Kasahara M (1996) Isolation of low molecular mass polypeptide complementary DNA clones from primitive vertebrates. Implications for the origin of MHC class I-restricted antigen presentation. J Immunol 156:4245–4253
Kasahara M, Hayashi M, Tanaka K, Inoko H, Sugaya K, Ikemura T, Ishibashi T (1996) Chromosomal localization of the proteasome Z subunit gene reveals an ancient chromosomal duplication involving the major histocompatibility complex. Proc Natl Acad Sci USA 93:9096–9101
Kasahara M, Nakaya J, Satta Y, Takahata N (1997) Chromosomal duplication and the emergence of the adaptive immune system. Trends Genet 13:90–92

Katsanis N, Fitzgibbon J, Fisher EMC (1996) Paralogy mapping: identification of a region in the human MHC triplicated onto human chromosomes 1 and 9 allows the prediction and isolation of novel PBX and NOTCH loci. Genomics 35:101–108

Litman GW, Anderson MK, Rast JR (1999) Evolution of antigen binding receptors. Annual Review of Immunology 17:109–147

Lundin LG (1993) Evolution of the vertebrate genome as reflected in paralogous chromosomal regions in man and the house mouse. Genomics 16:1–19

Mallatt J, Sullivan J (1998) 28 s and 18 s rDNA sequences support the monophyly of lampreys and hagfishes. Mol Biol Evol 15:1706–1718

Miles C, Elgar G, Coles E, Kleinjan DJ, van Heyningen V, Hastie N (1998) Complete sequencing of the Fugu WAGR region from WT1 to PAX6: dramatic compaction and conservation of synteny with human chromosome 11p13. Proc Natl Acad Sci USA 95:13068–13072

Murphy PM (1993) Molecular mimicry and the generation of host defense protein diversity. Cell 72: 823–826

Ohno S (1970) Evolution by Gene Duplication. Springer-Verlag, Heidelberg

Pancer Z, Rast JP, Davidson EH (1999) Origins of immunity: Transcription factors and homologs of effector genes of the vertebrate immune system expressed in sea urchin coelomocytes. Immunogenetics 49:773–786

Pebusque MJ, Coulier F, Birnbaum D, Pontarotti P (1998) Ancient large-scale genome duplications: phylogenetic and linkage analyses shed light on chordate genome evolution. Mol Biol Evol 15:1145–1159

Peterson KJ (1995) A phylogenetic test of the calcichordate scenario. Lethaia 28:25–38

Peterson KJ, Cameron RA, Davidson EH (1997) Set-aside cells in maximal indirect development: evolutionary and developmental significance. Bioessays 19:623–631

Smith LC, Britten RJ, Davidson EH (1995) Lipopolysaccharide activates the sea urchin immune system. Dev Comp Immunol 19:217–224

Smith LC, Chang L, Britten RJ, Davidson EH (1996) Sea urchin genes expressed in activated coelomocytes are identified by expressed sequence tags. Complement homologues and other putative immune response genes suggest immune system homology within the deuterostomes. J Immunol 156:593–602

Smith LC, Davidson EH (1992) The echinoid immune system and the phylogenetic occurrence of immune mechanisms in deuterostomes. Immunol Today 13:356–362

Smith LC, Davidson EH (1994) The echinoderm immune system. Characters shared with vertebrate immune systems and characters arising later in deuterostome phylogeny. Ann N Y Acad Sci 712: 213–226

Smith LC, Shih CS, Dachenhausen SG (1998) Coelomocytes express SpBf, a homologue of factor B, the second component in the sea urchin complement system. J Immunol 161:6784–6793

Stock DW, Whitt GS (1992) Evidence from 18 s ribosomal RNA sequences that lampreys and hagfishes form a natural group. Science 257:787–789

Tenen DG, Hromas R, Licht JD, Zhang DE (1997) Transcription factors, normal myeloid development, and leukemia. Blood 90:489–519

Turbeville JM, Schultz JR, Raff RA (1994) Deuterostome phylogeny and the sister group of the chordates: evidence from molecules and morphology. Mol Biol Evol 11:648–655

Wada H, Satoh N (1994) Details of the evolutionary history from invertebrates to vertebrates, as deduced from the sequences of 18 s rDNA. Proc Natl Acad Sci USA 91:1801–1804

Weiss MJ, Orkin SH (1995) GATA transcription factors: key regulators of hematopoiesis. Exp Hematol 23:99–107

The Antimicrobial Host Defense of *Drosophila*

M. Meister, C. Hetru, and J.A. Hoffmann

1 Introduction

It has long been known that insects are particularly resistant to bacteria. Early workers in the late 19th century attributed this resistance to phagocytosis and to encapsulation by hemocytes (Kowalevsky 1887; Cuénot 1896). Around 1920, a number of independent studies established that insects could be protected against the injection of lethal doses of bacteria by the prior administration of low doses (Metalnikow 1920; Paillot 1920; Glaser 1918) and that this induced protection was correlated to the appearance of a potent antibacterial activity in the cell-free hemolymph. These studies, and most of the subsequent investigations in the field of

Institut de Biologie Moléculaire et Cellulaire, UPR 9022 du CNRS, 15 Rue René Descartes, 67084 Strasbourg Cedex, France

insect immunity, were performed on large-sized insect species. It was only in 1972 that the problem of the inducible antibacterial activity was addressed in the small-sized *Drosophila*. Studies by Boman and associates (Boman et al. 1972) demonstrated that in *Drosophila*, as in other larger insect species, a primary infection can induce a protection against a secondary infection which otherwise would be lethal. In spite of the obvious interest of *Drosophila* as a model system, Boman and associates turned to the large pupae of the *Cecropia* moth for the first isolation of induced antibacterial molecules (cecropins; Steiner et al. 1981; attacins; Hultmark et al. 1983). Other groups subsequently worked on large-sized fly species, such as *Sarcophaga peregrina* (Okada and Natori 1985; Ando and Natori 1988; Matsuyama and Natori 1988) and *Phormia terranovae* (Dimarcq et al. 1988; Lambert et al. 1989). In the mid-eighties, two independent studies confirmed that *Drosophila* responds to the inoculation of bacteria within a few hours by the *de novo* synthesis of several peptides/polypeptides, some of which were presumed to be homologous to cecropins and attacins (Robertson and Postlethwait 1986; Flyg et al. 1987). However, at that time, the structure of the peptides remained elusive as a consequence of the limited amounts which could be extracted from these small-sized insects for sequence analysis. The situation evolved in the nineties, both through the advent of molecular biology techniques which allowed for homology screenings, and with the refinement of the analytical methods.

The current view is that the antimicrobial host defense of *Drosophila* is multifaceted process which involves the activation of proteolytic cascades, reactions by hemocytes, and the synthesis of antimicrobial peptides, as summarized in Fig. 1. We will review our present knowledge on these various aspects below.

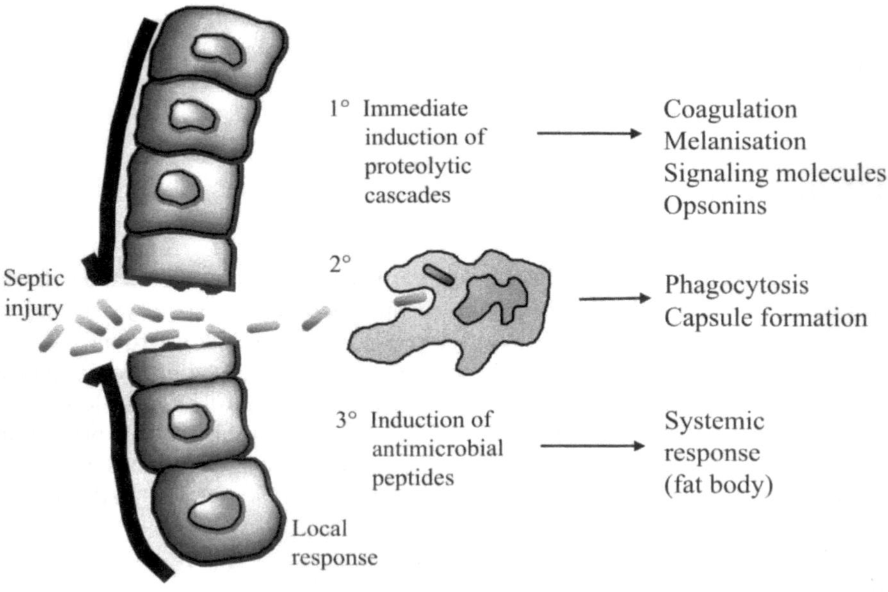

2 The Inducible Antimicrobial Peptides of *Drosophila*

Seven distinct molecules, plus several isoforms (see Fig. 2, Table 1), have been characterized from immune-challenged *Drosophila*. These are representatives of most of the antimicrobial peptide families found in Diptera and in other insect orders (reviewed in BOMAN 1995; HOFFMANN et al. 1996; HOFFMANN and REICHHART 1997). We first describe each of these peptides and then briefly discuss their modes of action.

2.1 Peptide Structures

2.1.1 Cecropins

The first information on the sequences of inducible antibacterial peptides of *Drosophila* was obtained from DNA cloning studies published in 1990. Using a cDNA clone corresponding to the major cecropin isolated by NATORI and colleagues from *Sarcophaga peregrina* (OKADA and NATORI 1985), HULTMARK and associates (KYLSTEN et al. 1990) characterized a compact gene cluster (mapping at 99 E) comprising three expressed cecropin genes plus two pseudogenes. Two of these genes (CecA1 and CecA2) code for a cecropin A with a deduced amino acid sequence identical to the major cecropin of *Sarcophaga peregrina*. The third expressed

Table 1. Amino acid sequences of the antimicrobial peptides identified in *Drosophila*

Drosophila cecropin A	GWLKKIGKKIERVGQHTRDATIQGLGIAQQAANVAATAR
Drosophila cecropin B	GWLRKLGKKIERIGQHTRDASIQVLGIAQQAANVAATAR
Drosophila cecropin C	GWLKKLGKRIERIGQHTRDATIQGLGIAQQAANVAATAR
Diptericin	DDMTMKPTPPPQYPLNLQGGGGGQSGDGFGFAVQGHQKVWTS- DNGRHEIGLNGGYGQHLGGPYGNSEPSWKVGSTYTYRFPNF
Drosocin	GKPRPYSPRPTSHPRPIRV
Drosophila defensin	ATCDLLSKWNWNHTACAGHCIAKGFKGGYCNDKAVCVCRN
Drosomycin	DCLSGRYKGPCAVWDNETCRRVCKEEGRSSGHCSPSLKCWCEGC
Metchnikowin A	HRHQGPIFDTRPSPFNPNQPRPGPIY
Metchnikowin B	HRRQGPIFDTRPSPFNPNQPRPGPIY
Drosophila attacin	QVLGGSLTSNPAGGADARLDLTKGIGNPNHNVVGQVFAAGNTQ- SGPVTTGGTLAYNNAGHGASLTKTHTPGVKDVFQQEAHANLF- NNGRHNLDAKVFASQNKLANGFEFQRNGAGLDYSHINGHGA- SLTHSNFPGIGQQLGLDGRANLWSSPNRATTLDLTGSASKWTS- GPFANQKPNFGAGLGLSHHFG

Fig. 1. The immune response to septic injury in *Drosophila*. The wounding, together with the invasion of the hemocoele by micro-organisms, triggers several defense processes, which include the induction of proteolytic cascades, a cellular response (the phagocyte in the figure represents a blood cell), and the synthesis by the fat body of antimicrobial peptides that are released into the hemolymph (systemic response). A local response is also observed in epithelial structures

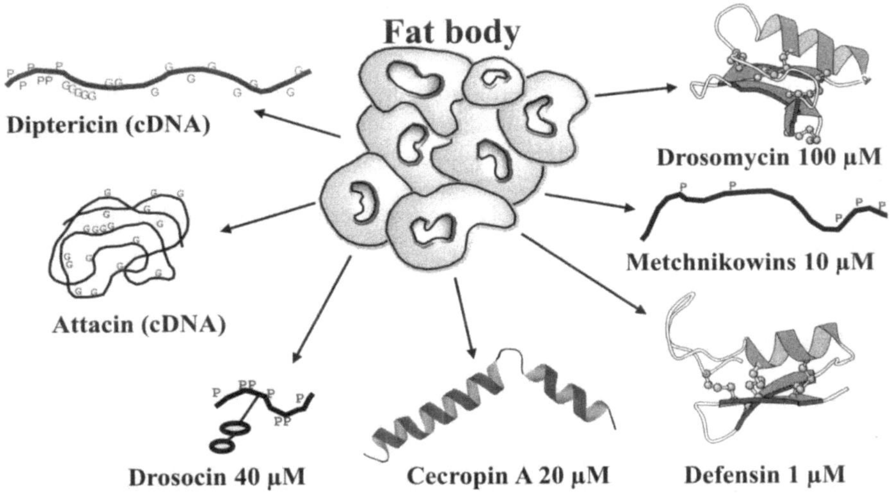

Fig. 2. Production of antimicrobial peptides by the fat body. In the hours that follow a septic injury in *Drosophila* adults, the fat body produces large amounts of antimicrobial peptides. These reach concentrations in the hemolymph which range from 1 to 100µM (P. Bulet, personal communication). In the case of diptericin and attacin the peptides have not yet been fully characterized

gene (CecB) codes for a cecropin which differs by five conservative amino acid replacements. Subsequently, a fourth cecropin gene (CecC), encoding a peptide which differs from cecropin A by three conservative amino acid replacements, was described (Tryselius et al. 1992). The four cecropin genes each contain a short intron and encode 63-residue prepropeptides. In the mid-1990s, Bulet and colleagues (personal communication) isolated cecropin A from challenged flies and observed that 50% of the native molecules were C-amidated. The isolated peptide was found to be active predominantly against gram-negative bacteria at the physiological concentration (20µM) found in immune-challenged adults.

2.1.2 Diptericin

A *Drosophila* cDNA encoding a member of the diptericin family of inducible antibacterial peptides was isolated by Wicker et al. in 1990. The main characteristic of this polypeptide is the simultaneous presence of a glycine-rich (67 residues) and a proline-rich (16 residues) domain. The cDNA was isolated with an oligonucleotide probe corresponding to the peptide sequence of diptericin identified in the fleshfly *Phormia terranovae* (Reichhart et al. 1989). The *Drosophila* gene encoding diptericin is intronless and is present in a single copy (mapping at 56 A). It encodes a 106-residue prepropeptide (Reichhart et al. 1992). We suspect *Drosophila* diptericin to be glycosylated, as with that of *Phormia*. However, the exact structure and activities of the mature *Drosophila* diptericin have not yet been established.

2.1.3 Drosocin

Drosocin is a 19-residue proline-rich peptide which was directly isolated and characterized from extracts of immune-challenged flies (BULET et al. 1993). This inducible cationic peptide carries an O-glycosylated substitution on a threonine residue. The substitution is mandatory for full biological activity of drosocin and corresponds to either the monosaccharide *N*-acetylgalactosamine or the *N*-acetylgalactosamine-galactose disaccharide. Drosocin is active mainly against gram-negative bacteria (0.2–2μM; BULET et al. 1996). It is processed from a 64-residue prepropeptide encoded by a single intronless gene (mapping at 51C1-6; CHARLET et al. 1996).

2.1.4 Defensin

Drosophila defensin was chemically isolated in the course of a large-scale purification of antibacterial peptides from immune flies. The N-terminal sequence obtained by Edman degradation served for the design of oligonucleotide probes which allowed the cloning of the corresponding cDNA and gene (DIMARCQ et al. 1994).

Drosophila defensin is a 40-residue peptide with activity directed against gram-positive bacteria. The single intronless defensin gene (mapping at 46CD) encodes a 92-residue prepropeptide from which mature defensin is processed. As is the case for all known insect defensins, the *Drosophila* molecule carries six cysteines engaged in three intramolecular disulfide bridges. Based on the significant sequence similarities with *Phormia* defensin, for which the 3D structure has been established (CORNET et al. 1995), it is assumed that *Drosophila* defensin consists of a central α-helix linked via two disulfide bridges to a C-terminal antiparallel β-sheet. The N-terminal residues of these molecules form a flexible loop linked via a disulfide bridge to the β-sheet.

The blood titers of defensin are usually very low (1μM) compared to those of the other inducible antibacterial peptides of *Drosophila*.

2.1.5 Drosomycin

In response to a septic injury, considerable amounts of a 44-residue peptide are synthesized (FEHLBAUM et al. 1994; Fig. 2). This peptide, named drosomycin, contains eight cysteines engaged in four intramolecular disulfide bridges and exhibits significant sequence similarity with plant defensins (BROEKAERT et al. 1995). It is active predominantly against filamentous fungi. The 3D structure of drosomycin shows a central α-helix linked to an antiparallel β-sheet via two disulfide bridges, as in the case of insect defensin (LANDON et al. 1997). Furthermore, drosomycin has an extended N-terminal sequence forming an additional β-sheet. This structure is very similar to that of plant defensins.

The *Drosophila* gene encoding drosomycin is unique and intronless and maps at position 63D1-2 (L. Michaut and E. Levashina, personal communication). In the hemolymph of immune challenged adults of *Drosophila*, drosomycin reaches a remarkably high concentration (100μM) equivalent to the sum of all of the other inducible antimicrobial peptides known in this species.

2.1.6 Metchnikowin

Metchnikowin is a 26-residue proline-rich immune-inducible peptide which exhibits both antibacterial (gram-positive) and antifungal activities (Levashina et al. 1995). The activity spectrum contrasts with that of other known proline-rich antimicrobial peptides which are active mainly against gram-negative bacteria. Peptide sequencing and cDNA cloning indicate the presence of two alleles (A and B, Table 1) in our *Drosophila* Oregon strain, which differ by one residue (His versus Arg), as a consequence of a single nucleotide change. The mature peptide is processed from a 52-residue prepropeptide encoded by a single intronless gene which maps at position 52 A1-2 (Levashina et al. 1998).

2.1.7 Attacin

Drosophila expresses a gene (mapping at 51A-B) encoding a 20-kDa attacin and there is evidence for several additional genes encoding attacins (Asling et al. 1995). The deduced peptide shows obvious homologies with the *Hyalophora* and *Sarcophaga* attacins, which are active against gram-negative bacteria (Hultmark et al. 1983; Ando and Natori 1988). The polypeptide has not yet been isolated from *Drosophila*.

2.2 Mode of Action of the Antimicrobial Peptides

As stated above, insect defensins are active against gram-positive bacteria, while cecropins are active mainly against gram-negative bacteria. Both target the cytoplasmic membrane and disrupt the permeability barrier of the membrane by a mechanism which is not yet fully established at molecular level. It has been proposed that cecropins and insect defensins induce the formation of channels (Christensen et al. 1988; Cociancich et al. 1993). More recently a "carpetlike" model has been proposed for the mode of action of polycationic antibacterial peptides on phospholipid membranes. The model suggests that the peptides bind randomly to the surface of the prokaryotic membrane through their hydrophobic face, and that above a given threshold concentration they generate transient holes which lead to the efflux of ions and solutes (Gazit et al. 1995).

Drosocin efficiently kills gram-negative bacteria. Potassium efflux experiments have shown that the membrane is not the direct target of this peptide (C. Hetru, unpublished).

Drosomycin exhibits a strong and selective activity against a broad range of filamentous fungi. High concentrations of the peptide completely inhibit spore germination, whereas low concentrations induce delayed growth of hyphae and abnormal morphology. The peptide also causes partial lysis of the hyphae

(Fehlbaum et al. 1994). The exact mechanism of action of drosomycin and of other antifungal peptides remains to be elucidated.

Attacins are bacteriostatic and act on the outer membrane of gram-negative bacteria (Engström et al. 1984). Nothing is known concerning the mode of action of metchnikowin and diptericin.

3 Regulation of Antimicrobial Peptide Gene Expression

A common feature of the genes encoding antimicrobial peptides is that they are silent under normal conditions. Upon experimental immune challenge (which consists of pricking larvae or flies with a needle coated with bacteria or fungi), all genes are rapidly and strongly induced in the fat body. Transcripts are detected as early as 30min after challenge and generally reach their highest levels after 12–24h. They return to basal levels after one to several days. The kinetics of induction are not identical for all genes (see Lemaitre et al. 1997).

3.1 Promoters of Antimicrobial Genes

The first data on the regulation of antimicrobial peptide gene expression were obtained through the analysis of the promoters of the cecropin A1 gene (Engström et al. 1993; Kadalayil et al. 1997; Roos et al. 1998) and the diptericin gene (Reichhart et al. 1992; Kappler et al. 1993; Georgel et al. 1993; Meister et al. 1994). These promoters contain several sequence motifs analogous to binding sites for transcriptional activators which are involved in the control of expression of mammalian immune-response genes. This is notably the case for NF-κB response elements (Fig. 3) which play a paramount role in the induction of many acute-phase response genes in mammals and are the targets of NF-κB, a DNA-binding protein that belongs to the family of inducible transactivators collectively referred to as Rel proteins (from the viral oncogene v-*rel*; reviewed in Ghosh et al. 1998). These Rel proteins share a conserved domain of ca. 300 amino acids, which is necessary for DNA binding, dimerization, and interaction with inhibitors (see below) and bind as homo- or heterodimers to the κB-response elements (also called Rel-sites).

In the cecropin A1 and diptericin genes, the role of the decameric Rel-sites was demonstrated by establishing transgenic fly lines or transfecting cell lines with reporter genes carrying wild-type promoter sequences or mutated κB sites within these sequences. These experiments established that the κB binding sites are mandatory for the immune-inducibility of cecropin and diptericin. Further experiments showed that additional regulatory sequences within the promoters are necessary to confer wild-type transcriptional levels. A functional dissection of the diptericin gene was performed by transgenic-based experiments, and the results are summarized in Fig. 4. Parallel investigations on the cecropin gene established that

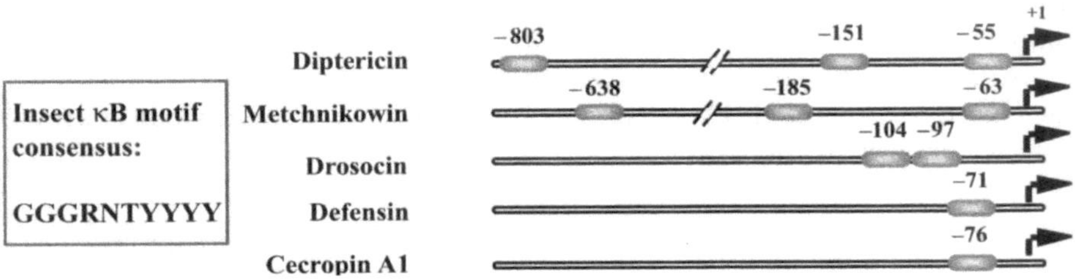

Fig. 3. The promoters of the genes encoding antimicrobial peptides contain sequence motifs similar to the mammalian NF-κB binding sites. An insect consensus κB motif is given. *Numerals*, the distance in basepairs from the transcription start site

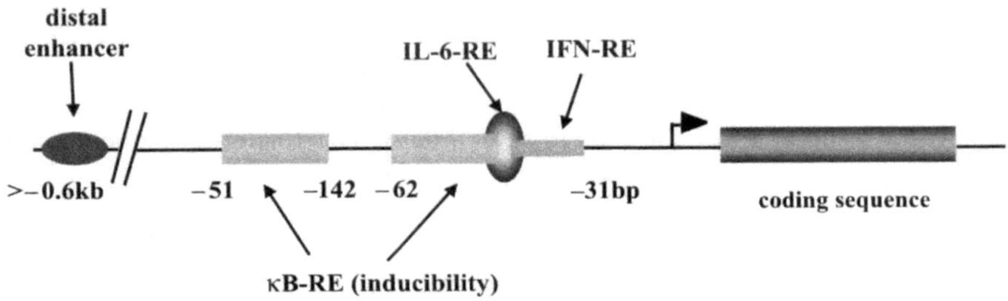

Fig. 4. Promoter of the diptericin gene. Two κB response elements (*κB-RE*) confer immune inducibility to the gene. Other regulatory sequences such as motifs homologous to mammalian IL-6 response elements (*IL-6-RE*) and interferon response elements (*IFN-RE*), together with a distal enhancer located upstream of −0.6kb, are responsible for the upregulation of the induced expression

GATA-sites proximal to the κB binding sites participate in the regulation of this gene during the immune response.

3.2 Immune Signaling Cascades in the Fat Body

At the time when the importance of the Rel-sites was demonstrated, the only Rel protein identified in *Drosophila* was Dorsal, which controls dorsoventral axis determination in early embryogenesis (Fig. 5; Anderson and Nüsslein-Volhard 1986; Stewart 1987). Dorsal is complexed in the cytoplasm of syncytial blastoderm stage embryos by binding to the inhibitory protein Cactus which is homologous to the mammalian NF-κB inhibitor I-κB. The dissociation of Cactus from Dorsal is regulated by an intracellular signaling cascade which is proposed to be initiated upon binding of a processed form of the cytokinelike protein Spätzle to the transmembrane receptor Toll. The processing of Spätzle results from the activation of an extracellular zymogen cascade, involving the products of the genes *gastrulation defective*, *snake*, and *easter*. Upon dissociation from Cactus, the Dorsal protein translocates to the nucleus and regulates the expression of zygotic genes that specify dorsoventral patterning (reviewed in Govind and Stewart

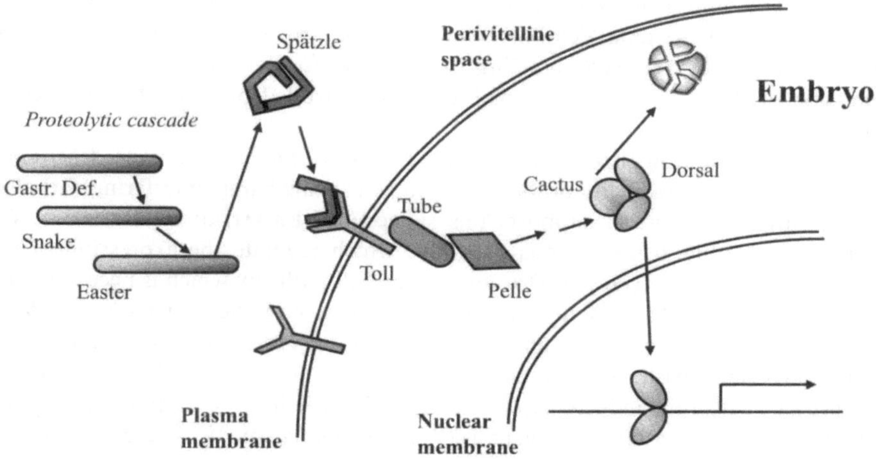

Fig. 5. The dorsoventral signal transduction pathway in the embryo. In the perivitelline space a proteolytic cascade involving successively the serine proteases Gastrulation defective (*Gastr. Def.*), Snake and Easter, locally produces a mature form of Spätzle which is thought to activate the intracellular transduction pathway upon its binding to the Toll receptor. This activation ultimately leads to the dissociation of the Dorsal/Cactus complex and to Dorsal nuclear import. Dorsal then regulates the restricted expression of zygotic genes responsible for the formation of the dorsoventral pattern

1991; Belvin and Anderson 1996). Strikingly, the *Drosophila* Toll receptor shares sequence homology in its intracellular domain with that of the interleukin-1 receptor. The serine-threonine kinase Pelle, which participates in the intracellular signaling pathway initiated by Toll, is related to kinases associated with mammalian cytokine signal transduction (i.e., IRAK for interleukin (interleukin-1) 1 receptor-associated kinase; Cao et al. 1996). In addition, the peptide sequence of the Spätzle protein strongly suggests structural analogies with the "cystine-knot" family (McDonald and Hendrickson 1993; DeLotto and DeLotto 1998) of cytokines/growth factors, such as platelet-derived growth factor, transforming growth factor-β, and nerve growth factor. The structural and functional similarities between the dorsoventral signaling cascade in the *Drosophila* embryo, and the cytokine-induced expression of a number of immune response genes in mammals, prompted us to investigate whether the induction of immune genes in *Drosophila* is regulated by similar mechanisms. A genetic analysis of the inducibility of all antimicrobial peptide genes was performed in flies mutant for the various genes of the embryonic dorsoventral signaling cascade (Lemaitre et al. 1996). Loss-of-function mutations in the genes encoding Spätzle, Toll, Tube, and Pelle resulted in a severe decrease in the inducibility of the drosomycin gene, the major antifungal peptide produced in *Drosophila*. Loss-of-function *cactus* mutations and gain-of-function *Toll* mutations induced high constitutive expression of the drosomycin gene. However, these various mutations did not affect the expression/inducibility of the diptericin gene. All other antimicrobial peptide genes were affected to varying degrees by the above mutations and could thus be placed between drosomycin and diptericin according to their level of dependence on the Toll pathway (see Fig. 6). These data demonstrated that the drosomycin gene is regulated predominantly by the Toll signaling cascade and further suggested that the expression of the various antimicrobial genes is controlled by at least two different pathways.

The serendipitous isolation of a new mutation, *imd* (for *immune deficiency*), that impairs the inducibility of the diptericin gene while barely interfering with that of the drosomycin gene, supports this suggestion (LEMAITRE et al. 1995a). We summarize in Fig. 6 the various pathways which regulate the expression of the antimicrobial peptide genes in *Drosophila*: the Toll pathway which is necessary and sufficient to regulate drosomycin gene expression, and the *imd* pathway which controls the expression of the antibacterial peptides. The data further suggest that some cross-talk exists between the two pathways. A number of mutations that reduce or abolish the inducibility of the dipericin gene have recently been obtained from a third chromosome mutagenesis (WU and ANDERSON 1998) and were named *ird* (for *immune response deficient*). The characterization of the corresponding genes will identify new members of pathways required for normal induction of antibacterial genes during the immune response.

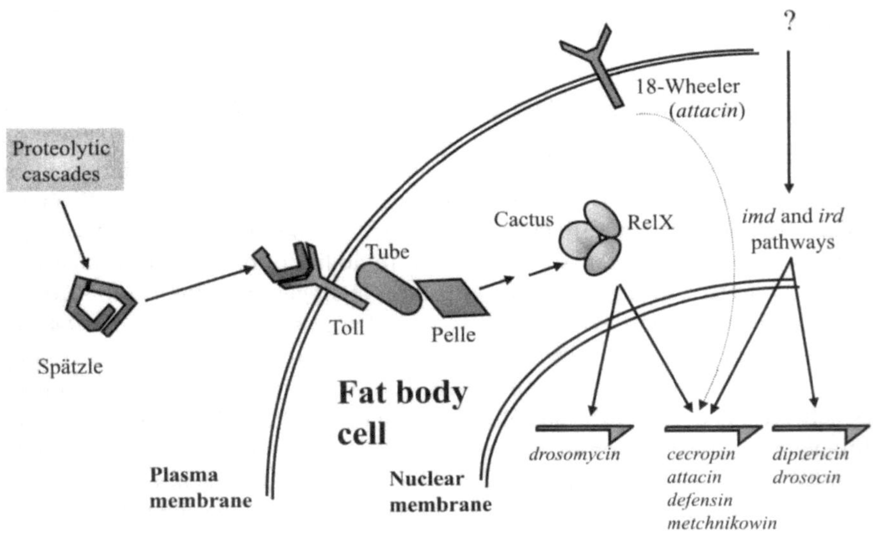

Fig. 6. Control of expression of genes encoding antimicrobial peptides in the fat body of *Drosophila*. Antimicrobial gene expression is controlled by at least three pathways. The Toll signaling pathway directs expression of the gene encoding the antifungal peptide drosomycin. The *imd* pathway controls expression of the antibacterial peptides diptericin and drosocin. Cecropin, attacin, and defensin genes are controlled by both pathways, and a contribution of the 18-Wheeler pathway to the control of attacin expression has been reported. Drosomycin is induced by a Rel protein that is retained in the cytoplasm by the *cactus* gene product in unchallenged *Drosophila*. The identity of RelX is not yet established, but at the larval stage both Dif and Dorsal appear to be good candidates. It is proposed that in the hemolymph, the product of the Spätzle gene, is cleaved to an active ligand form by a proteolytic cascade initiated upon injury or fungal infection

The identity of the Rel protein (designated RelX in Fig. 6) which is translocated into the nucleus upon activation of the Toll pathway has long remained elusive. Induction experiments in *dorsal* loss-of-function mutants yielded results similar to those of wild-type flies (LEMAITRE et al. 1995b, 1997), indicating that either Dorsal is not the Rel protein controlling antimicrobial peptide genes, or that functional redundancy allows another Rel protein to substitute for Dorsal in its absence. Two additional Rel proteins have since been identified in *Drosophila* (Fig. 7): Dif (for Dorsal-related immune factor; IP et al. 1993), and Relish (DUSHAY et al. 1996). Both are expressed in the fat body and their transcription is upregulated by immune challenge. Both Dif and Relish (D. Hultmark, personal communication) are translocated into the nucleus upon challenge. *Dif* and *relish* mutants have recently been generated (D. Ferrandon, D. Hultmark, personal communications) and genetic data should clarify the identity of RelX in the fat body after immune challenge. In an attempt to bypass the lack of *Dif* mutants, a recent analysis was based on the generation of mosaic larval fat body with clones of cells homozygous for a small deficiency uncovering both the *Dif* and *dorsal* genes, which are located in close proximity (MANFRUELLI et al. 1999). In these clones the drosomycin gene was not induced either after septic injury, or in the context of constitutively activated Toll. In both cases however, overexpression of either Dif or Dorsal could restore drosomycin expression, suggesting a redundancy of these Rel transactivators in the larval fat body for the regulation of the drosomycin gene.

The inhibitor of RelX proteins in the fat body is the product of the *cactus* gene, although the participation of other inhibitors cannot be ruled out. During embryonic development Dorsal nuclear translocation can be correlated with a signal-

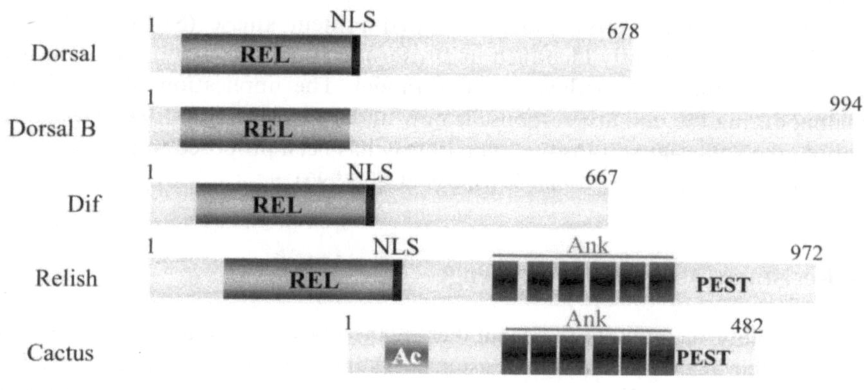

Fig. 7. The *Drosophila* homologues of mammalian Rel/NF-κB and I-κB proteins. Rel proteins share extensive homology within a 300 amino acid region (the Rel domain) responsible for DNA binding and protein dimerization, located at the N-terminus. Except in Dorsal B, a splice alternative of the *dorsal* gene (GROSS et al. 1999), they also contain a nuclear localization signal (*NLS*). I-κB proteins contain several ankyrin motifs (*Ank*) in the C-terminal region. Cactus contains an acidic region (*Ac*); Relish and Cactus contain a PEST domain associated with high proteolytic turnover. The structure of Relish is reminiscent of mammalian p105 and p100, which are precursors of p50 and p52, respectively. *Right*, total number of amino acids of each protein

dependent degradation of Cactus (Belvin et al. 1995). Rapid degradation of I-κBα was also reported in human T lymphocytes following various stimuli, including tumor-necrosis factor (TNF) α treatment of the cells (Brown et al. 1993). In the fat body of *Drosophila*, an immune challenge results in an increase in *cactus* transcription and in a rapid and transient degradation of the protein followed by a return to the equilibrium level within a few hours (Nicolas et al. 1998). The transcriptional regulation of *cactus* is controlled by the Toll pathway. These data reveal the striking functional similarities between transcriptional and posttranscriptional regulation of mammalian I-κB and *Drosophila cactus* in the immune response.

In *Drosophila* there are indications that several Toll homologues are present in the genome. One of these, *18-wheeler*, has been shown to play an important role in embryo morphogenesis (Eldon et al. 1994). It is also expressed at later stages, namely in the fat body cells, and participates in the control of several antibacterial peptide genes (Williams et al. 1997). In larvae mutant for the *18-wheeler* gene the induction of the attacin and cecropin genes is strongly reduced, and the analysis of nuclear translocation of Rel proteins points to an effect of the mutation on the dissociation of the Cactus-Rel complex (see Fig. 6). However, the precise integration of the 18-Wheeler pathway into the general scheme is not yet possible as the other members of the pathway have not been identified.

Mammalian signaling pathways leading from the binding of a ligand to the TNFα or the IL-1 receptors to NF-κB activation, include interaction of the receptor (or receptors aggregates) with a complex containing several distinct proteins. Among these are members of the TNF receptor associated factor (TRAF) family of signaling molecules (reviewed in Arch et al. 1998). Two TRAFs are involved in the response to TNFα and IL-1, they transmit the signal either to NF-κB or to the c-Jun N-terminal kinase (JNK)/stress-activated protein kinase (SAPK) pathway. Two TRAFs were recently identified in *Drosophila* (Liu et al. 1999), one of which was placed in the JNK pathway in this model. The implication of TRAFs in signaling during the immune response is now under investigation in *Drosophila*. In addition, the JNK signal transduction pathway has been proposed to participate in the immune response in *Drosophila* (Sluss et al. 1996).

3.3 Resistance to Immune Challenge

The availability of mutants for a number of the above genes allowed an in vivo study of their relevance for the resistance against bacterial or fungal infection (Lemaitre et al. 1996; Williams et al. 1997). Experiments performed in wild-type or mutant flies challenged with either gram-negative *Escherichia coli* or with the filamentous fungus *Aspergillus fumigatus* showed that: (a) *imd* mutants, in which the immune-inducibility of antibacterial peptides is dramatically reduced, exhibit a poor survival to injection with *E. coli*, in contrast to *Toll*-deficient and wild-type flies; their resistance to fungal infection is, however, similar to that of wild-type; (b) conversely, *Toll*-deficient flies with impaired drosomycin production do not resist an *A. fumigatus* injection but show a survival rate to *E. coli* close to that of wild-type flies. The same holds true for *spätzle-*, *tube-* and *pelle*-deficient flies. Finally larvae deficient for *18-wheeler* (the mutants do not survive to the adult stage) exhibit a significant reduction in survival to infection with *E. coli*. Thus, the Imd, Toll, and 18-Wheeler pathways are all three essential for a full antimicrobial resistance. The data establish a correlation between the impairment of antifungal gene induction and reduced resistance to fungal infection, and, equally, between the impairment of antibacterial gene induction and reduced resistance to bacterial infection. Evidently these results do not rule out the possibility that the various mutations compromise immune mechanisms other than the antimicrobial peptide synthesis by the fat body, which could contribute to defense against microbial infection. *Toll*, *tube*, *pelle*, *cactus* (Qiu et al. 1998), and *18-wheeler* (Williams et al. 1997) are also expressed in the hematopoietic system of *Drosophila* and therefore cellular defense reactions are likely to be altered in the mutants which were tested.

3.4 An Adapted Response

It has generally been assumed that the antimicrobial response of insects, including that of *Drosophila*, is nonspecific and that all effector genes are fully induced by septic injury whatever the infectious agent. A detailed analysis of the transcription of the known antimicrobial peptide genes after various immune stimuli has recently led us to reconsider this assumption. By taking advantage of the unique situation in *Drosophila*, where genes encoding antimicrobial peptides with distinct activity spectra are cloned, Lemaitre and coworkers (1997) showed that the genes encoding antibacterial and antifungal peptides are differentially induced by injection of various micro-organisms. This has led to the suggestion that *Drosophila* can discriminate between various classes of micro-organisms. This suggestion is further substantiated by the observation that flies which are naturally infected with an entomopathogenic fungus (i.e., in the absence of injury), exhibit an adapted response by producing only peptides with antifungal activities. This adapted response is mediated through the selective activation of the Toll pathway, and thereby provides a biological rationale to the existence of distinct regulatory cascades governing the humoral response. The recognition mechanisms that allow the discrimination between various micro-organisms, and their link to the subsequent activation of the regulatory pathways leading to the expression of the various effector genes, are current themes of research.

3.5 Local Immune Response

The data which have accumulated over the past decade on the production of antimicrobial peptides in *Drosophila*, have focused mostly on the production by the fat body (and to a minor extent the blood cells) of peptides that are released into the hemolymph. It has recently been shown in *Drosophila*, by using a drosomycin –

green fluorescent protein reporter gene, which allows direct observation of its expression in live animals, that in the absence of experimental septic injury a number of epithelia express the transgene (Ferrandon et al. 1998). In particular, when larvae or flies were raised on septic media, expression of the drosomycin reporter was frequently observed in structures such as tracheae, mouth parts, and salivary glands. These observations suggest the existence of a local immune response in *Drosophila* in addition to the classical systemic response. Interestingly, the local expression of the drosomycin reporter gene was not affected in mutants of the Toll-signaling cascade, and it therefore appears that expression of drosomycin in these tissues is regulated via a distinct pathway from that in fat body cells.

Thus insect epithelia, such as vertebrate epithelia, are certainly more than merely passive physical barriers and likely constitute an active component of host defense.

4 Proteolytic Cascades

In all arthropods, injury initiates two proteolytic cascades that lead to rapid clotting and melanization at the wound site. Black nodules or large melanized capsules also occur, for example, around encapsulated parasites. Extension of both cascades is prevented by specific protease inhibitors. Little information is available on the molecular bases underlying coagulation and melanization in *Drosophila*. The most complete data on coagulation in an arthropod have been obtained in the horseshoe crab (*Tachypleus tridentatus*) by the group of Iwanaga (reviewed in Iwanaga et al. 1998). In this species, lipopolysaccharide can induce the coagulation cascade, which involves three serine-protease zymogens, by binding to factor C, a 123-kDa serine-protease containing several complement control protein domains, an epithelial growth factor-like domain and a C-type lectin domain. Two of the proteases of this cascade, factor B and proclotting enzyme, contain small compact domains with three disulfide bridges, called CLIP-domains, which are also present in the *Drosophila* Snake and Easter serine-protease precursors active upstream of Spätzle in the embryonic dorsoventral patterning (see Fig. 5). Interestingly, recent studies on the 3D structure of coagulogen, the end target of the coagulation cascade in *Tachypleus*, point to unexpected structural similarities with Spätzle (Bergner et al. 1996). As the latter is required for induction of the gene encoding the antifungal peptide drosomycin (see above), it is tempting to draw a parallel between the two systems. The data on the *Tachypleus* coagulation cascade support the idea that the cascade can function as a system for nonself recognition.

A similar inference can also be made for the prophenoloxidase cascade from to studies performed on the crayfish *Pacifastacus leniusculus* (reviewed in Johansson and Söderhäll 1996) and on *Bombyx mori* (reviewed in Ashida and Brey 1997). In arthropods the formation of the black pigment melanin is catalyzed by the enzyme phenoloxidase, which is converted to its active form by a serine-protease cascade. A *Drosophila* gene for prophenoloxidase has been cloned (Fujimoto et al. 1995), and an activating enzyme for prophenoloxidase was recently isolated (Chosa et al. 1997).

5 Cellular Immunity

The body cavity of *Drosophila*, as with that of all arthropods, contains numerous blood cells or hemocytes, both circulating and sessile, which participate in the immune response. At the larval stage they are derived mainly from the lymph glands, which are paired organs located anteriorly along the dorsal vessel (Shrestha and Gateff 1982). In adults no hematopoietic organ can be identified, and it is commonly believed that blood cells multiply through division of circulating hemocytes. *Drosophila* hemocytes comprise three cell types (reviewed in Rizki and Rizki 1984): the majority of circulating cells are plasmatocytes which are small round cells with potent phagocytic capacities. Some 5%–10% of the hemocytes are crystal cells, characterized by conspicuous crystalline inclusions which are believed to correspond to the substrates of the melanization cascade (Rizki et al. 1980). It has been proposed that crystal cells contain both the substrates and the enzymes of the cascade. A third type of blood cells, the lamellocytes, are not normally found in circulation. These large flat cells differentiate following an immune challenge and participate in the encapsulation of large intruders such as parasitic wasp eggs, or in the formation of melanotic tumors.

The molecular mechanisms responsible for postembryonic hematopoiesis in *Drosophila* and for the differentiation of the various blood cell types are still poorly understood. Two pathways are, however, clearly linked to the process: the Toll signaling pathway and the Janus kinase (JAK)/signal transducer and activator of transcription protein (STAT) pathway. Qiu et al. (1998) have shown that mutations in *Toll* or *cactus* which hyperactivate the pathway cause blood cell hyperproliferation with abnormal differentiation into lamellocytes. Conversely, a reduced hemocyte count was found in loss-of-function *Toll*, *tube*, and *pelle* mutants. As *Toll*, *tube*, *pelle*, *cactus*, *dorsal*, and *dif* are expressed in the lymph glands, it has been proposed that the pathway plays a role in regulating hemocyte proliferation and differentiation in *Drosophila* larvae. This hematopoietic function of the Toll/Cactus pathway seems to be conserved in mammals as knock-out mice for members of the NF-κB/I-κB family exhibit defects in various blood cell lineages (reviewed in Ghosh et al. 1998).

JAK/STAT pathways participate in the response of mammalian blood cells to numerous cytokines and growth factors (reviewed in Ihle and Kerr 1995; Darnell 1997). In *Drosophila* a mutation of the JAK-kinase *hopscotch* which generates a constitutively activated protein leads to hyperproliferation of blood cells with abnormal differentiation into lamellocytes, and to lymph gland overgrowth (Hanratty and Dearolf 1993; Harrison et al. 1995; Luo et al. 1995). This phenotype is partially rescued by loss-of-function mutations of the *Drosophila* STAT gene as the latter suppress the abnormal differentiation but not the hyperproliferation phenotype (Luo et al. 1997). Thus it seems that diverse regulatory inputs such as those integrated by the Toll and the JAK/STAT pathways ensure controlled division and differentiation of the hematopoietic precursors in *Drosophila*.

Various roles of the hemocytes in *Drosophila* immunity such as phagocytosis, encapsulation, and melanization are well established, but many points concerning their functions are still unsolved. In particular, it is not known how the presence of micro-organisms in the body cavity is signaled to the fat body, thus inducing it to synthesize antimicrobial peptides. One hypothesis is that blood cells recognize and bind micro-organisms, thereafter releasing cytokinelike molecules that direct the fat body cells to produce these peptides. The recent isolation of a *Drosophila* mutant, named *domino* (Braun et al. 1997), in which homozygotes are totally devoid of blood cells and nevertheless exhibit good larval viability, has allowed us to test the role of hemocytes in the larval host defense (Braun et al. 1998). In this mutant a septic injury induces the genes encoding all antimicrobial peptides in the fat body at levels similar to wild-type. This indicates that hemocytes are not mandatory for this process. The body cavity of *domino* mutant larvae contains numerous micro-organisms, the presence of which induces only a weak antimicrobial response in the fat body. A full response is observed only after septic injury. These data lead to the proposition that fat body cells can be activated both by the presence of microbes and by injury, and that injury potentiates the effect of micro-organisms. Survival experiments show that *domino* larvae devoid of blood cells maintain a wild-type resistance to septic injury. This is also valid for *imd* larvae in which the humoral antibacterial response is impaired (*imd* larvae are less sensitive to bacterial infection than *imd* adults), and for mutant larvae that are deficient for humoral melanization (carrying the *Black cells* mutation; Rizki et al. 1980). However, when *domino* is combined with either the *imd* or the *Black cells* mutation, the resistance to septic injury is dramatically reduced. These results establish the relevance of the three immune reactions: phagocytosis, synthesis of antibacterial peptides, and melanization: by working in synergy, they provide *Drosophila* with a highly effective defense against injury and/or infection.

6 Perspectives

The field of *Drosophila* immunity has experienced exciting developments over the past decade. Probably the most unexpected result was the discovery of significant similarities with mammalian innate immunity. As this review emphasizes, both insects and mammals essentially rely on short cationic antimicrobial peptides to fend off invading micro-organisms. The intracellular signaling pathways which control the immune-induced synthesis of these peptides in insects are remarkably similar to those of cytokine-induced expression of inflammatory and acute-phase response genes in mammals. In this particular context a striking follow-up of the studies on the role of Toll in the *Drosophila* host defense was the recent discovery that mammals express Toll-like receptors and that they participate in the lipopolysaccharide-induced activation of NF-κB (Medzhitov et al. 1997; Yang et al. 1998).

In spite of the obvious progress in our knowledge on *Drosophila* immunity, wide gaps and broad uncertainties still plague the field. For example, no receptor molecules of infectious nonself have yet been identified. Equally, the proteolytic cascades initiated by septic injury have not yet been deciphered in molecular terms. In particular, the mechanisms of activation of these cascades upon interaction of upstream receptors with micro-organisms remains to be elucidated. In this respect, information in *Drosophila* lags far behind that in the horseshoe crab. Furthermore, wide gaps persist in our understanding of the intracellular signaling pathways which control the expression of the immune response genes.

The full genome sequence of *Drosophila* should be available by the time this review is published. This invaluable information, taken in conjunction with the fact that a rapidly increasing number of scientists are joining the field of *Drosophila* immunity, should guarantee that the next decade will be as exciting as the preceding decade.

Acknowledgements. The authors' research is funded by the Center National de la Recherche Scientifique, the Université Louis Pasteur, the Association pour la Recherche sur le Cancer, the Ligue Nationale contre le Cancer, Human Frontiers in Science, Training and Mobility in Research, and Rhône-Poulenc.

References

Anderson KV, Nüsslein-Volhard C (1986) Dorsal-group genes of *Drosophila.* In: Gametogenesis and the early embryo. Ed J Gall (New York: Alan R. Liss) pp 177–194

Ando K, Natori S (1988) Molecular cloning, sequencing, and characterization of cDNA for SarcotoxinIIA, an inducible antibacterial protein of *Sarcophaga peregrina* (flesh fly). Biochem 27:1715–1721

Arch RH, Gedrich RW, Thompson CB (1998) Tumor necrosis factor receptor-associated factors (TRAFs) – a family of adaptor proteins that regulates life and death. Genes & Devel 12:2821–2830

Ashida M, Brey P (1997) Recent advances in research on the insect prophenoloxidase cascade. In: Molecular Mechanisms of Immune Response in insects. (Ed) Brey PT, Hultmark D (Chapman & Hall, London) pp 135–172

Åsling B, Dushay MS, Hultmark D (1995) Identification of early genes in the *Drosophila* immune response by PCR-based differential display: the Attacin A gene and the evolution of attacin-like proteins. Insect Biochem Mol Biol 25:511–518

Belvin MP, Jin Y, Anderson KV (1995) Cactus protein degradation mediates *Drosophila* dorsal-ventral signaling. Genes Dev 9:783–793

Belvin MP, Anderson KV (1996) A conserved signaling pathway: the *Drosophila* Toll-Dorsal pathway. Annu Rev Cell Dev Biol 12:393–416

Bergner A, Oganessyan V, Muta T, Iwanaga S, Typke D, Huber R, Bode W (1996) Crystal structure of coagulogen, the clotting protein from horseshoe crab: a structural homologue of nerve growth factor. EMBO J 15:6789–6797

Boman HG, Nilsson I, Rasmuson B (1972) Inducible antibacterial defence system in *Drosophila.* Nature 237:232–235

Boman HG (1995) Peptide antibiotics and their role in innate immunity. Ann Rev Immunol 13:61–92

Braun A, Lemaitre B, Lanot R, Zachary D, Meister M (1997) *Drosophila* immunity: analysis of larval hemocytes by P-element-mediated enhancer trap. Genetics 147:623–634

Braun A, Hoffmann JA, Meister M (1998) Analysis of the *Drosophila* host defense in *domino* mutant larvae, which are devoid of hemocytes. Proc Natl Acad Sci USA 95:14337–14342

Broekaert WF, Terras FRG, Cammue BPA, Osborn RW (1995) Plant defensins: novel antimicrobial peptides as components of the host defense system. Plant Physiol 108:1353–1358

Brown K, Park S, Kanno T, Franzoso G, Siebenlist U (1993) Mutual regulation of the transcriptional activator NF-κB and its inhibitor, IκB-α Proc Natl Acad Sci USA 90:2532–2536

Bulet P, Dimarcq JL, Hetru C, Lagueux M, Charlet M, Hégy G, Van Dorsselaer A, Hoffmann JA (1993) A novel inducible antibacterial peptide of *Drosophila* carries an O-glycosylated substitution. J Biol Chem 268:14893–14897

Bulet P, Urge L, Ohresser S, Hetru C, Otvos L Jr (1996) Enlarged scale chemical synthesis and range of activity of drosocin, an O-glycosylated antibacterial peptide of *Drosophila*. Eur J Biochem 238:64–69

Cao Z, Henzel WJ, Gao X (1996) IRAK: a kinase associated with the Interleukin-1 receptor. Science 271:1128–1131

Charlet M, Lagueux M, Reichhart JM, Hoffmann D, Braun A, Meister M (1996) Cloning of the gene encoding the antibacterial peptide drosocin involved in *Drosophila* immunity. Expression studies during the immune response. Eur J Biochem 241:699–706

Chosa N, Fukumitsu T, Fujimoto K, Ohnishi E (1997) Activation of prophenoloxidase A1 by an activating enzyme in *Drosophila melanogaster*. Insect Biochem Molec Biol 27:61–68

Christensen B, Fink J, Merrifield RB, Mauzerall D (1988) Channel-forming properties of cecropins and related model compounds incorporated into planar lipid membranes. Proc Natl Acad Sci USA 85:5072–5076

Cociancich S, Ghazi A, Hetru C, Hoffmann JA, Letellier L (1993) Insect defensin, an inducible antibacterial peptide, forms voltage-dependent channels in *Micrococcus luteus*. J Biol Chem 268: 19239–19245

Cornet B, Bonmatin JM, Hetru C, Hoffmann JA, Ptak M Vovelle F (1995) Refined three-dimensional solution structure of insect defensin A. Structure 3:435–448

Cuénot L (1896) Etudes physiologiques sur les orthoptères. Arch Biol Liège 14:293–341

Darnell JE (1997) Stats and gene regulation. Science 277:1630–1635

DeLotto Y, DeLotto R (1998) Proteolytic processing of the *Drosophila* Spätzle protein by Easter generates a dimeric NGF-like molecule with ventralising activity. Mechanisms of Devel 72:141–148

Dimarcq JL, Hoffmann D, Meister M, Bulet P, Lanot R, Reichhart JM, Hoffmann JA (1994) Characterization and transcriptional profiles of a *Drosophila* gene encoding an insect defensin. A study in insect immunity. Eur J Biochem 221:201–209

Dimarcq JL, Keppi E, Dunbar B, Lambert J, Reichhart JM, Hoffmann D, Rankine SM, Fothergill JE, Hoffmann JA (1988) Insect Immunity. Purification and characterization of a family of novel inducible antibacterial proteins from immunized larvae of the dipteran *Phormia terranovae* and complete amino-acid sequence of the predominant member, diptericin A. Eur J Biochem 171:17–22

Dushay M, Asling B, Hultmark D (1996) Origins of immunity: relish, a compound Rel-like gene in the antibacterial defense of *Drosophila*. Proc Natl Acad Sci USA 93:10343–10347

Eldon E, Kooyer S, D'Evelyn D, Duman M, Lawinger P, Botas J, Bellen H (1994) The *Drosophila 18-wheeler* is required for morphogenesis and has striking similarities to Toll. Development 120:885–899

Engström P, Carlsson A, Engström A, Tao ZJ, Bennich H (1984) The antibacterial effect of attacins from the silk moth *Hyalophora cecropia* is directed against the outer membrane of *Escherichia coli*. EMBO J 3:3347–3351

Engström Y, Kadalayil L, Sun SC, Samakovlis C, Hultmark D, Faye I (1993) κB-like motifs regulate the induction of immune genes in *Drosophila* J Mol Biol 232:327–333

Fehlbaum P, Bulet P, Michaut L, Lagueux M, Broekaert WF, Hetru C, Hoffmann JA (1994) Insect immunity: septic injury of *Drosophila* induces the synthesis of a potent antifungal peptide with sequence homology to plant antifungal peptides. J Biol Chem 269:33159–33163

Ferrandon D, Jung A, Criqui MC, Michaut L, Reichhart JM, Hoffmann JA (1998) A drosomycin-GFP reporter transgene reveals a local immune response in *Drosophila* that is not dependent on the Toll pathway. EMBO J 17:1217–1227

Flyg C, Dalhammar G, Rasmuson B, Boman HG (1987) Insect Immunity. Inducible antibacterial activity in *Drosophila*. Insect Biochem 17:153–160

Fujimoto K, Okino N, Kawabata S, Iwanaga S, Ohnishi E (1995) Nucleotide sequence of the cDNA encoding the proenzyme of phenol oxidase A_1 of *Drosophila melanogaster*. Proc Natl Acad Sci USA 92:7769–7773

Gazit E, Boman A, Boman HG, Shai Y (1995) Interaction of the mammalian antibacterial peptide cecropin P1 with phospholipid vesicles. Biochemistry 34:11479–11488

Georgel P, Meister M, Kappler C, Lemaitre B, Reichhart JM, Hoffmann JA (1993) Insect immunity: the diptericin promoter contains multiple functional regulatory sequences homologous to mammalian acute-phase response elements. Biochem Biophys Res Commun 197:508–517

Georgel P, Kappler C, Langley E, Gross I, Nicolas E, Reichhart JM, Hoffmann JA (1995) *Drosophila* immunity. A sequence homologous to mammalian interferon consensus response element enhances the activity of the diptericin promoter. Nucleic Acids Res 23:1140–1145

Ghosh S, May MJ, Kopp EB (1998) NF-κB and Rel proteins: evolutionarily conserved mediators of immune responses. Annu Rev Immunol 16:225–260

Glaser RW (1918) On the existence of immunity principles in insects. Psyche 25:38–46

Govind S, Steward R (1991) Dorsoventral pattern formation in *Drosophila*: signal transduction and nuclear targeting. Trends in Genetics 7:119–125

Gross I, Georgel P, Oertel-Buchheit P, Schnarr M, Reichhart JM (1999) Dorsal-B, a splice variant of the *Drosophila* factor dorsal, is a novel Rel/NF-κB transcriptional activator. Gene 228:233–242

Hanratty WP, Dearolf CR (1993) The *Drosophila Tumorous-lethal* hematopoietic oncogene is a dominant mutation in the *hopscotch* locus. Mol Gen Genet 238:33–37

Harrison DA, Binari R, Stines Nahreini T, Gilman M, Perrimon N (1995) Activation of a *Drosophila* Janus kinase (JAK) causes hematopoietic neoplasia and developmental defects. EMBO J 14:2857–2865

Hoffmann JA, Reichhart JM, Hetru C (1996) Innate immunity in higher insects. Current Opinion in Immunol 8:8–13

Hoffmann JA, Reichhart JM (1997) *Drosophila* immunity. Trends in Cell Biol 7:309–316

Hultmark D, Engström A, Andersson K, Steiner H, Bennich H, Boman HG (1983) Insect immunity. Attacins, a family of antibacterial proteins from *Hyalophora cecropia*, EMBO J 2:571–576

Ihle JN, Kerr IM (1995) Jaks and Stats in signaling by the cytokine receptor superfamily. Trends in Genetics 11:69–74

Ip TY, Reach M, Engström Y, Kadalayil L, Cai H, Gonzalez-Crespo S, Tatei K, Levine M (1993) *Dif*, a *dorsal*-related gene that mediates an immune response in *Drosophila*. Cell 75:753–763

Iwanaga S, Kawabata SI, Muta T (1998) New types of clotting factors and defense molecules found in horseshoe crab hemolymph: their structures and functions. J Biochem 123:1–15

Johansson MW, Söderhäll K (1996) The prophenoloxidase activating system and associated proteins in invertebrates. Prog Mol Subcell Biol 15:46–66

Kadalayil L, Petersen UM, Engström Y (1997) Adjacent GATA and κB-like motifs regulate the expression of a *Drosophila* immune gene. Nucl. Acids Res 25:1233–1239

Kappler C, Meister M, Lagueux M, Gateff E, Hoffmann JA, Reichhart JM (1993) Insect Immunity. Two 17bp repeats nesting a κB-related sequence confer inducibility to the diptericin gene and bind a polypeptide in bacteria-challenged *Drosophila*. EMBO J 12:1561–1568

Kowalevsky A (1887) Ein Beiträge zur Kenntniss des Exkretions Organen. Biol Zentralbl 6:125–144

Kylsten P, Samakovlis C, Hultmark D (1990) The cecropin locus in *Drosophila*; a compact gene cluster involved in the response to infection. EMBO J 9:217–224

Lambert J, Keppi E, Dimarcq JL, Wicker C, Reichhart JM, Dunbar B, Lepage P, van Dorsselaer A, Hoffmann JA, Fothergill J (1989) Insect immunity: isolation from immune blood of the dipteran *Phormia terranova* of two insect antibacterial peptides with sequence homology to rabbit lung macrophage bactericidal peptides. Proc Natl Acad Sci USA 86:262–266

Landon C, Sodano P, Hetru C, Hoffmann JA, Ptak M (1997) Solution structure of drosomycin, the first inducible antifungal protein from insects. Protein Science 6:1878–1884

Lemaitre B, Kromer-Metzger E, Michaut L, Nicolas E, Meister M, Georgel P, Reichhart JM, Hoffmann JA (1995a) A recessive mutation, *immune deficiency* (*imd*), defines two distinct control pathways in the *Drosophila* host defense. Proc Natl Acad Sci USA 92:9465–9469

Lemaitre B, Meister M, Govind S, Georgel P, Steward R, Reichhart JM, Hoffmann JA (1995b) Functional analysis and regulation of nuclear import of *dorsal* during the immune response in *Drosophila*. EMBO J 14:536–545

Lemaitre B, Nicolas E, Michaut L, Reichhart JM, Hoffmann JA (1996) The dorsoventral regulatory gene cassette *spaetzle/Toll/cactus* controls the potent antifungal response in *Drosophila* adults. Cell 86: 973–983

Lemaitre B, Reichhart JM, Hoffmann JA (1997) *Drosophila* host defense: differential induction of antimicrobial peptide genes after infection by various classes of microorganisms. Proc Natl Acad Sci USA 94:14614–14619

Levashina EA, Ohresser S, Bulet P, Reichhart JM, Hetru C, Hoffmann JA (1995) Metchnikowin, a novel immune-inducible proline-rich peptide from *Drosophila* with antibacterial and antifungal properties. Eur J Biochem 233:694–700

Levashina EA, Ohresser S, Lemaitre B, Imler JL (1998) Two distinct pathways can control expression of the gene encoding the *Drosophila* antimicrobial peptide metchnikowin. J Mol Biol 278:515–527

Liu H, Su YC, Becker E, Treisman J, Skolnik EY (1999) A *Drosophila* TNF-receptor-associated factor (TRAF) binds the Ste20 kinase Misshapen and activates Jun kinase. Current Biology 9:101–104

Luo H, Hanratty WP, Dearolf CR (1995) An amino acid substitution in the *Drosophila* hop^{Tum-l} Jak kinase causes leukemia-like hematopoietic defects. EMBO J 14:1412–1420

Luo H, Rose P, Barber D, Hanratty WP, Lee S, Roberts TM, D'Andrea AD, Dearolf CR (1997) Mutation in the Jak kinase JH2 domain hyperactivates *Drosophila* and mammalian Jak-Stat pathways. Molec Cell Biol 1997:1562–1571

Manfruelli P, Reichhart JM, Steward R, Hoffmann JA, Lemaitre B (1999) A mosaic analysis in *Drosophila* fat body cells of the control of antimicrobial peptide genes by the Rel proteins Dorsal and DIF. EMBO J 18:3380–3391

Matsuyama K, Natori S (1988) Molecular cloning of cDNA for sapecin and unique expression of the sapecin gene during the development of *Sarcophaga peregrina*. J Biol Chem 263:17117–17121

McDonald NQ, Hendrickson WA (1993) A structural superfamily of growth factors containing a cystine knot motif. Cell 73:421–424

Medzhitov R, Preston-Hurlburt P, Janeway Jr CA (1997) A human homologue of the *Drosophila* Toll protein signals activation of adaptive immunity. Nature 388:394–397

Meister M, Braun A, Kappler C, Reichhart JM, Hoffmann JA (1994) Insect immunity. A transgenic analysis in *Drosophila* defines several functional domains in the diptericin promoter. EMBO J 13:5958–5966

Metalnikow S (1920) Immunité naturelle ou acquise des chenilles de *Galleria mellonella*. CR Acad Sci Paris 83:817–820

Nicolas E, Reichhart JM, Hoffmann JA, Lemaitre B (1998) In vivo regulation of the IκB homologue cactus during the immune response of *Drosophila*. J Biol Chem 273:10463–10469

Okada M, Natori S (1985) Primary structure of sarcotoxin I, an antibacterial protein induced in the hemolymph of *Sarcophaga peregrina* (flesh fly) larvae. J Biol Chem 260:7174–7177

Paillot A (1920) L'immunité acquise chez les insectes. CR, Acad Sci Paris 83:278–280

Qiu P, Pan PC, Govind S (1998) A role for the *Drosophila* Toll/cactus pathway in larval hematopoiesis. Development 125:1909–19520

Reichhart JM, Meister M, Dimarcq JL, Zachary D, Hoffmann D, Ruiz C, Richards G, Hoffmann JA (1992) Insect immunity: developmental and inducible activity of the *Drosophila* diptericin promoter. EMBO J 11:1469–1477

Reichhart JM, Essrich M, Dimarcq JL, Hoffmann D, Hoffmann JA, Lagueux M (1989) Insect Immunity. Isolation of cDNA clones corresponding to diptericin, an inducible antibacterial peptide from *Phormia terranovae* (Diptera). Eur J Biochem 182:423–427

Rizki TM, Rizki RM (1984) The cellular defense system of *Drosophila melanogaster*. In: King RC, Akai H (eds) Insect ultrastructure Vol 2. Plenum Publishing Corporation, pp 579–604

Rizki TM, Rizki RM, Grell EH (1980) A mutant affecting the crystal cells in *Drosophila melanogaster*. Wilhelm Roux's Arch 188:91–99

Robertson M, Postlethwait JH (1986) The humoral antibacterial response of *Drosophila* adults. Dev Comp Immunol 10:167–179

Roos E, Björklund G, Engström Y (1998) In vivo regulation of tissue-specific and LPS-inducible expression of the *Drosophila* Cecropin genes. Insect Molec Biol 7:51–62

Shrestha R, Gateff E (1982) Ultrastructure and cytochemistry of the cell types in the larval hematopoietic organs and hemolymph of *Drosophila melanogaster*. Dev Growth Differ 24:65–82

Sluss HK, Han Z, Barett T, Davis RJ, Ip YT (1996) A JNK signal transduction pathway that mediates morphogenesis and an immune response in *Drosophila*. Genes & Dev10:2745–2758

Steiner H, Hultmark D, Engström A, Bennich H, Boman HG (1981) Sequence and specificity of two antibacterial proteins involved in insect immunity. Nature 292:246–248

Steward R (1987) Dorsal, an embryonic polarity gene in *Drosophila*, is homologous to the vertebrate proto-oncogene, c-rel. Science 238:692–694

Tryselius Y, Samakovlis C, Kimbrell DA, Hultmark D (1992) *CecC*, a cecropin gene expressed during metamorphosis in *Drosophila* pupae. Eur J Biochem 204:395–399

Wicker C, Reichhart JM, Hoffmann D, Hultmark D, Samakovlis C, Hoffmann JA (1990) Insect immunity. Characterization of a *Drosophila* cDNA encoding a novel member of the diptericin family of immune peptides. J Biol Chem 265:22493–22498

Williams MJ, Rodriguez A, Kimbrell DA, Eldon E (1997) The *18-wheeler* mutation reveals complex antibacterial gene regulation in *Drosophila* host defense. EMBO J 16:6120–6130

Wu LP, Anderson KV (1998) Regulated nuclear import of Rel proteins in the *Drosophila* immune response. Nature 392:93–97

Yang RB, Mark MR, Gray A, Huang A, Xie MH, Zhang M, Goddard A, Wood WI, Gurney AL, Godowski PJ (1998) Toll-like receptor-2 mediates lipopolysaccharide-induced cellular signalling. Nature 395:284–288

Origin and Evolution of the Complement System

M. NONAKA

1 Introduction

The vertebrate immune system is composed of two parts, innate immunity which recognizes the invading microbes using germline-encoded molecules, and adaptive immunity, which depends on recognition molecules generated by somatic mechanisms during the ontogeny of each individual organism (MEDZHITOV and JANEWAY 1997). All data available to date indicate that adaptive immunity became established at the early stage of vertebrate evolution around the time of cartilaginous fish emergence. Thus the genes which encode the pivotal elements of adaptive immunity, such as immunoglobulin (LITMAN et al. 1993), T-cell receptor (RAST et al. 1997), major histocompatibility complex (MHC) class I (HASHIMOTO et al. 1992) and class II molecules (KASAHARA et al. 1992; BARTLE and WEISSMAN 1994) and recombination activating gene (GREENHALGH and STEINER 1995) have been identified in cartilaginous fish and higher vertebrates. None of the attempts to isolate these genes from the most primitive extant vertebrates, cyclostomes, has yet succeeded. In contrast, vertebrate innate immunity is believed to have a more ancient

Department of Biological Sciences, Graduate School of Science, University of Tokyo, Hongo, Tokyo 113-0033, Japan
e-mail: mnonaka@biol.s.u-tokyo.ac.jp

origin, and an apparently primitive complement system has been found in lamprey (Nonaka et al. 1983). However, it was not clear until recently whether the origin of the complement system can be traced back to invertebrate. Identification of C3/C4/C5-like expressed sequence tag (Est) from sea urchin coelomocytes (Smith et al. 1996) and molecular studies of the complement system in sea urchin and ascidian established the presence of the multicomponent, opsonic complement system in invertebrates. Therefore the complement system seems to have a more ancient origin than adaptive immunity, although its name, which is based on the fact that it was found as a factor to complement the bacteriocidal activity of antibody, implies the reverse. This chapter discusses recent progress in understanding the evolution of the complement system mainly in invertebrates and lower vertebrates. Evolution of the complement system in vertebrates has previously been reviewed (Dodds and Day 1993; Nonaka 1998).

2 Mammalian Complement System

As a background to understanding complement evolution, I briefly summarize our present knowledge on the mammalian complement system. For more details and references, refer to the recent review (Volanakis 1998). The mammalian complement system, as revealed by extensive studies in human, consists of about 30 serum and cell surface proteins and plays a role in host defense by mediating acute inflammatory reactions, clearance of foreign cells, and killing of pathogenic microorganisms. The complement activation reaction is divided into four parts, merging at the central component C3. These four parts consist of three activation pathways named the classical pathway (CP), the alternative pathway (AP), and the lectin pathway (LeP) and one effector, the lytic pathway (LyP), resulting in the formation of a membrane attack complex (MAC). Activation of the CP is usually triggered by antibodies bound to antigens. C1 complex, composed of C1q, C1r, and C1s is then activated by binding to antibodies which in turn activates C4 and C2 by cleaving a single peptide bond. The bimolecular complex formed between C4 and C2 is the CP C3 convertase, in which the catalytic moiety is C2. Activation of AP is believed to be initiated by spontaneous hydrolysis of the thioester bond of C3. Hydrolyzed C3 binds to factor B, making the latter susceptible to proteolytic attack by factor D and leading to the formation of AP C3 convertase from C3 and B. Under normal conditions AP C3 convertase is immediately inactivated by regulatory proteins with no physiological consequences. However, should an activating surface protect the AP C3 convertase from regulatory proteins, AP C3 convertase activates C3, which starts a positive feedback amplification loop of the AP. Activation of the LeP is initiated by the binding of mannan binding lectin (MBL) to certain sugar on the surface of foreign particles. Although molecular details need to be clarified, MBL-associated serine protease (MASP) 1 and -2 are believed to circulate as a complex with MBL similar to the CP C1 complex, and binding of MBL to sugars induces activation of MASP-1 and MASP-2. MASPs are serine proteases, which activate C2, C4, and C3. There are several pairs or sets of components of the complement system believed to be generated from a common ancestor by gene duplications. These are C1q (CP) and MBL (LeP); C1r (CP), C1s (CP), MASP-1 (LeP), and MASP-2 (LeP); C2 (CP) and B (AP); C3 (general), C4 (CP), and C5 (LyP); C8α and C8β (LyP). The presence of these homologous pairs or sets of genes indicates that gene duplication played an important role in increasing the number of components and expanding the functional repertory of the complement system. The important questions to be addressed by this phylogenetic study are to substantiate when the original complement system was established, and how it looked, and also when the critical gene duplications occurred enabling development of the complement system. Molecular evidence on the phylogenetic aspect of the LyP and regulatory proteins is limited, and the only nonmammalian cDNA clones reported so far are trout C9 (Stanley and Herz 1987) and barred sand bass H (Dahmen et al. 1994). Therefore I focus below on the evolution of the apparently redundant three activation pathways.

3 Invertebrate Complement System

For some time the complement system was considered to be the unique property of vertebrates. However, this changed when an Est, isolated from sea urchin, *Strongylocentrotus purpuratus*, coelomocyte cDNA library, was found to show significant sequence homology to C3/C4/C5 (Smith et al. 1996). The primary structure of the corresponding protein deduced by cDNA analysis showed sequence similarity to C3/C4/C5 throughout its entire length (Al-Sharif et al. 1998). This protein was termed SpC3, and it showed structural similarities to vertebrate C3 such as an α-β subunit chain structure and an intrachain thioester bond. Phylogenetic analysis of SpC3 compared with other members of the thioester protein family, including C3, C4, C5, and α_2-macroglobulin, showed SpC3 as the first divergent complement protein, falling at the base of the complement protein clade. This indicated that the C3/C4/C5 gene duplication occurred in the chordate lineage after divergence of echinoderm. Another Est from the same sea urchin cDNA library proved to encode B and was termed SpBf (Smith et al. 1998). SpBf had the same domain structure as vertebrate B/C2, except for two additional SCR domains at the N-terminus. Phylogenetic analysis of SpBf indicated that it is the most ancient member of Bf/C2, again suggesting that the B/C2 gene duplication occurred in the chordate lineage after divergence of echinoderm. The identification of C3 and B strongly suggests the presence of the AP of the complement system in sea urchin, although the presence of additional components, especially those of the LeP (see below) still need to be analyzed. The sea urchin is the lowest animal to date in which the complement has been found. It is important to establish by complement phylogeny whether it is present in more primitive animals.

Another invertebrate which possesses the complement system is the ascidian, *Halocynthia roretzi*. C3 (Nonaka et al. 1999), B (Ji et al., unpublished data), and two MASPs (Ji et al. 1997) have been identified and termed, AsC3, AsB, AsMASPa and AsMASPb, respectively. From the structural point of view, AsC3 and AsB are similar to the sea urchin counterparts in that they are considered to represent the ancestral state of C3/C4/C5 and B/C2, respectively. Similarly, AsMASPa and AsMASPb seemed to represent the ancestral state of MASP-1/MASP-2/C1r/C1s. Since MASPs are components of the LeP, this strongly suggests that the primitive complement system contains both the AP and LeP. Although these pathways are recognized as separate activation pathways in the mammalian complement system, they could also be two parts of a single activation pathway in the ancient complement system. This hypothetical activation pathway, consisting of MBL, MASP, B, and C3, shows a close similarity to the CP. Interestingly, this suggests the possibility of the CP being generated by the duplication of a set containing the AP plus LeP (Nonaka et al. 1998). Ascidian coelomocytes show phagocytic activity against yeast, enhanced by pretreating yeast with ascidian body fluid. Antibody against AsC3 or EDTA blocked this opsonic activity in body fluid, indicating that the opsonic factor is C3 activated through a complement cascade (Nonaka et al. 1999). Further details of the activation mechanism are still under investigation.

4 Agnathan Complement System

The previous functional study of the lamprey complement system indicated the presence of an apparently primitive system, which is activated through the AP and shows an opsonic activity (Nonaka et al. 1984). Trials to detect the CP or cytolytic activity failed, suggesting that the lamprey complement system is simpler than that of mammals. C3 protein and cDNA clones have been isolated from both lamprey (Nonaka et al. 1984; Nonaka and Takahashi 1992) and hagfish (Fujii et al. 1992; Ishiguro et al. 1992; Hanley et al. 1992), two extant representatives of agnatha. Both lamprey and hagfish C3 retained all basic characteristics of the vertebrate C3, except that lamprey C3 has three subunit chains, as in the mammalian C4, rather than two subunit chains, as in the mammalian C3. Although the functional study indicating the absence of the CP and LyP in agnatha suggested that they lack C4 and C5, the phylogenetic tree analysis of C3, C4, and C5 showed that the agnathan C3s form a monophyletic group with vertebrate C3s which is supported by a high bootstrap value (Nonaka et al. 1998). Thus we cannot exclude the possibility that the gene duplications to generate C4 and C5 predated the emergence of agnatha, but agnatha lost the C4 and C5 genes secondarily. In addition to C3, B (Nonaka et al. 1994) and MASP (Endo et al. 1998) cDNA clones have been isolated from lamprey, making the situation very similar to ascidians. These results suggest that the original opsonic complement system was composed of the LeP and APs, and that this primitive state was conserved for a long time at least from the emergence of echinoderms to the emergence of cyclostomes.

5 Unique Features of the Teleost Complement System

Recent progress in phylogenetic analysis of the complement system has not only revealed the ancient origin of the system but also the uniqueness of teleost complement. Molecular cloning of C3 has been reported from all major classes of vertebrate and two invertebrate species; mammals, man (DE BRUIJN and FEY 1985), mouse (FEY et al. 1984), rat (MISUMI et al. 1990), guinea pig (AUERBACH et al. 1990), rabbit (KUSANO et al. 1986); bird, chicken (MAVROIDIS et al. 1995); reptile, cobra (FRITZINGER et al. 1992); amphibia, toad (LAMBRIS et al. 1995); bony fish, trout (LAMBRIS et al. 1993), carp (M. Nakao, personal communication), Medaka fish (Kuroda et al., 2000); cartilaginous fish, shark (Terado et al., unpublished); cyclostome, lamprey (NONAKA and TAKAHASHI 1992), hagfish (ISHIGURO et al. 1992); protochordate, ascidian (NONAKA et al. 1999); echinoderm, sea urchin (AL-SHARIF et al. 1998). In all cases studied so far, except for cobra and bony fish, C3 is a single copy gene. Cobra has an additional gene, although it is expressed only in the venom gland and is considered to be a specialized form for venom (FRITZINGER et al. 1995). On the other hand, multiple bony fish C3 genes are all expressed in liver and their products are present in the serum (SARRIAS et al. 1998; Nakao et al., personal communication; Kuroda et al., 2000). Multiple C3 proteins are also identified in sea bream (SUNYER et al. 1997), indicating that the C3 multiplication is a rule rather than exception in bony fish. Some interesting points were noted with the multiple C3 of bony fish. (a) In all three species analyzed so far at least one of the multiple C3 lacks the catalytic His residue which determines the reactivity of the thioester bond. (b) A functional study performed in trout indicated that multiple forms of C3 show various binding preference to various complement activating surfaces. (c) Phylogenetic analysis of multiple C3 forms of trout, carp, and Medaka fish with other C3 indicated that multiplication of C3 in these species occurred independently. It is proposed that C3 multiplication is a unique strategy taken by teleost to increase its repertory to recognize foreign molecules by innate immunity where adaptive immunity is still in an underdeveloped state (SUNYER et al. 1998a). Another interesting point with the teleost complement is the organization of mammalian MHC class III complement genes. As reviewed recently (FLAJNIK et al. 1999), the class IA genes are not linked to the class IIA and B genes in all teleost species analyzed so far. Only genes showing almost the same degree of similarity to both B and C2 has been identified from zebrafish (SEEGER et al. 1996) and Medaka fish (KURODA et al. 1996). In both species these genes are linked neither to class I nor to class II genes (BINGULAC-POPOVIC et al. 1997; Kuroda et al., unpublished). Moreover, the Medaka fish C4 is not linked to B or to class II (Kuroda et al., 2000). Although it is not clear at present whether dispersed MHC found in teleost is ancestral or derived character, it is intriguing to speculate that it is related to underdeveloped state of adaptive immunity in teleost.

6 Evolution of B and C2

Mammalian B and C2 play a pivotal role in complement activation as a catalytic subunit of the C3 and C5 convertase of the AP and CP, respectively. B and C2 share a unique domain structure consisting of three SCR domains, a von Willebrand factor type A domain and a serine protease domain, indicating that they have arisen from a common ancestor by gene duplication (Ishikawa et al. 1990). Recent progress in both the structural analysis of the serine protease domain of human B and the phylogenetic analysis in lower vertebrates and invertebrates has revealed an interesting evolutionary history of B and C2. The serine protease domain of human B and C2 has been recognized to be unique among serine proteases with trypsinlike specificity in the respect that it lacks an Asp-189 (trypsin numbering) residue at the bottom of the specificity pocket. A recent crystallographic study of B has indicated that the Asp-715 (B numbering) near the C-terminus forms a part of a wall of the substrate binding pocket, probably giving a trypsinlike substrate specificity to B instead of Asp-189 (Narayana et al. 1998). This conclusion is supported by a functional study using the site-directed mutagenized proteins (Hourcade et al. 1998). Substitution of Asp-715 with Asn, Glu, Ala, Ser, or Tyr almost completely abrogated C3-cleaving activity without affecting convertase assembly with C3. Thus the serine protease domain of B and C2 seems to have specialized during evolution from the trypsinlike prototype. As far as the Asp residue involved in interaction with P1 Arg of substrate is concerned, phylogenetic studies clearly showed when this specialization occurred. Invertebrate B from sea urchin (Smith et al. 1998) and ascidian (Ji et al., unpublished). As well as lamprey B (Nonaka et al. 1994) have Asp-189 but not Asp-715 (Fig. 1).

In contrast, all B from cartilaginous fish, Japanese dog fish (Terado et al., unpublished) and nurse shark (S. Smith et al., personal communication); bony fish, Medaka fish (Kuroda et al. 1996), zebrafish (Seeger et al. 1996), carp (Nakao et al. 1998), and trout (Sunyer et al. 1998b); amphibia, *Xenopus* (Kato et al. 1994); mammals, man (Mole et al. 1984), and mouse (Ishikawa et al. 1990) have Asp-715 but not Asp-189. Therefore the change in the serine protease domain of B from a trypsinlike prototype to a specialized one should have occurred in the main line of vertebrate evolution after the divergence of cyclostomes but before the divergence of cartilaginous fish. It is interesting to note that the CP and LyP of the complement system and adaptive immunity are also believed to have been established around the time of cartilaginous fish emergence (see Introduction). It is possible therefore that the change in serine protease domain structure has some functional relevance for example to C5 convertase formation or CP activation. Further structure/function correlation study of mammalian B, C2, and primitive B found in lamprey, ascidian, and sea urchin will extend our understandings of the functional development of the complement system.

Another interesting question regarding the B and C2 genes is the timing of the duplication during evolution. The human and mouse C2 and B genes locate in the class III region of the MHC in a head to tail configuration (Carroll et al. 1984;

```
        631                                   !     #                                                $   721
HosaB   CERDAQYAP-GYDKVKDISEVVTPRFLCTGGVSPYADPNTCRGDSGGPLIVHKRS-RFIQVGVISWGVVDVCK----NQKRQKQVPA---HARDFHINLF
MumuB   CERDATKAQ-GYEKVKDASEVVTPRFLCTGGVDPYADPNTCKGDSGGPLIVHKRS-RFIQVGVISWGVVDVCR----DQRRQQLVPS---YARDFHINLF
XelaB1  CLEAAKKAP-ELKNVTNIEDAVTDQFLCTGGIVPVADPPVCKGDSGGPLLIQVKR-RYVQVGIISWGTVDHCD----KGTRIKQTQK---NARDFYQDIF
XelaB2  CLDAAKKTP-ELKDVTNIEDAVSDQFLCTGGLIPVVDPPVCKGDSGGPLLVQVKR-RYVQVGIISWGTVDHCE----KGKRVKQTKS---NARDFYQDIF
OrlaB   CIKKAL--KAKGITTTDPKVPVTDNFLCTGGD---RDHIACTGDSGGAVFKNYES-RTIQIALVSWGTQEICTGGG---MRETTPES-----RDFHINLF
CycaB1  CVEDAK--KAKGINVENAREAVTDNFLCSGGIEPETDDVACKGESGGASYVNKKG-RVFQVGIISWGVKDICKGS--SKKFTSDADS-----RDYHSNLF
CycaB2  CIEEAR-STIPEHSTASLSEVITDRFMCTGGTEKNRVQLTCKGDSGGPLFLRKRM-RYFQVAVVSWGTKQICDS--QTDVKEWPQDA-----RDFHISVF
OnmyB1  CIEMAVT-EVEGITPLNLKDIVTDNFLCTGGKQPTRDNVACKGDSGGAVFMDYDKHRTIQVGVISWGTKDLCPGGNSDIKQESSEKS-----RDFHINLF
OnmyB2  CIELAT--KVDGITSDNLKDVVTENFLCSGGRQPTRDHVACKGDSGGALFKNYDDYRTIQVGLISWGTKNLCPDGDNDIQQESSDDS-----RDFHINLF
DareB   CVEDAK--KAKESKWQMRRRQLQKISCGSGGNQPQRDDVSCKGESGGATHVDKYG-RLIQIGVVSWGVKNLCSKKR-NLMQFSVSDS-----RDYHINPF
TrscB   CEANALQAP-EYKNMSHVHLVVTDRFLCTGGHEPVVEKVSCKGDSGGALFIQKKR-RYIQVGVLSWGVINSCDN-----VALKEKYA-----RDFHINLF
LajaB   CLKDIM--ARFNVTEEKAEKHITENMLCAWNAT----ADTCRGDSGGPLVLQKNR-RWIQVGIVAGGVAQHCG---------KNIKS------SFYTNVA
StpuB   CSIAMS---------AHGIAVDTTTELCAGIER----KDSCQGDSGGPLVVQRNN-KYRQIGIVSYGIG--CG---------VTYG-------VYTRVP
HosaC2  CAEVVSQEKTMFPNLTDVREVVTDQFLCSGTQE---DESPCKGESGGAVFLERRF-RFFQVGLVSWGLYNPCLGSADKNSRKRAPRSKVPPPRDFHINLF
MumuC2  CIQAVSQNKNIFPSLTNVSEVVTDQFLCSGMEEE--DDNPCKGESGGAVFLGRRY-RFFQVGLVSWGLFDPCHGSSNKNLRKKPPRG--VLPRDFHISLF
        *                                       * * ***            *      *    *
```

Fig. 1. Amino acid alignment of part of the serine protease domain of B and C2. The entire amino acid sequences of B and C2 of various species are aligned using the Clustal w software (Thompson et al. 1994). See text for references for the aligned sequences. Only part of the serine protease domain is shown here. #, The catalytic serine residue. The aspartic acid residues are the putative determinant of the trypsinlike specificity of the prototype (!) and B/C2-specific ($) serine protease domains, respectively; *asterisks*, the completely conserved residues; *at top*, residue numbers of the human B sequence. *Hosa*, *Homo sapiens* (human); *Mumu*, *Mus musculus* (mouse); *Xela*, *Xenopus laevis* (toad); *Orla*, *Oryzias latipes* (Medaka fish); *Cyca*, *Cyprinus carpio* (carp); *Onmy*, *Onchorhynchus mykiss* (trout); *Dare*, *Danio rerio* (zebrafish); *Trsc*, *Triakis scyllia* (shark); *Laja*, *Lampetra japonica* (lamprey); *Stpu*, *Strongylocentrotus purpuratus* (sea urchin)

Chaplin et al. 1983), and the distance from the poly-A addition site of the C2 gene to the transcriptional start site of the B gene is less than 500bp (Wu et al. 1987; Nonaka et al. 1989). This may reflect tandem duplication of the ancestral single gene. From a functional point of view, the presence of both the CP, defined as an antibody-dependent, Ca^{2+}- and Mg^{2+}-requiring pathway, and the AP, antibody-independent and requiring only Mg^{2+}, has been demonstrated in all jawed vertebrate classes (Dodds and Day 1993). However, the evidence of B/C2 gene duplication is not clear at the lower vertebrate stage. A single B/C2-like sequence has been isolated from sea urchin (Smith et al. 1998), ascidian (Ji et al., unpublished), lamprey (Nonaka et al. 1994), Japanese dogfish (Terado et al., unpublished), Medaka fish (Kuroda et al. 1996), and zebrafish (Seeger et al. 1996), and all of these sequences show almost the same similarity to the mammalian B and C2 sequences.

In contrast, two B/C2-like sequences were identified from trout (Sunyer et al. 1998b), carp (Nakao et al. 1998), and *Xenopus* (Kato et al. 1995). The amino acid sequence identity between the two genes in trout, carp, and *Xenopus* is 75%, 26%, and 82%, respectively. Thus duplication of trout and *Xenopus* seems to be a recent event, whereas duplication of carp seems to be quite ancient. Since all these species are tetraploid, it seems likely that duplication of trout and *Xenopus* is the result of tetraploidization. However, this possibility is not supported by the results of linkage analysis performed in *Xenopus*. Two *Xenopus* B/C2-like genes show a close linkage, indicating that tandem duplication rather than tetraploidization is a probable molecular basis (Kato et al. 1995). Two trout and *Xenopus* sequences show a higher amino acid identity to mammalian B than to mammalian C2, whereas two carp sequences show an almost identical similarity to mammalian B and C2. The phylogenetic tree analysis shown in Fig. 2 suggests that the putative B/C2 gene duplication (indicated by the filled circle) is located after the divergence of cyclostomes but before the divergence of cartilaginous fish. The conclusion that the divergence of cyclostomes predated this gene duplication is supported by a high bootstrap percentage. On the other hand, bootstrap percentages to support the later branching pattern are not high enough, and it is still not established that this gene duplication predated the divergence of cartilaginous fish. In this context, it is interesting to note that the major one of the two trout B/C2-like molecule plays a dual role in both the CP and AP (Sunyer et al. 1998b). In contrast, it is established that the C3/C4 gene duplication event occurred before the divergence of cartilaginous fish, since both C3 and C4 have been identified in shark (Dodds et al. 1998) and Medaka fish (Kuroda et al., 2000).

7 Evolution of MASP-1, MASP-2, C1r, and C1s

Mammalian MASP-1 (Sato et al. 1994), MASP-2 (Thiel et al. 1997), C1r (Leytus et al. 1986), and C1s (Tosi et al. 1987) constitute another serine protease family

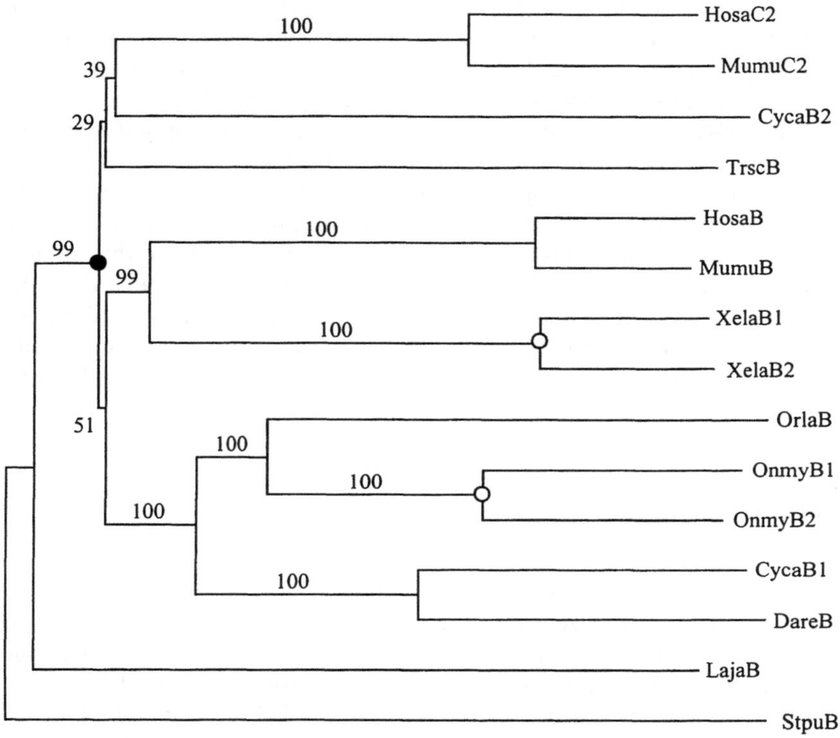

Fig. 2. Phylogenetic tree of B and C2. The phylogenetic tree is constructed using the Neighbor-joining method (SAITOU and NEI 1987) based on the alignment made by Clustal w. *Numerals on the branches*, bootstrap percentages to support the given partitioning; *filled circle*, a putative B/C2 gene duplication; *open circles*, gene duplication within the species

specific to the complement system. The domain structure of this family member is from the N-terminus; CUB, EGF-like, CUB, SCR, and serine protease domain. Functionally, C1r and C1s belong to the CP, and MASP-1 and MASP-2 to the LeP. From the structure of the serine protease domain, however, C1r, C1s, and MASP-2 belong to a specialized group, and MASP-1 belongs to another common group. Unique features of the serine protease domain of the former group compared to other serine proteases are (a) the entire serine protease domain is encoded by a single exon, (b) the disulfide bond called "histidine loop" is missing, and (c) the catalytic serine is encoded by the AGY codon instead of the more common TCN codon. Since these characteristics are considered to be a derived feature, and it is highly unlikely that all these three changes occurred independently twice during evolution, C1r, C1s, and MASP-2 most probably shared a common ancestor after diverging from MASP-1. Phylogenetic studies revealed the complicated evolutionary history of these proteins. Two ascidian clones, termed AsMASPa and AsMASPb, had the TCN-type serine protease domain and both AsMASPa and AsMASPb were ancestral molecules before the divergence of MASP-1 from C1r/C1s/MASP-2 (Ji et al. 1997). On the other hand, MASP molecules identified from lamprey, shark and carp had AGY-type serine protease domain (Endo et al. 1998). Both MASP-1 and MASP-2 were identified from *Xenopus*, indicating that the divergence between MASP-1 and MASP-2 predated the divergence between amphibia and mammals (Endo et al. 1998). The unusual point about this phylogenetic study is that the TCN-type molecules, found in ascidian and considered to be a prototype, are missing between the cyclostomes and bony fish, and appear again in amphibia. Since it is highly unlikely that previously lost introns are reinserted, and that the histidine loop and the TCN codon for catalytic serine are recovered at the same time, the TCN-type serine protease domain should be present in cyclostome, cartilaginous fish, and bony fish in some form. These genes could simply be difficult to identify due to low expression, or these domain might be present as a part of pseudogenes.

8 Conclusion

The hypothetical story of the origin and evolution of the complement system has been composed from data discussed above (Fig. 3). The birth of the complement system is located somewhere during evolution of invertebrate phyla, although the exact stage at which the complement system was established must still be clarified. The original complement system most probably resembled the LeP and AP of the mammalian complement system. Lectin-based recognition by MBL, followed by the activation of MASP, B, and then C3, covalent binding of C3 to foreign particles, and finally phagocytosis of C3-tagged foreign particles by phagocytes with C3 receptors on their surface seem to be the most probable constitution of the original complement system. Although MBL and C3 receptor are still to be identified, MASP, B, and C3 have been identified from ascidian and lamprey, and B and C3 have been identified from sea urchin. Thus it is conceivable that the original complement system, once established, continued until the Agnatha stage without major changes. This hypothesis is supported not only by the composition of the components elucidated by cDNA cloning study but also by the structure of the serine protease domain of the B protein.

The appearance of cartilaginous fish was landmark event in the history of the immune system since all aspects of adaptive immunity started at this stage. The complement system also experienced a major change between Agnatha and cartilaginous fish. Both the CP and the cytolytic pathway seem to have emerged at this stage, and the B protein seems to acquire its specialization at this stage too. The reason why such a change in the immune system occurred at this stage is not fully understood yet. However, duplication of the entire genome by tetraploidization seemed to have played a major role in this immunological revolution. MHC may have had a role in developing the immune system, although this point would need to be clarified by future studies.

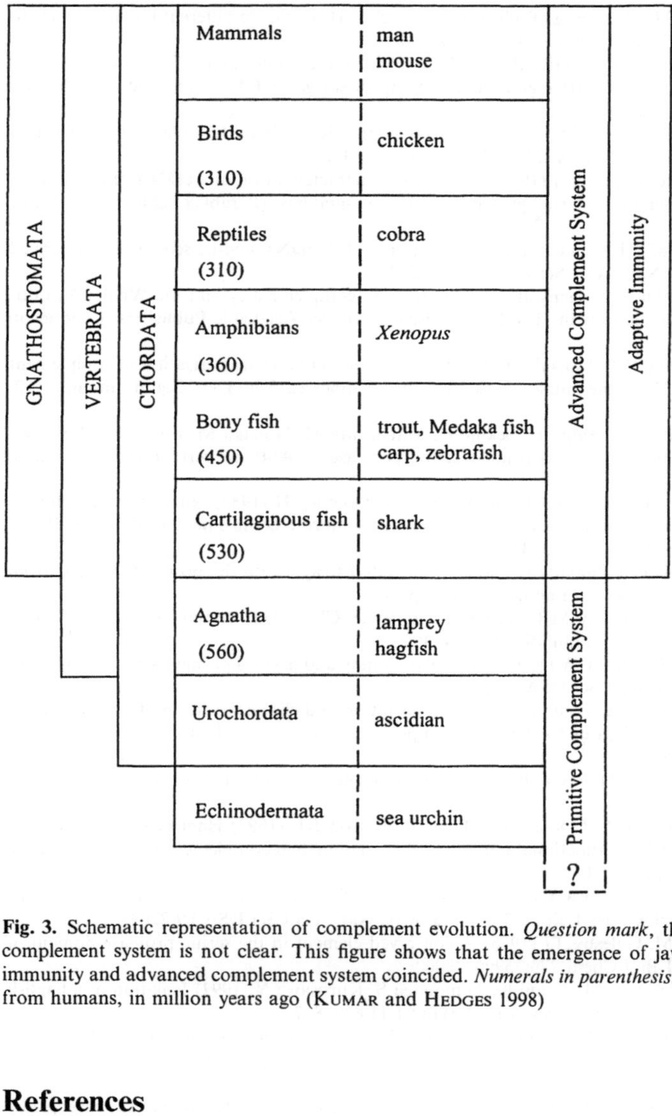

Fig. 3. Schematic representation of complement evolution. *Question mark*, the origin of the primitive complement system is not clear. This figure shows that the emergence of jawed vertebrates, adaptive immunity and advanced complement system coincided. *Numerals in parenthesis*, separation of each group from humans, in million years ago (KUMAR and HEDGES 1998)

References

Al-Sharif WZ, Sunyer JO, Lambris JD, Smith LC (1998) Sea urchin coelomocytes specifically express a homologue of the complement component C3. J Immunol 160:2983–2997

Auerbach HS, Burger R, Dodds A, Colten HR (1990) Molecular basis of complement C3 deficiency in guinea pigs. J Clin Invest 86:96–106

Bartl S, Weissman IL (1994) Isolation and characterization of major histocompatibility complex class IIB genes from the nurse shark. Proc Natl Acad Sci USA 91:262–266

Bingulac-Popovic J, Figueroa F, Sato A, Talbot WS, Johnson SL, Gates M, Postlethwait JH, Klein J (1997) Mapping of *Mhc* class I and class II regions to different linkage groups in the zebrafish, *Danio rerio*. Immunogenetics 46:129–134

Postlethwait JH, Klein J (1997) Mapping of mhc class I and class II regions to different linkage groups in the zebrafish, *Danio rerio*. Immunogenetics 46:129–134

Carroll MC, Campbell RD, Bentley DR, Porter RR (1984) A molecular map of the human major histocompatibility complex class III region linking complement genes C4, C2 and factor B. Nature 307:237–241

Chaplin DD, Woods DE, Whitehead AS, Goldberger G, Colten HR, Seidman JG (1983) Molecular map of the murine S region. Proc Natl Acad Sci USA 80:6947–6951

Dahmen A, Kaidoh T, Zipfel PF, Gigli I (1994) Cloning and characterization of a cDNA representing a putative complement-regulatory plasma protein from barred sand bass (Parablax neblifer). Biochem J 301:391–397

de Bruijn MH, Fey GH (1985) Human complement component C3: cDNA coding sequence and derived primary structure. Proc Natl Acad Sci USA 82:708–712

Dodds AW, Day AJ (1993) The phylogeny and evolution of the complement system. In: Whaley K, Loos M, Weiler JM (eds) Complement in Health and Disease. Kluwer Academic Publishers, London pp 39–88

Dodds AW, Smith SL, Levine RP, Willis AC (1998) Isolation and initial characterisation of complement components C3 and C4 of the nurse shark and the channel catfish. Dev Comp Immunol 22: 207–216

Endo Y, Takahashi M, Nakao M, Saiga H, Sekine H, Matsushita M, Nonaka M, Fujita T (1998) Two lineages of mannose-binding lectin-associated serine protease (MASP) in vertebrates. J Immunol 161:4924–4930

Fey GH, Lundwall A, Wetsel RA, Tack BF, de Bruijn MH, Domdey H (1984) Nucleotide sequence of complementary DNA and derived amino acid sequence of murine complement protein C3. Philos Trans R Soc Lond B Biol Sci 306:333–344

Faljnik MF, Ohta Y, Namikawa-Yamada C, Nonaka M (1999) Insight into the primordial MHC from studies in ectothermic vertebrates. Immunol Rev in press

Fritzinger DC, Petrella EC, Connelly MB, Bredehorst R, Vogel CW (1992) Primary structure of cobra complement component C3. J Immunol 149:3554–3562

Fritzinger DC, Bredehorst R, Vogel CW (1995) Molecular cloning and derived primary structure of cobra venom factor. Proc Natl Acad Sci USA 91:12775–12779

Fujii T, Nakamura T, Sekizawa A, Tomonaga S (1992) Isolation and characterization of a protein from hagfish serum that is homologous to the third component of the mammalian complement system. J Immunol 148:117–123

Greenhalgh P, Steiner LA (1995) Recombination activating gene 1 (Ragl) in zebrafish and shark. Immunogenetics 41:54–55

Hanley PJ, Hook JW, Raftos DA, Gooley AA, Trent R, Raison RL (1992) Hagfish humoral defense protein exhibits structural and functional homology with mammalian complement components. Proc Natl Acad Sci USA 89:7910–7914

Hashimoto K, Nakanishi T, Kurosawa Y (1992) Identification of a shark sequence resembling the major histocompatibility complex class I alpha 3 domain. Proc Natl Acad Sci USA 89:2209–2212

Hourcade DE, Mitchell LM, Oglesby TJ (1998) A conserved element in the serine protease domain of complement factor B. J Biol Chem 273:25996–26000

Ishiguro H, Kobayashi K, Suzuki M, Titani K, Tomonaga S, Kurosawa Y (1992) Isolation of a hagfish gene that encodes a complement component. EMBO J 11:829–837

Ishikawa N, Nonaka M, Wetsel RA, Colten HR (1990) Murine complement C2 and factor B genomic and cDNA cloning reveals different mechanisms for multiple transcripts of C2 and B. J Biol Chem 265:19040–19046

Ji X, Azumi K, Sasaki M, Nonaka M (1997) Ancient origin of the complement lectin pathway revealed by molecular cloning of mannan binding protein-associated serine protease from a urochordate, the Japanese ascidian, *Halocynthia roretzi*. Proc Natl Acad Sci USA 94:6340–6345

Kasahara M, Vazquez M, Sato K, McKinney EC, Flajnik MF (1992) Evolution of the major histocompatibility complex: isolation of class II A cDNA clones from the cartilaginous fish. Proc Natl Acad Sci USA 89:6688–6692

Kato Y, Salter-Cid L, Flajnik MF, Kasahara M, Namikawa C, Sasaki M, Nonaka M (1994) Isolation of the Xenopus complement factor B complementary DNA and linkage of the gene to the frog MHC. J Immunol 153:4546–4554

Kato Y, Salter-Cid L, Flajnik MF, Namikawa C, Sasaki M, Nonaka M (1995) Duplication of the MHC-linked Xenopus complement factor B gene. Immunogenetics 42:196–203

Kumar S, Hedges SB (1998) A molecular timescale for vertebrate evolution. Nature 392:917–920

Kuroda N, Wada H, Naruse K, Simada A, Shima A, Sasaki M, Nonaka M (1996) Molecular cloning and linkage analysis of the Japanese medaka fish complement Bf/C2 gene. Immunogenetics 44:459–467
Kuroda N, Naruse K, Shima A, Nonaka M, Sasaki M, Nonaka M (2000) Molecular cloning and linkage analysis of complement C3 and C4 genes of the Japanese medaka fish. Immunogenetics in press
Kusano M, Choi NH, Tomita M, Yamamoto K, Migita S, Sekiya T, Nishimura S (1986) Nucleotide sequence of cDNA and derived amino acid sequence of rabbit complement component C3 alpha-chain. Immunol Invest 15:365–378
Lambris JD, Lao Z, Pang J, Alsenz J (1993) Third component of trout complement. cDNA cloning and conservation of functional sites. J Immunol 151:6123–6134
Lambris JD, Pappas J, Mavroidis M, Wang Y, Manzone H, Schwager J, Du Pasquier L, Silibovsky R (1995) The third component of Xenopus complement: cDNA cloning, structural and functional analysis, and evidence for an alternate C3 transcript. Eur J Immunol 25:572–578
Leytus SP, Kurachi K, Sakariassen KS, Davie EW (1986) Nucleotide sequence of the cDNA coding for human complement C1r. Biochemistry 25:4855–4863
Litman GW, Rast JP, Shamblott MJ, Haire RN, Hulst M, Roess W, Litman RT, Hinds-Frey KR, Zilch A, Amemiya CT (1993) Phylogenetic diversification of immunoglobulin genes and the antibody repertoire. Mol Biol Evol 10:60–72
Mavroidis M, Sunyer JO, Lambris JD (1995) Isolation, primary structure, and evolution of the third component of chicken complement and evidence for a new member of the alpha 2-macroglobulin family. J Immunol 154:2164–2174
Medzhitov R, Janeway CA Jr (1997) Innate immunity: the virtues of a nonclonal system of recognition. Cell 91:295–298
Misumi Y, Sohda M, Ikehara Y (1990) Nucleotide and deduced amino acid sequence of rat complement C3. Nucleic Acids Res 18:2178
Mole JE, Anderson JK, Davison EA, Woods DE (1984) Complete primary structure for the zymogen of human complement factor B. J Biol Chem 259:3407–34012
Nakao M, Fushitani Y, Fujiki K, Nonaka M, Yano T (1998) Two diverged complement factor B/C2-like cDNA sequences from a teleost, the common carp (*Cyprinus carpio*). J Immunol 161:4811–4818
Narayana SVL, Jing H, Moore D, Macon KJ, Ma YX, Volanakis JE (1998) Structure of factor B serine protease domain. Mol Immunol 35:398
Nonaka M (1988) Phylogeny of the complement system. In: Volanakis JE, Frank MM (eds) The Human Complement System in Health and Disease. Marcel Dekker, Inc. New York pp 203–215
Nonaka M, Takahashi M (1992) Complete complementary DNA sequence of the third component of complement of lamprey. Implication for the evolution of thioester containing proteins. J Immunol 148:3290–3295
Nonaka M, Fujii T, Kaidoh T, Natsuume-Sakai S, Nonaka M, Yamaguchi N, Takahashi M (1984) Purification of a lamprey complement protein homologous to the third component of the mammalian complement system. J Immunol 133:3242–3249
Nonaka M, Gitlin JD, Colten HR (1989) Regulation of human and murine complement: comparison of 5′ structural and functional elements regulating human and murine complement factor B gene expression. Mol Cell Biochem 89:1–14
Nonaka M, Takahashi M, Sasaki M (1994) Molecular cloning of a lamprey homologue of the mammalian MHC class III gene, complement factor B. J Immunol 152:2263–2269
Nonaka M, Kuroda N, Naruse K, Shima A (1998) Molecular genetics of the complement C3 convertases in lower vertebrates. Immunol Rev 166:59–65
Nonaka M, Azumi K, Ji X, Namikawa-Yamada C, Sasaki M, Saiga H, Dodds AW, Sekine H, Homma MK, Matsushita M, Endo Y, Fujita T (1999) Opsonic complement component C3 in the solitary ascidian, *Halocynthia roretzi*. J Immunol 162:387–391
Rast JP, Anderson MK, Strong SJ, Luer C, Litman RT, Litman GW (1997) alpha, beta, gamma, and delta T cell antigen receptor genes arose early in vertebrate phylogeny. Immunity 6:1–11
Saitou N, Nei M (1987) The neighbor-joining method: a new method for reconstructing phylogenetic trees. Mol Biol Evol 4:406–425
Sarrias MR, Zarkadis I, Sunyer JO, Lambris JD (1998) Cloning of three trout C3 isoforms: structural, functional, and phylogenetic analysis. Mol Immunol 35:370
Sato T, Endo Y, Matsushita M, Fujita T (1994) Molecular characterization of a novel serine protease involved in activation of the complement system by mannose-binding protein. Int Immunol 6: 665–669
Seeger A, Mayer WE, Klein J (1996) A complement factor B-like cDNA clone from the zebrafish (*Brachydanio rerio*). Mol Immunol 33:511–520

Smith LC, Chang L, Britten RJ, Davidson EH (1996) Sea urchin genes expressed in activated coelomocytes are identified by expressed sequence tags. Complement homologues and other putative immune response genes suggest immune system homology within the deuterostomes. J Immunol 156:593–602

Smith LC, Shin C-S, Dachenhausen SG (1998) Coelomocytes express SpBf, a homologue of factor B, the second component in the sea urchin complement system. J Immunol 161:6784–6793

Stanley KK, Herz J (1987) Topological mapping of complement component C9 by recombinant DNA techniques suggests a novel mechanism for its insertion into target membranes. EMBO J 6:1951–1957

Sunyer JO, Tort L, Lambris JD (1997) Structural C3 diversity in fish: characterization of five forms of C3 in the diploid fish Sparus aurata. J Immunol 158:2813–2821

Sunyer JO, Zarkadis IK, Lambris JD (1998a) Complement diversity: a mechanism for generating immune diversity? Immunol Today 19:519–523

Sunyer JO, Zarkadis I, Sarrias MR, Hansen JD, Lambris JD (1998b) Cloning, structure, and function of two rainbow trout Bf molecules. J Immunol 161:4106–4114

Thiel S, Vorup-Jensen T, Stover CM, Schwaeble W, Laursen SB, Poulsen K, Willis AC, Eggleton P, Hansen S, Holmskov U, Reid KB, Jensenius JC (1997) A second serine protease associated with mannan-binding lectin that activates complement. Nature 386:506–510

Thompson JD, Higgins DG, Gibson TJ (1994) CLUSTAL W: improving the sensitivity of progressive multiple sequence alignment through sequence weighting, position-specific gap penalties and weight matrix choice. Nucleic Acids Res 22:4673–4680

Tosi M, Duponchel C, Meo T, Julier C (1987) Complete cDNA sequence of human complement Cls and close physical linkage of the homologous genes Cls and Clr. Biochemistry 26:8516–8524

Volanakis JE (1998) Overview of the complement system. In: Volanakis JE, Frank MM (eds) The Human Complement System in Health and Disease. Marcel Dekker, Inc. New York pp 9–32

Wu LC, Morley BJ, Campbell RD (1987) Cell-specific expression of the human complement protein factor B gene: evidence for the role of two distinct 5′-flanking elements. Cell 48:331–342

Major Vertebrate Evolutionary Issues

Genome Paralogy: A New Perspective on the Organization and Origin of the Major Histocompatibility Complex

M. Kasahara

1 Introduction

The coordinated international project to sequence the entire human major histocompatibility complex (MHC) is approaching its end. We will probably see the complete nucleotide sequence of the mouse MHC in the near future. Thus, for the time being, the MHC is likely to remain one of the best characterized genetic regions in the mammalian genome.

The MHC of nonmammalian vertebrates also enjoys such a privileged status. Thanks to the efforts of a number of laboratories we now know the basic genomic organizations of the MHC in most classes of jawed vertebrates (Kaufman et al. 1995; Trowsdale 1995; Du Pasquier and Flajnik 1998; Klein and Sato 1998; Flajnik et al. 1999). In only a few other genetic systems is such extensive information encompassing phylogenetically diverse organisms available. Therefore the

Department of Biosystems Science, School of Advanced Sciences, Graduate University for Advanced Studies, and Core Research for Evolutionary Science and Technology, JST, Hayama 240-0193, Japan
e-mail: kasahara@soken.ac.jp

MHC holds the promise of serving as a paradigm for understanding the long-term evolutionary dynamics of the megabase-scale region.

Until recently, however, this potential had not been fully exploited. Almost 3 years ago two groups of investigators (Kasahara et al. 1996; Katsanis et al. 1996) discovered independently that the mammalian genome contains regions paralogous to the MHC. This discovery has offered a new clue for understanding the genome dynamics of the MHC (Kasahara 1999). At the same time it has opened up new avenues for exploring the origin of the adaptive immune system and genome evolution in vertebrates (Kasahara et al. 1997; Kasahara 1998). In this chapter I discuss the structure, evolution, and origin of the MHC from the viewpoint of genome paralogy.

2 What Is Genome Paralogy?

The term "paralogous" is used to denote genes within the same species that are descended from a common ancestral gene. For example, the human genome contains four structurally related *NOTCH* genes designated *NOTCH1–NOTCH4* which are paralogous to one another. The term contrasting with "paralogous" is "orthologous", with the latter referring to genes that diverged by speciation events. Thus the human *NOTCH4* gene is orthologous to the mouse *Notch4* gene.

Scrutiny of the vertebrate genome occasionally reveals that closely linked sets of paralogous genes are located on more than two chromosomal segments (Nadeau and Kosowsky 1991; Lundin 1993). This phenomenon is referred to as genome paralogy, and such chromosomal segments are called paralogous regions.

A classic example of genome paralogy is provided by the *HOX* gene clusters and their adjacent regions located on human chromosomes 2, 7, 12, and 17 (Ruddle et al. 1994). Here not only are the *HOX* gene clusters quadruplicated, but many of the adjacent genes have paralogous copies on chromosomes 2, 7, 12, and 17. It is generally thought that this paralogous group emerged early in vertebrate evolution by genome-wide or large-scale chromosomal duplications (Holland et al. 1994; Ruddle et al. 1994; Finnerty and Martindale 1998). However, not all paralogous regions in the vertebrate genome are of such ancient origin. Recent evidence indicates that pericentromeric and subtelomeric regions of human chromosomes occasionally contain the paralogous segments of relatively small size (approx. 50kb) that are of recent origin (Eichler 1998).

3 MHC and Genome Paralogy

Kasahara et al. (1996) and Katsanis et al. (1996) noticed independently that the human and mouse genomes contain regions paralogous to the MHC. We came to this conclusion from the study of the interferon-γ (IFN-γ) inducible subunits of the proteasome, a multisubunit protease responsible for the production of peptides presented by MHC class I molecules (Tanaka and Kasahara 1998). Katsanis et al. (1996) reached the same conclusion from the analysis of *PBX* (pre-B cell leukemia transcription factor) and *NOTCH* genes. Initially only a handful of paralogous genes were identified. Since then the number of the members constituting the MHC paralogous group has increased steadily. The current listing (Fig. 1) contains more than 20 gene families. It is now clear that the three regions in the human genome, 9q33–q34, 1q21–q25/1p11–p32, and 19p13.1–p13.3, are paralogous to the MHC located on 6p21.3. Because the currently available database does not contain full sequence information on the MHC class I and class III regions, the membership of the MHC paralogous group is likely to increase further when the full gene contents of these regions become available.

Traditionally, the human MHC has been estimated to span approx. 4Mb, of which the class II, class III, and class I regions are thought to occupy approx. 800kb, 1100kb, and 1800kb, respectively (Campbell and Trowsdale 1997). Recently a nonclassical class I gene, designated *HFE*, was discovered about 4Mb telomeric from *HLA-F* (Feder et al. 1996), the class I gene located at the telomeric end of the traditionally defined class I region. Thus, if we define the class I region as a contiguous DNA segment containing class I genes, the size of the MHC increases almost twofold. The replication-dependent histone gene cluster in the vicinity of the *HFE* gene (*H1*, *H2A*, *H2B*, *H3*, and *H4*) has a paralogous cluster on 1q21. Likewise, the *HFE* region-encoded zinc finger genes *ZNF184* and *KIAA0426* have closely related paralogous copies designated *ZNF85* and *ZNF91* in the 19p12–p13.1 region. Thus the paralogous segment on human chromosome 6 appears to extend to the *HFE* region. On the centromeric side of the MHC, *RPL12L* has a paralogous copy designated *RPL12* on 9q34 (UniGene Hs. 119590). Similarly, *BING1* has a paralogous copy termed *ZNF-X* (*KIAA0414*) on 9q34 (UniGene Hs. 127649). Therefore the paralogous segment on human chromosome 6 appears to cover the entire region of the extended MHC and to span at least some 8Mb.

4 Origin of the MHC Paralogous Group

Most of the gene families which belong to the MHC paralogous group are of relatively small size, normally with two to four members. Furthermore, in many cases all of the members map to the paralogous regions. These observations indicate that the four regions MHC, 9q33–q34, 1q21–q25/1p11–p32, and 19p13.1–p13.3 share ancestry and have diverged by chromosomal duplications. At least two rounds of duplication are required to create four paralogous regions. Phylogenetic analysis of the genes coding for tenascins (*TNX*, *TNR*, *HXB*), PBX, type V/XI collagens (*COL*), retinoid X receptors (*RXR*), and NOTCH indicates that the copies located on chromosomes 1 and 9 are more closely related to each other

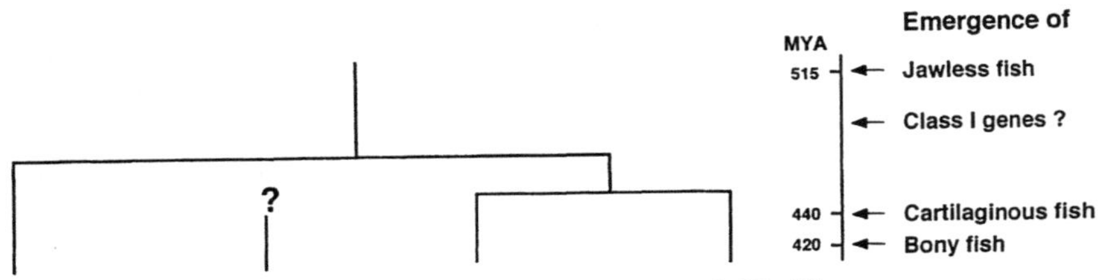

6p21.3 (*HLA*)	19p13.1-p13.3	1q21-q25/ p11-p32	9q33-q34	
RPL12L	-	-	*RPL12*	Class II region
BING1	-	-	*ZNF-X*	
RALGDS2	-	-	*RALGDS*	
RXRB	-	*RXRG*	*RXRA*	
COL11A2	-	*COL11A1*	*COL5A1*	
RING3	*HUNKI*	*BRDT*	*RING3L*	
LMP	-	-	*PSMB7*	
TAP	-	-	*ABC2?*	
NOTCH4	*NOTCH3*	*NOTCH2*	*NOTCH1*	Class III region
PBX2	-	*PBX1*	*PBX3*	
LPAATA	-	-	*LPAATB*	
PPT2	-	*PPT*	-	
CREBL1	-	*ATF6*	-	
TNX	-	*TNR*	*HXB*	
C4	*C3*	-	*C5*	
G9A	-	*KIAA0067*	-	
HSPA1B *HSPA1A* *HSPA1L*	-	*HSPA6?* *HSPA7?*	*HSPA5?*	
BAT2	-	-	*BAT2-like*	
AIF1	-	-	*AIF1-like*	
LTB *TNF* *LTA*	*CD70* *CD137L*	*APT1LG1* *OX40L*	*CD30L*	
BAT1	*BAT1-like*	-	-	
NTRK4	-	*NTRK1*	*NTRK2*	Class I region
TUBB	*TUBB5*	-	*TUBB2*	
MHC class I *HFE*	*FCGRT?*	*CD1* *HLALS?*	-	
H1, H2A, H2B, *H3, H4*	-	*H2A, H2B, H3, H4*	-	
ZNF184, KIAA0426	*ZNF85, ZNF91*	-	-	
OLFR2	*OLFR*	-	*OLFR3*	
-	*CACNL1A4*	*CACNL1A6*	*CACNL1A5*	Paralogous genes with no copy in the MHC
-	*VAV1*	*VAV3*	*VAV2*	

than they are to the corresponding copies on 6p21.3 (KASAHARA 1999). Thus it appears that the paralogous regions on chromosomes 1 and 9 shared an immediate common ancestral region (Fig. 1). The relationship of 19p13.1–p13.3 to the other paralogous regions is less clear (KASAHARA 1999). If 19p13.1–p13.3 and 6p21.3 share an immediate common ancestral region, as suggested by some of the gene families, the existence of the four paralogous regions would be accounted for by postulating two rounds of chromosomal duplication. If this is not the case, one must postulate a more complex scenario: at least three rounds of duplication and cluster loss events.

There is now considerable evidence that the duplication involving the MHC and the presumed common ancestral region of 1q21–q25/1p11–p32 and 9q33–q34 took place most likely in a common ancestor of jawed vertebrates after its divergence from the lineage leading to the extant jawless fishes (KASAHARA 1999). Analysis of the *CACNL1A* gene family with copies on chromosomes 1, 9, and 19 (Fig. 1) indicates that the paralogous regions on chromosomes 1 and 9 diverged before the emergence of the cartilaginous fish (KASAHARA 1999). Thus it appears likely that the two rounds of duplication in Fig. 1 took place in a common ancestor of jawed vertebrates.

5 Chromosomal Duplication and the Emergence of Adaptive Immunity

Of particular interest in this regard is the fact that the adaptive immune system characterized by the MHC and antigen-specific rearranging receptors has been identified only in jawed vertebrates, thus in the organisms that are thought to have experienced the chromosomal duplications discussed above (KASAHARA et al. 1995; THOMPSON 1995; LITMAN 1996; SCHLUTER et al. 1997; DU PASQUIER and FLAJNIK

Fig. 1. The MHC paralogous group in the human genome and the presumed relationship of the paralogous regions. The upper part of the diagram shows the presumed relationship of the regions paralogous to the MHC. The estimated time point when each class of vertebrates emerged is indicated in million years ago (*MYA*). The genes in the *HLA* complex are arranged from centromere (*top*) to telomere (*bottom*). *Hyphen*, the paralogous genes have not been identified; *question mark*, genes which are paralogous but might not have diverged as a result of the duplications that took place in a common ancestor of jawed vertebrates (ENDO et al. 1997; KASAHARA 1997; HUGHES 1998; KASAHARA 1999). *HLALS* is the gene originally described as *MR1* (HASHIMOTO et al. 1995). The class I gene coding for Zn-α_2-glycoprotein (*AZGP1*) maps to 7q22.1 (UEYAMA et al. 1993), thus outside the four paralogous regions. For detailed discussion on the class I genes encoded outside the MHC, see KASAHARA (1999). At least ten gene families share copies only in the paralogous regions other than the MHC (KASAHARA 1999). For lack of space, only two of them, *VAV* and *CACNL1A*, are shown. Note that human chromosome 1 contains paralogous regions on both arms, presumably because this chromosome has experienced a pericentric inversion. Detailed information on the genes can be obtained from The Online Mendelian Inheritance in Man (http://www3.ncbi.nlm.nih.gov/Omim/) or The UniGene Database (http://www.ncbi.nlm.nih.gov/UniGene/Hs.Home.html). Some of the gene symbols used in this figure are not officially approved. (Modified from KASAHARA 1999)

1998). A large number of accessory molecules employed by the adaptive immune system also appear to exist only in jawed vertebrates (Kasahara 1998).

Phylogenetic analysis indicates that several accessory molecules of the adaptive immune system owe their existence to the chromosomal duplications that formed the MHC paralogous group (Kasahara 1997, 1999). These include LMP2 and LMP7, the IFN-γ inducible proteasomal subunits that facilitate the production of class I binding peptides (Kandil et al. 1996; Nonaka et al. 1997b); C4 and C5, the components of the classical and lytic pathways of complement activation, respectively (Sunyer et al. 1998); the NOTCH1 molecule that plays a critical role in thymic development of T cells by affecting the CD4 versus CD8 as well as the αβ versus γδ lineage choice (Robey 1997); and the RXRB molecule that binds to the MHC class I enhancer A region and upregulates gene expression (Ting and Baldwin 1993).

The other gene families of immunological interest, the members of which might have diverged by the chromosomal duplications that formed the MHC paralogous group, are *TNF* (Spring 1997) and *VAV*. The *TNF* family has members on all of the four paralogous regions (gene symbols in parentheses): tumor-necrosis factor-α (*TNF*), lymphotoxin-α (*LTA*), and lymphotoxin-β (*LTB*) on 6p21.3; the CD30 ligand (*CD30L*) on 9q33; the Fas ligand (*APT1LG1*) and the OX40 ligand (*OX40L*) on 1q23; and the CD27 ligand (*CD70*) and the CD137 ligand (*CD137L*, also known as 4-1BBL) on 19p13 (Fig. 1). Of the ten known family members only the CD40 ligand and TWEAK (TNF-related weak inducer of apoptosis) are encoded outside the paralogous regions. VAV is a guanine nucleotide exchange factor that selectively activates the Rho family GTPase Rac1 (Cantrell 1998). Recent evidence indicates that the VAV1 molecule encoded in the 19p13.3–p13.2 region is essential for T- and B-cell development and activation (Fischer et al. 1995; Tarakhovsky et al. 1995; Zhang et al. 1995).

6 The Prototype of the MHC Inferred from the Ancestral Syntenic group

6.1 The Ancestral Syntenic Group

Figure 1 suggests that before the first round of chromosomal duplication the genetic region was formed that contained the precursors of the genes belonging to the MHC paralogous group. We call the constellation of such precursor genes the ancestral syntenic group. It is reasonable to assume that the prototype of the MHC was made up of the members of the ancestral syntenic group. The evidence accumulated thus far is largely consistent with this assumption. For example, Fig. 1 predicts that the MHCs of jawed vertebrates contain the genes coding for RING3 and LMP. This is indeed the case with the MHCs of the bony fish (Takami et al. 1997) and *Xenopus laevis* (Namikawa et al. 1995; Salter-Cid et al. 1996). Of course, genes can be lost from, inserted into, or even assembled in the MHC and its paralogous regions after the chromosomal duplications. The absence of the *LMP* genes in the chicken *B* complex presumably represents an example of gene loss (S. Milne, J. Kaufman, S. Beck; GenBank accession number AL023516). Secondary alterations of this kind must have also occurred in the human MHC. Full elucidation of the ancestral syntenic group therefore requires the information on the structures of the MHC in various classes of jawed vertebrates.

The concept of the ancestral syntenic group and genome paralogy has practical value in that it helps to identify new paralogous copies of a gene family as previously emphasized (KATSANIS et al. 1996). For example, only two *VAV* genes are currently known: *VAV1* on 19p13.3–p13.2 and *VAV2* on 9q34. While searching for new members of the MHC paralogous group, we found that a novel *VAV* gene, which we tentatively named *VAV3*, is located in the paralogous region on human chromosome 1 (Fig. 1).

6.2 Origin of the Class III Region

TRACHTULEC et al. (1997) found that the genes coding for the RXR-like, PBX-like, NOTCH-like, and tenascin-like molecules are located within the distance of 5.8Mb on chromosome III of the nematode *Caenorhabditis elegans*. In man these gene families have copies in the regions paralogous to the MHC, and three of them have copies in the MHC class III region (Fig. 1). Thus a primitive form of the ancestral syntenic group defined above appears to exist also in invertebrates. Because *C. elegans* has no MHC class I or class II genes, this finding indicates that the class III like region was assembled long before the emergence of class I and class II genes. Presumably, a considerable number of genes that now reside in the class III region were the original inhabitants of the chromosomal segment that eventually gave rise to the MHC.

6.3 The Prototype of the MHC

When we originally proposed the chromosomal duplication model (KASAHARA et al. 1996), it was not clear whether the preduplicated region contained class I or class II genes. Subsequently we noted that *CD1*, a member of the class I gene family, maps to one of the regions paralogous to the MHC and proposed that MHC class I and *CD1* diverged as a result of the chromosomal duplications that formed the MHC paralogous group (KASAHARA et al. 1997). One of the observations supporting this suggestion is that the membrane proximal domain of CD1 is equidistant to the corresponding domains of MHC class I and class II molecules (MARTIN et al. 1986). Because the cartilaginous fish have typical class I and class II genes, *CD1* is likely to have diverged from MHC class I before the emergence of this class of vertebrates. Failure to identify class I genes in jawless fishes suggests that this divergence probably occurred in a common ancestor of jawed vertebrates, thus at the time the duplications in Fig. 1 are predicted to have taken place. Therefore I presume that a class I gene emerged in the preduplicated region at some point in the stage of a common ancestor of jawed vertebrates (Fig. 1). What function this primitive class I gene performed is currently a matter of pure conjecture. It might have had the ability to bind peptides similar to MHC class I molecules, or to bind glycolipids similar to CD1 molecules. Alternatively, it might have had a totally different function. Whichever is the case, it appears likely that authentic MHC class I molecules emerged only after the chromosomal duplication.

Figure 2 shows the hypothetical model of MHC evolution deduced from the currently available information. One of the major uncertainties in this model is whether the preduplicated region contained class II genes. The apparent absence of class II genes in the regions paralogous to the MHC leaves three possibilities open. First, class II genes might have been assembled from class I genes within the MHC after the chromosomal duplication. Second, both class I and class II genes might have existed before the chromosomal duplication, with the latter then lost from the paralogous regions other than the MHC. Third, non-MHC-encoded class II genes might still exist in the regions paralogous to the MHC or elsewhere but have evaded detection thus far.

Figure 2 also shows the secondary alterations postulated to have taken place in the MHC in each major class of vertebrates. In mammals the *LMP* and *TAP* (transporter associated with antigen processing) genes appear to have been translocated from the class I to the class II region (Takami et al. 1997), as we predicted previously (Kasahara et al. 1995). Furthermore, the studies in *X. laevis* (Nonaka et al. 1997a) and chickens (Kaufman and Salomonsen 1997) suggest that, unlike the organization in the human and mouse MHCs, the class I and class II regions were adjacent to each other in the prototypic MHC. In the chicken the MHC is split into the two arms of the chromosome (Miller et al. 1996) and seems to have lost many resident genes, possibly because it resides on a microchromosome (Kaufman et al. 1995). In the amphibian *X. laevis* tetraploidization and subsequent differential silencing of duplicated genes appear to have taken place, thus placing some of the duplicated genes in the functional MHC chromosome and others in an inactivated MHC chromosome (Flajnik et al. 1999). In the teleost fish such as zebrafish, MHC class I and class II genes are fragmented to different chromosomes (Bingulac-Popovic et al. 1997; Graser et al. 1998), possibly because of an additional tetraploidization and subsequent gene silencing event (see below for further discussion). Currently there is little information on the genomic organization of the MHC in the cartilaginous fish. Preliminary evidence suggests that its class I and class II genes are linked to each other as in mammals (Flajnik et al. 1999). The overall impression gained from these discussion is that the major factors which made the MHCs of *X. laevis* and zebrafish look different from those of mammals are tetraploidization events presumed to have taken place exclusively in these animals.

Finally, there remains an important, but not well understood, issue concerning the origin of the MHC. Our model predicts that the accessory molecules such as LMP, VAV1, NOTCH1, TNF, and C4 and the Fas ligand emerged only after the

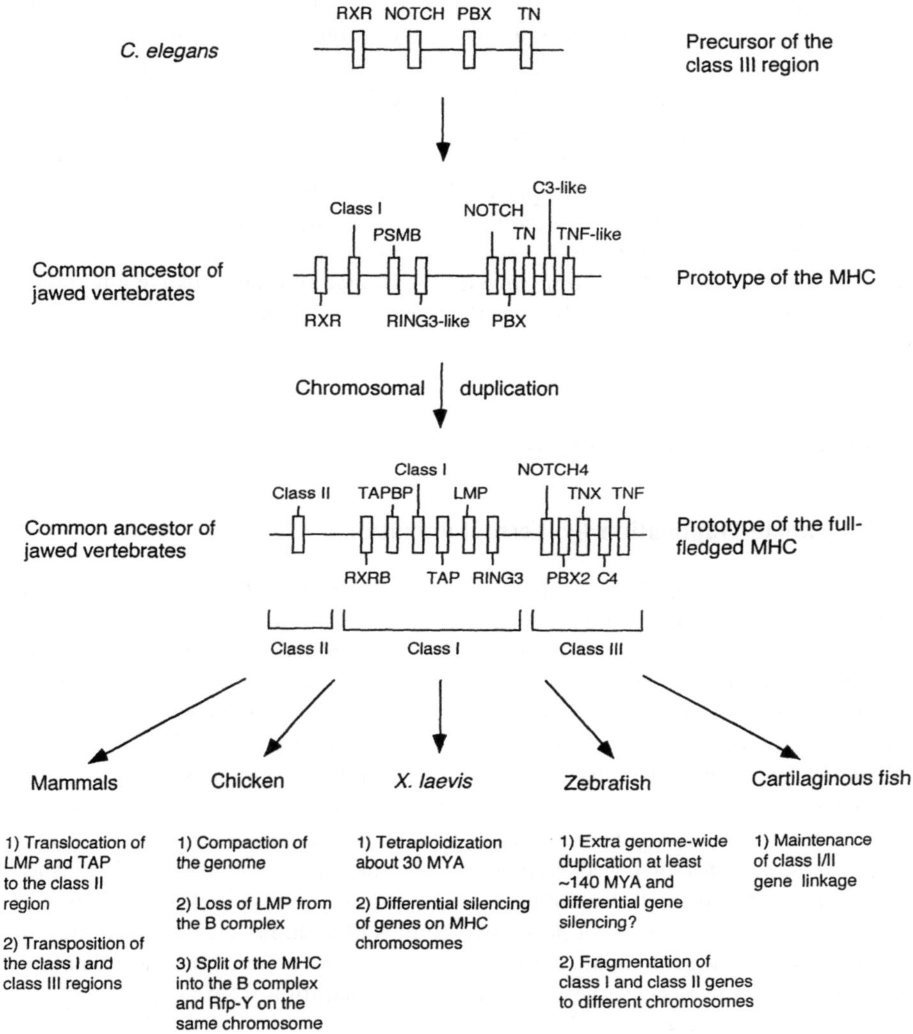

Fig. 2. The chromosomal duplication model of the MHC. The ancestral syntenic group contains more than 20 gene families (Fig. 1). For lack of space, this figure shows only some of them. No data are available on the MHC precursor region predicted to exist in the genome of jawless fishes. This region probably contains no class I, class II, *TAP*, or *TAPBP* (tapasin) genes, but otherwise might have an organization similar to that depicted for the prototype of the MHC. The genome of jawless fishes presumably did not experience the two rounds of duplication that led to the formation of the MHC paralogous group. However, this assumption does not rule out the possibility of lineage-specific duplications. The prototype of the MHC in its strict sense was formed probably in a common ancestor of jawed vertebrates. It presumably contained class I genes. Whether this region contained the class II, *TAP*, and/or *TAPBP* genes is not clear. We presume that the prototype of the full-fledged, functionally mature MHC, equipped with almost complete antigen-presentation machinery, emerged in a common ancestor of jawed vertebrates after its genome underwent the two rounds of chromosomal duplication. For gene symbols, see Fig. 1 and KASAHARA (1999). The order of genes shown in this figure is arbitrary. *PSMB*, the genes coding for the constitutively expressed β-type subunits (PSMB5, PSMB6, PSMB7) of the proteasome; *MYA*, million years ago

chromosomal duplication (Figs. 1, 2). This implies that the full-fledged adaptive immune system probably did not exist before the chromosomal duplication (Kasahara et al. 1997; Kasahara 1998). If so, it appears likely that the clustering of the genes in the prototypic MHC occurred by chance, or, if there were any reasons, for those unrelated to adaptive immunity. This explanation appears reasonable for the majority of the MHC-encoded genes that have copies in the paralogous regions. Less clear is the origin of the genes whose presence in the MHC is thought to be selectively advantageous. One such example is *LMP*. In this case the genes coding for the constitutively expressed proteasome subunits and primordial class I molecules might have come together by chance and their linkage might have gained selective advantages only after the chromosomal duplications.

7 Origin of Genome Paralogy

7.1 Genome Duplication by Tetraploidization

Nearly 30 years ago, the hypothesis was put forward by Ohno (1970) that the vertebrate genome underwent one or two rounds of tetraploidization at the stage of fish or amphibians. Recently this hypothesis was refined taking into account the rapidly accumulating gene mapping data and has become quite popular (Aparicio 1998). There is, however, considerable controversy as to the nature, number, and timing of the presumed genome doubling events (Skrabanek and Wolfe 1998). Lundin (1993) suggested that two or three rounds of genome-wide duplication took place early in vertebrate evolution. Sharman and Holland (1996) proposed that widespread gene duplication took place close to vertebrate origins and again in a common ancestor of jawed vertebrates after its separation from jawless fishes; they proposed that the latter duplication presumably involved tetraploidization of the genome. Sidow (1996) proposed two rounds of genome-wide duplication, one in a common ancestor of all jawed and jawless vertebrates, and the other in a common ancestor of jawed vertebrates after its separation from jawless fishes.

From the viewpoint of the origin of the MHC, it is noteworthy that both Sharman and Holland (1996) and Sidow (1996) proposed one round of genome-wide duplication in a common ancestor of jawed vertebrates. These proposals raise the possibility that the duplications responsible for the formation of the MHC paralogous group took place as part of the genome-wide duplication. As discussed above, our analysis of the MHC paralogous group suggests at least two rounds of chromosomal duplication in a common ancestor of jawed vertebrates (Fig. 1). Thus, if the proposals of Sidow (1996) and Sharman and Holland (1996) are correct, one of the chromosomal duplications in Fig. 1 might have been a regional one (Kasahara 1997). Alternatively, if both duplications in Fig. 1 were genome-wide ones, at least two rounds of tetraploidization rather than one round might have taken place in a common ancestor of jawed vertebrates.

7.2 An Additional Tetraploidization Event in the Teleost Fish?

The MHC of zebrafish, and presumably more generally of the teleost fish, is unique in that class I and class II genes map to different chromosomes (BINGULAC-POPOVIC et al. 1997). Until recently it was not known whether this type of organization reflects an ancestral or a derived one (KLEIN and SATO 1998). Preliminary evidence reported by FLAJNIK and coworkers (1999) suggests that class I and class II genes are linked in the cartilaginous fish. Thus it appears reasonable to assume that the two classes of MHC genes, which were linked in a common ancestor of jawed vertebrates, became fragmented to separate chromosomes in the lineage leading to the teleost fish.

Of interest in this regard is the recent observation by AMORES et al. (1998) that zebrafish have seven *HOX* clusters and hence three more clusters than do mammals. These authors suggest that this is the outcome of the chromosome doubling event that took place after the divergence of ray-finned and lobe-finned fishes but before the teleost radiation. They also suggest that this chromosome doubling event probably occurred as part of genome-wide duplication. If this is the case, fragmentation of class I and class II genes in zebrafish might have occurred as a result of the lineage-specific genome-wide duplication followed by the differential loss of duplicated genes.

8 Jaws and the Adaptive Immune System

In addition to the key constituent molecules, such as the rearranging receptors and MHC molecules, the adaptive immune system requires cooperative interactions of a large number of accessory molecules for its proper operation. Among them are LMP, proteasome activators, various coreceptors, transcription factors, and signal transducing molecules involved in lymphocyte activation and development. Accumulating evidence suggests that the complex network of protein interactions constituting the adaptive immune system emerged rather abruptly within about 100 million years in a common ancestor of jawed vertebrates. (We consider the possibility quite unlikely that this system was lost from jawless fishes because of their peculiar ways of life). Figure 1 suggests that the two rounds of chromosomal duplication, presumed to have taken place in a common ancestor of jawed vertebrates, enabled the emergence of several key accessory molecules, and presumably the formation of authentic MHC class I molecules. If these duplications occurred as part of genome-wide duplication, it is not difficult to imagine that they provided unparalleled opportunities for the emergence of the adaptive immune system.

Because the presence or absence of the jaw appears to make much difference to the make-up of the immune system, this organ has been given a special symbolic status (MATSUNAGA and RAHMAN 1998). However, it seems more reasonable to assume that the emergence of jaws is merely one manifestation of many organizational changes triggered by the presumed genome-wide duplication event(s).

The jawless fish appear to be the immediate extant predecessor to the organisms with the adaptive immune system. For this very simple reason we can expect that the study of the immune system in this organism will provide valuable information on the origin of adaptive immunity. Furthermore, information on the architecture of its genome is likely to help resolve current controversies surrounding vertebrate genome evolution.

Acknowledgements. I thank my colleagues and collaborators for discussions and Ms. Atsuko Usami for her expert secretarial assistance. This work was supported in part by Grants-in-Aid for Scientific Research from The Ministry of Education, Science, Sports, and Culture of Japan. This article is contribution no. 11 from the Department of Biosystems Science, Graduate University for Advanced Studies, Hayama 240-0193, Japan.

References

Amores A, Force A, Yan Y-L, Joly L, Amemiya C, Fritz A, Ho RK, Langeland J, Prince V, Wang Y-L, Westerfield M, Ekker M, Postlethwait JH (1998) Zebrafish *hox* clusters and vertebrate genome evolution. Science 282:1711–1714

Aparicio S (1998) Exploding vertebrate genomes. Nature Genet 18:301–303

Bingulac-Popovic J, Figueroa F, Sato A, Talbot WS, Johnson SL, Gates M, Postlethwait JH, Klein J (1997) Mapping of *Mhc* class I and class II regions to different linkage groups in the zebrafish, *Danio rerio*. Immunogenetics 46:129–134

Campbell D, Trowsdale J (1997) A map of the human major histocompatibility complex. Immunol Today 18:43

Cantrell D (1998) Lymphocyte signalling: A coordinating role for Vav? Curr Biol 8:R535–R538

Du Pasquier L, Flajnik MF (1998) Origin and evolution of the vertebrate immune system. In: Paul WE (ed) Fundamental Immunology. Lippincott Williams & Wilkins, Philadelphia, PA, pp 605–650

Eichler EE (1998) Masquerading repeats: paralogous pitfalls of the human genome. Genome Res 8: 758–762

Endo T, Imanishi T, Gojobori T, Inoko H (1997) Evolutionary significance of intra-genome duplications on human chromosomes. Gene 205:19–27

Feder JN, Gnirke A, Thomas W, Tsuchihashi Z, Ruddy DA, Basava A, Dormishian F, Domingo RJ, Ellis MC, Fullan A, Hinton LM, Jones NL, Kimmel BE, Kronmal GS, Lauer P, Lee VK, Loeb DB, Mapa FA, McClelland E, Meyer NC, Mintier GA, Moeller N, Moore T, Morikang E, Prass CE, Quintana L, Starnes SM, Schatzman RC, Brunke KJ, Drayna DT, Risch NJ, Bacon BR, Wolff RK (1996) A novel MHC class I-like gene is mutated in patients with hereditary haemochromatosis. Nature Genet 13:399–408

Finnerty JR, Martindale MQ (1998) The evolution of the Hox cluster: insights from outgroups. Curr Opin Genet Dev 8:681–687

Fischer KD, Zmuldzinas A, Gardner S, Barbacid M, Bernstein A, Guidos C (1995) Defective T-cell receptor signalling and positive selection of Vav-deficient CD4+ CD8+ thymocytes. Nature 374:474–477

Flajnik MF, Ohta Y, Namikawa-Yamada C, Nonaka M (1999) Insights into the primordial MHC from studies in ectothermic vertebrates. Immunol Rev 167:in press

Graser R, Vincek V, Takami K, Klein J (1998) Analysis of zebrafish *Mhc* using BAC clones. Immunogenetics 47:318–325

Hashimoto K, Hirai M, Kurosawa Y (1995) A gene outside the human MHC related to classical HLA class I genes. Science 269:693–695

Holland PWH, Garcia-Fernandez J, Williams NA, Sidow A (1994) Gene duplications and the origins of vertebrate development. Development 1994 Suppl.:125–133

Hughes AL (1998) Phylogenetic tests of the hypothesis of block duplication of homologous genes on human chromosomes. 6:9, and 1. Mol Biol Evol 15:854–870

Kandil E, Namikawa C, Nonaka M, Greenberg AS, Flajnik MF, Ishibashi T, Kasahara M (1996) Isolation of low molecular mass polypeptide complementary DNA clones from primitive vertebrates: Implications for the origin of MHC class I-restricted antigen presentation. J Immunol 156: 4245–4253

Kasahara M (1997) New insights into the genomic organization and origin of the major histocompatibility complex: Role of chromosomal (genome) duplication in the emergence of the adaptive immune system. Hereditas 127:59–65

Kasahara M (1998) What do the paralogous regions in the genome tell us about the origin of the adaptive immune system? Immunol Rev 166:159–175

Kasahara M (1999) The chromosomal duplication model of the major histocompatibility complex. Immunol Rev 167:17–32

Kasahara M, Flajnik MF, Ishibashi T, Natori T (1995) Evolution of the major histocompatibility complex: a current overview. Transplant Immunol 3:1–20

Kasahara M, Hayashi M, Tanaka K, Inoko H, Sugaya K, Ikemura T, Ishibashi T (1996) Chromosomal localization of the proteasome Z subunit gene reveals an ancient chromosomal duplication involving the major histocompatibility complex. Proc Natl Acad Sci USA 93:9096–9101

Kasahara M, Nakaya J, Satta Y, Takahata N (1997) Chromosomal duplication and the emergence of the adaptive immune system. Trends Genet 13:90–92

Katsanis N, Fitzgibbon J, Fischer EMC (1996) Paralogy mapping: Identification of a region in the human MHC triplicated onto human chromosomes 1 and 9 allows the prediction and isolation of novel *PBX* and *NOTCH* loci. Genomics 35:101–108

Kaufman J, Salomonsen J (1997) The "Minimal Essential MHC" revisited: Both peptide-binding and cell surface expression level of MHC molecules are polymorphisms selected by pathogens in chickens. Hereditas 127:67–73

Kaufman J, Völk H, Wallny H-J (1995) A "minimal essential MHC" and an "unrecognized MHC": two extremes in selection for polymorphism. Immunol Rev 143:63–88

Klein J, Sato A (1998) Birth of the major histocompatibility complex. Scand J Immunol 47:199–209

Litman GW (1996) Sharks and the origins of vertebrate immunity. Sci Amer 275:47–51

Lundin LG (1993) Evolution of the vertebrate genome as reflected in paralogous chromosomal regions in man and the house mouse. Genomics 16:1–19

Martin LH, Calabi F, Milstein C (1986) Isolation of CD1 genes: A family of major histocompatibility complex-related differentiation antigens. Proc Natl Acad Sci USA 83:9154–9158

Matsunaga T, Rahman A (1998) What brought the adaptive immune system to vertebrates? – The jaw hypothesis and the seahorse. Immunol Rev 166:177–186

Miller MM, Goto RM, Taylor RL Jr, Zoorob R, Auffray C, Briles RW, Briles WE, Bloom SE (1996) Assignment of Rfp-Y to the chicken major histocompatibility complex/NOR microchromosome and evidence for high-frequency recombination associated with the nuclear organizer region. Proc Natl Acad Sci USA 93:3958–3962

Nadeau JH, Kosowsky M (1991) Mouse map of paralogous genes. Mamm Genome 1:S433–S460

Namikawa C, Salter-Cid L, Flajnik MF, Kato Y, Nonaka M, Sasaki M (1995) Isolation of *Xenopus LMP-7* homologues. Striking allelic diversity and linkage to MHC. J Immunol 155:1964–1971

Nonaka M, Namikawa C, Kato Y, Sasaki M, Salter-Cid L, Flajnik MF (1997a) Major histocompatibility complex gene mapping in the amphibian *Xenopus* implies a primordial organization. Proc Natl Acad Sci USA 94:5789–5791

Nonaka M, Namikawa-Yamada C, Sasaki M, Salter-Cid L, Flajnik MF (1997b) Evolution of proteasome subunits δ and LMP2: complementary DNA cloning and linkage analysis with MHC in lower vertebrates. J Immunol 159:734–740

Ohno S (1970) Evolution by gene duplication. Springer-Verlag, New York

Robey E (1997) Notch in vertebrates. Curr Opin Genet Dev 7:551–557

Ruddle FH, Bentley KL, Murtha MT, Risch N (1994) Gene loss and gain in the evolution of the vertebrates. Development 1994 Suppl.:155–161

Salter-Cid L, Du Pasquier L, Flajnik MF (1996) *RING3* is linked to the *Xenopus* major histocompatibility complex. Immunogenetics 44:397–399

Schluter SF, Bernstein RM, Marchalonis JJ (1997) Molecular origins and evolution of immunoglobulin heavy-chain genes of jawed vertebrates. Immunol Today 18:543–549

Sharman AC, Holland PWH (1996) Conservation, duplication, and divergence of developmental genes during chordate evolution. Netherlands J Zool 46:47–67

Sidow A (1996) Gen(om)e duplications in the evolution of early vertebrates. Curr Opin Genet Dev 6: 715–722

Skrabanek L, Wolfe KH (1998) Eukaryote genome duplication – where's evidence? Curr Opin Genet Dev 8:694–700

Spring J (1997) Vertebrate evolution by interspecific hybridization – are we polyploid? FEBS Lett 400:2–8

Sunyer JO, Zarkadis IK, Lambris JD (1998) Complement diversity: a mechanism for generating immune diversity? Immunol Today 19:519–523

Takami K, Zaleska-Rutczynska Z, Figueroa F, Klein J (1997) Linkage of *LMP*, *TAP*, and *RING3* with *MHC* class I rather than class II genes in the zebrafish. J Immunol 159:6052–6060

Tanaka K, Kasahara M (1998) The MHC class I ligand-generating system: roles of immunoproteasomes and the interferon-γ-inducible proteasome activator PA28. Immunol Rev 163:161–176

Tarakhovsky A, Turner M, Schaal S, Mee PJ, Duddy LP, Rajewsky K, Tybulewicz VL (1995) Defective antigen receptor-mediated proliferation of B and T cells in the absence of Vav. Nature 374:467–470

Thompson CB (1995) New insights into V(D)J recombination and its role in the evolution of the immune system. Immunity 3:531–539

Ting JP-Y, Baldwin AS (1993) Regulation of MHC gene expression. Curr Opin Immunol 5:8–16

Trachtulec Z, Hamvas RMJ, Forejt J, Lehrach HR, Vincek V, Klein J (1997) Linkage of TATA-binding protein and proteasome subunit C5 genes in mice and humans reveals synteny conserved between mammals and invertebrates. Genomics 44:1–7

Trowsdale J (1995) "Both man & bird & beast": Comparative organization of MHC genes. Immunogenetics 41:1–17

Ueyama H, Han-Xiang D, Ohkubo I (1993) Molecular cloning and chromosomal assignment of the gene for human Zn-α2-glycoprotein. Biochemistry 32:12968–12976

Zhang R, Alt FW, Davidson L, Orkin SH, Swat W (1995) Defective signalling through T- and B-cell antigen receptors in lymphoid cells lacking the Vav proto-oncogene. Nature 374:470–473

Phylogeny of Lower Vertebrates and Their Immunological Structures

A. Zapata[1] and C.T. Amemiya[2]

1 Introduction

Vertebrate organisms have evolved sophisticated immunological mechanisms for the recognition of self versus nonself. The immune system is functionally characterized by specificity, diversity, and memory but is based at the molecular level on highly diverse antigen receptors (T-cell receptor, TcR; immunoglobulin, Ig), which together with major histocompatibility complex (MHC) molecules allow the acti-

[1] Department of Cell Biology, Faculty of Biology, Complutense University, Madrid, Spain
[2] Center for Human Genetics, Boston University School of Medicine, 700 Albany Street, W408, Boston, MA 02118, USA

vation of the so-called immunocompetent cells (T- and B-lymphocytes) during the immune response. Development and functional activation of immune cells occur in specialized tissue microenvironments provided by the primary and/or secondary lymphoid organs. These lymphoid organs then, represent a higher-order morphological structure culminating from complex molecular, cellular and tissue interactions, and are an invariable feature of the vertebrate immune system.

For many years the patterns of organization of the lymphoid tissues of lower vertebrates were inferred or established by analogy to observations in mammals. Recently, numerous comparative histological and ultrastructural studies on the lymphoid organs of lower vertebrates have allowed the reconstruction of a more realistic picture of the emergence and evolutionary organization of vertebrate lymphoid organs (see reviews Zapata et al. 1995, 1996a,b; Galindez and Aggio 1997). [Note: we use the term "lower vertebrates" strictly to indicate representatives of taxa that diverged before the emergence of endothermic (warm-blooded) vertebrates. Likewise, the term "primitive vertebrates" is meant to indicate those vertebrates (or representatives thereof) that diverged very early in the vertebrate clade.]

This chapter is divided into two sections. First, we briefly review the phylogeny of vertebrates, with special emphasis on the lower vertebrates, particularly fishes. This overview is not meant to be a comprehensive treatise but merely serves to outline major extant taxa and salient departure points during vertebrate evolution, and, importantly, to familiarize the reader with the positions of many of the taxa that are discussed throughout this volume. Second, we describe the phylogenetic development and acquisition of immune (lymphoid) tissues, as revealed through histological and ultrastructural observation. Where relevant, evolutionary trends in the immune system that are correlated with phylogenetic advancement are presented and discussed.

2 Phylogeny of Lower Vertebrates and Vertebrate-like Taxa

The revolutionary thinker Ohno (1973) once wrote: "When one attempts to look at the overall picture of vertebrate evolution, it soon becomes evident that the genetic foundation which assured the eventual emergence of mammals and of man was laid at the very beginning; i.e., at the stage of fish." One could certainly paraphrase this statement in the context of the immune system, since it is during the stage of fishes where adaptive immunity appears to have arisen. In this section we review lower vertebrate (chordate) phylogeny and discuss interrelationships and some of the systematic characteristics that separate the key phylads from one another. [Readers interested in more in-depth phylogenetic and evolutionary analyses of fishes should consult the primary reference material, particularly Nelson (1994), Stiassney et al. (1996), and Carroll (1988). An excellent source for obtaining broad phylogenetic information of any form of life can be found at the Tree of Life website, http://ag.arizona.edu/tree.]

2.1 Cladistic Methodology: Brief Overview

Before reviewing lower vertebrate relationships, a brief discussion of "cladistics" (HENNIG 1966; CARROLL 1988) and systematic nomenclature is warranted. A phylogenetic tree, or "cladogram," is an inference of interrelationships whose cohesiveness or "monophyly" is based exclusively on "shared-derived" characters (HENNIG 1966). [Shared-derived characters are also referred to as "synapomorphies," and can consist of morphological or molecular features. Characters can also be defined as shared-primitive ("symplesiomorphies") or convergences ("homoplasies"); however, according to cladistic rules such characters cannot be used to define monophyletic assemblages.] Thus, a cladogram depicts a series of successively smaller and smaller monophyletic groups that are defined by subsets of unique characters not shared by the so-called "outgroups." A partial cladogram of the metazoans (animals) without invertebrates is given in Fig. 1. This tree corresponds to what is commonly referred to as phylum Chordata. Notably, a character that separates the chordates from the rest of metazoans is their possession of a notochord. The notochord defines the phylum Chordata, is shared by members of the chordates but not by lineages (clades) that diverged earlier, such as the echinoderms and arthropods.

One other term that deserves clarification is "paraphyly". As mentioned above, the goal of cladistics is to establish a nested series of monophyletic groups. In Fig. 1, monophyletic assemblages are listed horizontally on the bottom of the figure, so that Chordata, Craniata, Vertebrata, etc., define successively smaller monophyletic groups. A paraphyletic group, however, would be the Agnatha, since two separate clades, Mixini (M) and Petromyzontiformes (P), emerge from the main stem. Thus, the term Agnatha is not a valid systematic unit but is used to denote a paraphyletic group. Likewise, we use the term "lower vertebrates" to denote fishes and fish-like vertebrates; however, this is a paraphyletic grouping which is used for the sake of convenience and is not a valid systematic unit. This contradiction in terms is not meant to confuse the reader, but rather is presented so that the reader can recognize when, for example, Sarcopterygii *sensu stricto* refers to lobe-finned fishes (i.e., paraphyly) and when it refers *sensu lato* to the monophyletic assemblage of lobe-finned fishes plus tetrapods (Fig. 1). The literature is rife with paraphyletic designations, which are usually employed for the sake of convenience. In this review for the most part avoids using paraphyletic groups in our discussions, with exceptions where noted.

2.2 Overview of Chordate Phylogeny

In this section, we specifically discuss the higher-level taxonomic relationships comprising Fig. 1, and briefly highlight their respective phylogenetic positions and relevant biological aspects. For those readers interested in divergence times between the various taxa discussed here and elsewhere in this volume, Table 1 lists approximate times. It should be noted that these figures are expressed as "time since

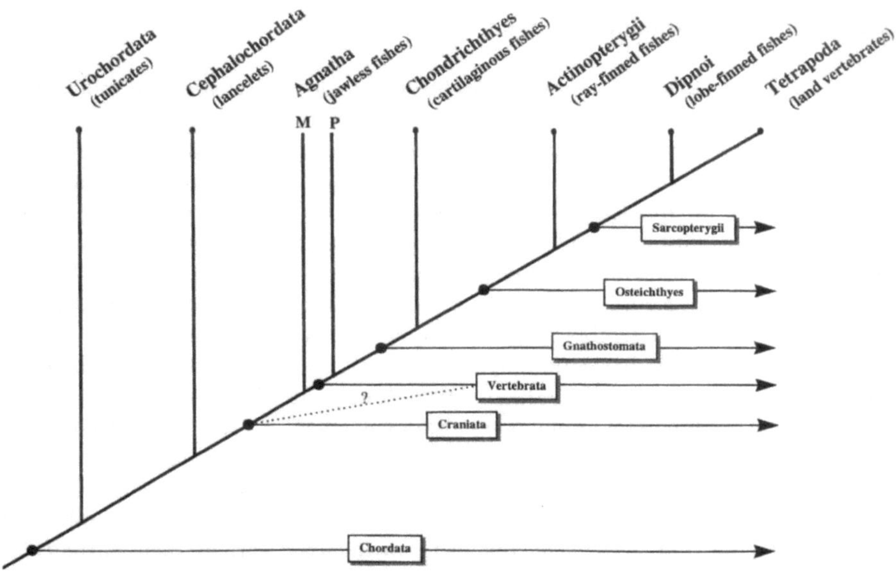

Fig. 1. The proposed relationships of the phylum Chordata. The cladogram represents a consensus of chordate relationships according to several sources (Romer 1966; Greenwood et al. 1966; Janver 1981; Carroll 1988; Nelson 1994; Helfman et al. 1997), with well-defined monophyletic assemblages listed below the cladogram. Briefly, the Chordata is defined by the presence of a notochord, a rodlike, skeletal structure lying dorsal to the gut and ventral to the nerve cord that is imperative for proper tissue induction and embryonic development. Craniata is defined by possession of a cranium (skull), at least a three-part brain, and advanced sensory structures. Vertebrata is defined by the presence of a vertebral column and neural crest cells. Gnathostomata is defined by the possession of true jaws and a horizontal semicircular canal in the inner ear. Osteichthyes (*sensu lato*) comprises the ray- and lobe-finned fishes plus tetrapods, and is defined by an internal bony skeleton, unique musculature of the jaw and gill (pharyngeal pouch) regions, and a lung/swim bladder connected to the gut. Sarcopterygii (*sensu lato*) comprises the lobe-finned fishes and tetrapods. The tetrapods (all land vertebrates including amphibians, reptiles, birds and mammals) are defined by the possession of four limbs. Note, each individual offshoot (e.g., tunicates or cartilaginous fishes) also has characters that define their own monophyletic cohesiveness. This cladogram merely provides a framework for the major chordate clades, and taxa of questionable lineage (e.g., coelacanths) have been omitted. With regard to the immune system, salient features of immunological import have notably been superimposed on phylogenies by Du Pasquier and Flajnik (1999) and Litman et al. (1999). Lastly, it is important to note that the Agnatha, a group that includes the extant cyclostomes (*M*) Myxini (hagfishes) and (*P*) Petromyzontiformes (lampreys), is thought to be a paraphyletic group. Several advanced anatomical characters exist in lampreys that are absent in hagfishes, although the presence or absence of neural crest cells (which, in part, define the Vertebrata) has not been adequately assessed in hagfishes due to the paucity of embryos

the taxa last shared a common ancestor". Thus, while mammals, for example, emerged in the fossil record in the Jurassic (approx. 150 million years ago, MYA), the divergence time between mammals and lampreys would be the point where the lineages that led to the two taxa diverged in the Cambrian (> 500 MYA). The fossil records for certain vertebrate taxa are very sketchy (Janvier 1981; Foote and Sepkoski 1999) and the estimates provided in Table 1 should be interpreted with some degree of caution.

Table 1. Approximate divergence times for selected taxa

	Hagfish	Lamprey	Chimeras	Horned shark	Nurse shark	Sandbar shark	Spiny dogfish	Sting rays	Skates	Bichir	Chondrosteans	Gar	Bowfin	Osteoglossids	Elops	Clupeids	Catfish, goldfish	Salmonids	Perciforms	Coelacanth	Lungfish	Amphibians	Reptiles	Birds	Marsupials	Mouse
Lamprey	550																									
Chimeras	550	530																								
Horned shark	550	530	415																							
Nurse shark	550	530	415	175																						
Sandbar shark	550	530	415	175	160																					
Spiny dogfish	550	530	415	200	200	200																				
Sting rays	550	530	415	200	200	200	155																			
Skates	550	530	415	200	200	200	155	90																		
Bichir	550	530	470	470	470	470	470	470	470																	
Chondrosteans	550	530	470	470	470	470	470	470	470	380																
Gar	550	530	470	470	470	470	470	470	470	380	380															
Bowfin	550	530	470	470	470	470	470	470	470	380	380	270														
Osteoglossids	550	530	470	470	470	470	470	470	470	380	380	270	250													
Elops	550	530	470	470	470	470	470	470	470	380	380	270	250	200												
Clupeids	550	530	470	470	470	470	470	470	470	380	380	270	250	200	200											
Catfish, goldfish	550	530	470	470	470	470	470	470	470	380	380	270	250	200	200	180										
Salmonids	550	530	470	470	470	470	470	470	470	380	380	270	250	200	200	180	150									
Perciforms	550	530	470	470	470	470	470	470	470	380	380	270	250	200	200	180	150	150								
Coelacanth	550	530	470	470	470	470	470	470	470	400	400	400	400	400	400	400	400	400	400							
Lungfish	550	530	470	470	470	470	470	470	470	400	400	400	400	400	400	400	400	400	400	400						
Amphibians	550	530	470	470	470	470	470	470	470	400	400	400	400	400	400	400	400	400	400	400	400					
Reptiles	550	530	470	470	470	470	470	470	470	400	400	400	400	400	400	400	400	400	400	400	400	380				
Birds	550	530	470	470	470	470	470	470	470	400	400	400	400	400	400	400	400	400	400	400	400	380	200			
Marsupials	550	530	470	470	470	470	470	470	470	400	400	400	400	400	400	400	400	400	400	400	400	380	300	300		
Mouse	550	530	470	470	470	470	470	470	470	400	400	400	400	400	400	400	400	400	400	400	400	380	300	300	190	
Human	550	530	470	470	470	470	470	470	470	400	400	400	400	400	400	400	400	400	400	400	400	380	300	300	190	100

These estimates were taken from several sources (Romer 1966; Carroll 1987, 1997; Helfman et al. 1997; Nelson 1994; Shirai 1996; Kumar and Hedges 1998) and have relatively large margins of error (± 50 MYA).

2.2.1 Tunicates and Lancelets

These chordate taxa are thought to be the immediate predecessors to the craniates and vertebrates (Fig. 1; MAISEY 1986; CARROLL 1988; SATOH and JEFFERY 1995; JEFFERY 1997). There are roughly 1600 extant species of tunicates and 22 extant species of lancelets (NELSON 1994; HELFMAN et al. 1997). Tunicates are relatively sessile as adults; however, they do exist as free-swimming tadpoles as larvae. Amphioxus resemble fishes superficially but lack many structures, including a cranium, brain and hemoglobin. While these organisms are clearly quite distinct from modern vertebrates, their overall characteristics are apparently much more fish-like than invertebrate-like, prompting hypothetical scenarios by various investigators as to how genetic contingencies in developmental regulation could have led to the ultimate emergence of vertebrates (MAISEY 1986; SATOH and JEFFERY 1995; HELFMAN et al. 1997; HOLLAND and GARCIA-FERNANDEZ 1996).

Phylogenetically, the lancelets are considered to be the archetypal protovertebrate since it has a relatively simple genome with limited redundancy (PEBUSQUE et al. 1998; BOEDDRICH et al. 1999). The latter point is best exemplified by the Hox clusters, which serve as a barometer for whole genome duplications. As determined by the number and composition of Hox clusters, the genomes of all vertebrates studied thus far (including hagfishes and lampreys) have apparently undergone at least one round of genome duplication since the lancelets and craniates last shared a common ancestor (PENDLETON et al. 1993; GARCIA-FERNANDEZ and HOLLAND 1994; HOLLAND et al. 1994; HOLLAND and GARCIA-FERNANDEZ 1996; BAILEY et al. 1997; AMORES et al. 1998; F. Ruddle, unpublished data). Neither the lancelets or tunicates show indications of an adaptive immune system, however, the latter clearly possess elements of cellular immunity (SCHLAKE et al. 1997; reviewed in DU PASQUIER and FLAJNIK 1999). Surprisingly, Amphioxus appears to possess whn, a transcription factor intimately involved in thymus formation in mammals (SCHLAKE et al. 1997), although to what extent it functions in an immune capacity is yet to be determined.

2.2.2 Jawless Fishes

There are roughly 43 extant species of hagfish and 41 extant species of lampreys (HELFMAN et al. 1997). The similarity in eel-like body forms of the hagfish and lamprey is superficial and earlier efforts to group these two taxa together have since been shown to be flawed. Thus Agnatha (cyclostomes) is considered a paraphyletic assemblage (Fig. 1; ROMER 1966; JANVIER 1981; FOREY and JANVIER 1993). These taxa are thought to have emerged in the Cambrian (Table 1) although their fossil records are rather poor. Several morphological characters suggest that hagfishes are older than lampreys despite the paucity of fossils. Both lampreys and hagfish possess a well-defined notochord structure, however, neither have constricted vertebrae or paired fins. Lampreys clearly possess neural crest cells, a feature that suggests it is a true vertebrate.

At this stage of phylogenetic development, hematopoietic structures and lymphocytes are clearly observed as discussed below. However, neither of these taxa appear to possess hallmarks of an immune system, where the receptors are generated somatically: Ig and TcR (reviewed in Du Pasquier and Flajnik 1999). This suggests that either these taxa are lacking the components altogether or that they simply have not been discovered to this point. Lastly, lampreys, as mentioned above for Amphioxus, appear to possess whn transcription factor genes (Schlake et al. 1997).

2.2.3 Jawed Fishes

Any discussion of vertebrate evolution must make mention of the Placodermi. These fishes, all of which are now extinct, existed in the Cambrian (550 MYA) and are thought to have been the immediate predecessors of all vertebrates (Romer 1966; Carroll 1988; Joseph 1994). The placoderms had thick external plates (armor), and have been systematically placed in-between the jawless fishes (Agnathans) and the rest of the vertebrates. Thus they are a presumed primitive sister-group to the rest of the vertebrates. Unfortunately, because this group was extinct by the end of the Devonian (approx. ~360 MYA), most characteristics germane to its biology (including a presumptive immune system that gave rise to that observed in all extant vertebrates) have been lost. It is unknown whether placoderm fossils exist whose internal organs were preserved such that ultrastructural analyses could be performed. This would be one way, albeit indirect, for assessing whether this key taxon already possessed a primordial immune system 360 MYA reminiscent of what we observe in modern vertebrates.

2.2.3.1 Cartilaginous Fishes

The cartilaginous fishes are thought to have arisen from a placoderm ancestor (Romer 1966; Carroll 1988). Figure 2 gives a cladogram of the 11 orders of chondrichthyans generally recognized. The old age of the group together with difficulties associated with character definition have made systematic relationships within and among the chondrichthyans somewhat difficult to establish. There are roughly 850 extant species, over half of which are batoids and around 30 of which are chimeroids (Compagno 1973; Nelson 1994; Helfman et al. 1997). The two major clades, the holocephans and the elasmobranchs, diverged some 415 MYA, and in general, have a rich, but incomplete fossil record (Romer 1966; Carroll 1988; Foote and Sepkoski 1999).

The chondrichthyans are the first vertebrates to show a humoral immune response. They possess Ig, TcR and MHC class I and II, and their Ig system is marked by multiple IgH and IgL isotypes (reviewed in Du Pasquier and Flajnik 1999). Coincident with these molecular findings, they possess well-defined immunological structures and some unique tissues that are not found in any other vertebrate taxa (discussed in Sect. 3).

2.2.3.2 Ray-Finned Fishes

The ray-finned (actinopterygian) fishes are the largest single group of vertebrates and consist of well over 22,000 species in some 40 orders and 400 families

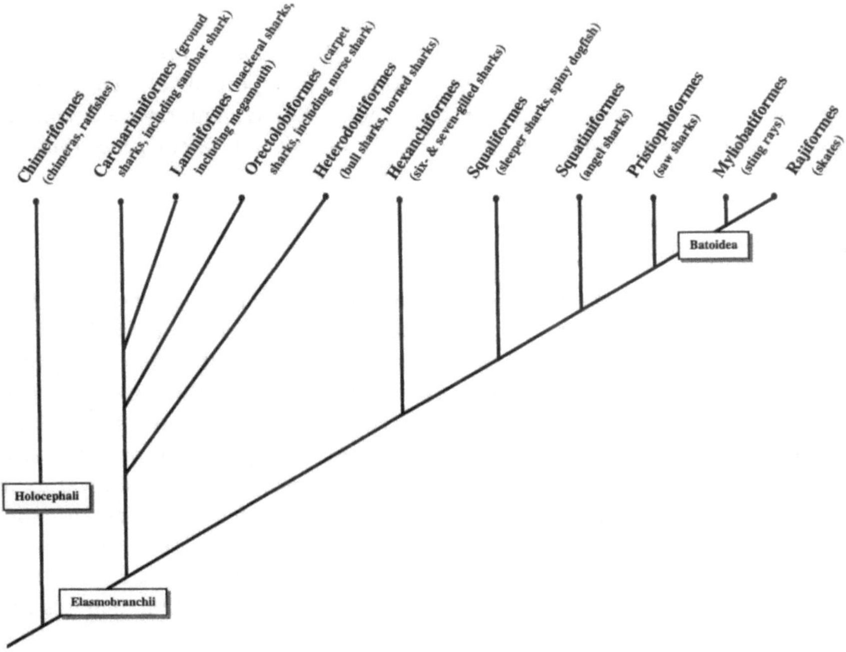

Fig. 2. Interrelationships of chondrichthyans. This cladogram is based on several sources (COMPAGNO 1973, 1977; MAISEY 1984; MIYAKE and MCEACHRAN 1991; NELSON 1994; MCEACHRAN et al. 1996; CARVALHO 1994; SHIRAI 1994). The branching order is based primarily on CARVALHO (1994) and SHIRAI (1994). The chondrichthyans are defined from other gnathostomes by having a calcified cartilaginous endoskeleton that is seldom ossified, and copulatory pelvic organs known as "claspers" in the males. They are generally divided into two clades: (a) Holocephali (ratfish, chimeras); and (b) Elasmobranchii (sharks and batoid fishes)

(COHEN 1970; NELSON 1994). The actinopterygians, the vast majority of which are teleosts, comprise over one-half the extant species of vertebrates. The group is divided here into two cladograms: Fig. 3A lists the basal ray-finned fishes, and Fig. 3B lists several of the taxa that comprise the teleost assemblage. It should be noted that in Fig. 3A, the age-old subdivision of actinopterygian fishes into chondrostean-holostean-teleostean is no longer valid. At present, polypterids (bichirs) are excluded from the chondrosteans *proper*, and the gars and bowfins, taxa that formerly comprised the holosteans, have been shown to be paraphyletic.

In this stage of phylogenetic development, the immune systems are similar to those of tetrapods, with notable exceptions or variations. Studies on basal and advanced ray-finned fishes have yielded much information as to the manner in which the immune tissues are organized (discussed below), as well as the ways in which immune receptor genes are both similar and different from those of cartilaginous fishes and tetrapods (LITMAN et al. 1999).

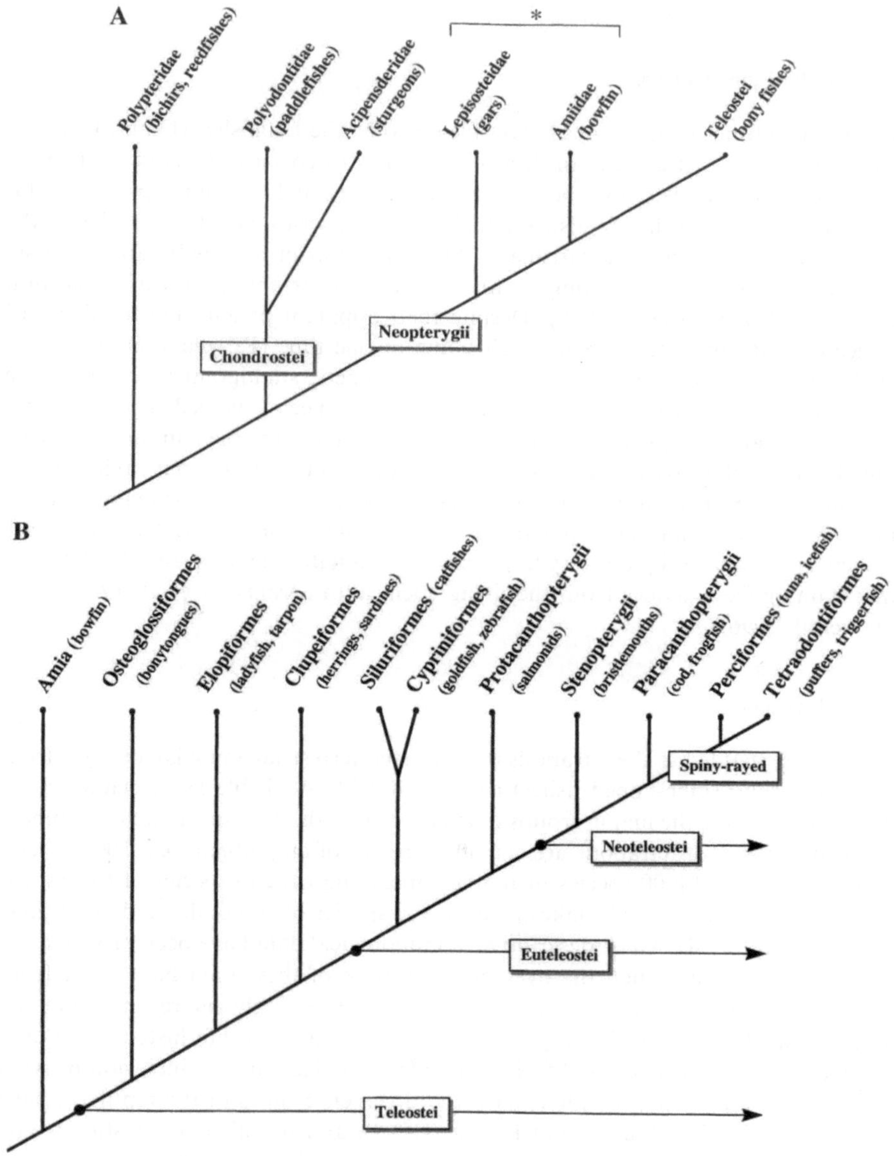

Fig. 3A,B. Interrelationships of ray-finned fishes (Actinopterygii). These cladograms are based primarily on Greenwood et al. (1966); Nelson (1994), Grande and Bemis (1996), Helfman et al. (1997), Gardiner et al. (1996), De Pinna (1996), Fink and Fink (1996), and Cloutier and Ahlberg (1996). **A** Interrelationships of basal actinopterygian fishes. This cladogram is a fairly consensual view of the basal actinopterygian fishes. The polyterids (Brachiopterygii) have historically presented problems in their placement, and some systematists have considered this group as part of the "chondrosteans" (Denton and Howell 1972; Nelson 1994; Helfman et al. 1997). Note, the * denotes two taxa that were formerly considered a monophyletic assemblage (Holostei) but which have since been split into two discrete lineages (Nelson 1994; Gardiner et al. 1996). **B** Interrelationships of teleostean fishes. Note, only a select number of taxa were chosen for this cladogram; several taxa were omitted due to space limitations. *Amia* is considered the sister taxon to all of the teleosts, which form a putative monophyletic group based on a suite of morphological characters (De Pinna 1996)

2.2.4 Lobe-Finned Fishes

Lobe-finned fishes (Figs. 1, 4) are considered to be the lungfishes (Dipnoi) and the coelacanths. The actual relationship of lungfishes to coelacanths is unclear; however, most systematists now believe the two represent long and separate evolutionary lineages with the lungfishes actually being closer to the tetrapods than is the coelacanth (Cloutier and Ahlberg 1996; Helfman et al. 1997). There are six currently extant species of lungfish and one or two extant species of coelacanth (Nelson 1994; Erdmann 1998). Despite their apparent paucity the fossil record indicates that both species had a rich fauna at one time (Romer 1966; Carroll 1988). The interest in these taxa from a phylogenetic standpoint is that they are representative of forms whose ancestors may have been involved in the fin-limb transition and subsequent vertebrate invasion of land. Are their immune systems preadapted to the environmental constraints imposed by a terrestrial environment? Immune system structural organization has not been examined thoroughly in these taxa. Moreover, molecular genetic studies on their immune system genes are somewhat hampered by the very large size of the lungfish genome and the difficulty in obtaining fresh tissues from the living coelacanth (Amemiya et al. 1993, 1996; Ota et al. 1996).

2.2.5 Tetrapods

The lineage that led to the tetrapods (four-limbed terrestrial animals) diverged from its sister group (lobe-finned fishes) roughly 400 MYA (Table 1). A cladogram is given in Fig. 4 for the major groups of extant tetrapods. The approximate numbers of extant species of tetrapods are: ca. 4000 species of amphibians, ca. 5000 species of mammals, ca. 14000 species of reptiles (including ca. 240 species of turtles, ca. 5100 species of lizards and snakes, and ca. 900 species of crocodiles and birds; Zug 1993; Nelson 1994). Much descriptive immunological data have been procured for tetrapod species, although the overwhelming bulk of these data have come from mouse or man. As a group, the tetrapods are exceedingly interesting from the standpoint of its overall heterogeneity in biology, bauplan, life history, and evolutionary rates. This is exemplified by the reptilian clade and its inclusion of Aves (Fig. 4). Most recently, the phylogeny and divergence times of the reptilians have been challenged by Hedges and Poling (1999), and it will be interesting to ascertain whether characteristics of the immune system have followed the presumed phylogenetic trends.

3 Evolutionary Development of Immune Tissues

The previous section outlines the phylogenetic and evolutionary history of extant taxa that comprise chordates. This section now deals with the origins and histo-

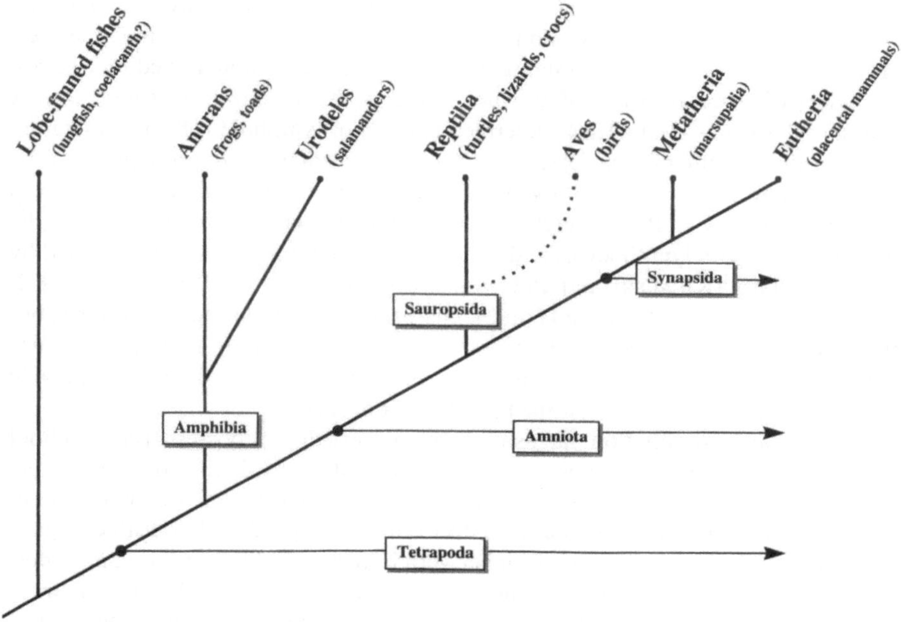

Fig. 4. Interrelationships of tetrapods. This cladogram depicts the major tetrapod taxa as a monophyletic group with the lobe-finned fishes as an outgroup (CARROLL 1988). The tetrapods are defined by limbs that bear digits (rather than fins). All the major phylads are well-supported cladistically: Amniota refers to the distinctive attribute of an amniotic egg, and Sauropsida and Synapsida refer to defined conditions in the temporal openings in the skull. The stippled line in the Aves clade indicates that the birds are thought to have evolved from a reptilian ancestor (CRACRAFT 1981; FEDUCCIA 1996), presenting a quandary in terms of their taxonomy and systematics

logical characterization of immune tissues throughout the chordate radiation. We do not document the molecular evolution of immune molecules, as this is dealt with in-depth in several other chapters in this volume and elsewhere. Rather, we describe the phylogenetic emergence and appearance of lymphoid tissue and discuss their implications with regard to the evolution of the immune response.

3.1 Agnathans Do Not Have True Lymphoid Organs

Modern Agnathans (cyclostomes) are a paraphyletic group that contains the most primitive living craniates (see Sect. 2.2.2 above). The immunological structures of Agnatha are striking. Although pioneering studies assumed capacity to mount weak immune responses to both T-dependent and T-independent antigens in both myxinoids and lampreys, attempts to find specific antigen receptors (i.e., TcR and Ig) as well as MHC molecules have failed repeatedly (LITMAN et al. 1992; DU PASQUIER 1993; KRONENBERG et al. 1994). In contrast, a member of the C3-like complement protein family identified in the serum of Pacific hagfish, *Eptatretus stoutii* (VARNER et al. 1991; Hanley et al. 1992; Ishiguro et al. 1992), seems to be involved in innate responses which resemble the nonspecific defense mechanisms of invertebrates. However, circulating cells resembling lymphocytes have been identified by electron microscopy in several species of myxinoids (Fig. 5; reviewed in Fänge and Zapata 1985), and Zapata et al. (1984) described clusters of lymphoid cells, including developing and mature plasma cells, in the so-called central mass of pronephros of the Atlantic hagfish, *Myxine glutinosa*. Other authors have failed to identify plasma cells in various species of Pacific hagfish (Fänge and Zapata 1985). In contrast, the occurrence of both lymphocytes and plasma cells in lampreys has been reported by several authors (Kilarski and Plytycz 1981; Zapata et al. 1981a; Fujii 1982; Hagen et al. 1983). Putative Ig H and L chains had been reported in lampreys, but the nature of L chains needs confirmation and the two subunits are apparently not covalently joined together as in other Ig molecules. Correspondingly, lamprey Ig genes, if any, are yet to be identified (Litman et al. 1992).

Upon close examination, hagfishes and lampreys do not possess true lymphoid organs. Neither species possesses a morphologically identifiable thymus or peripheral lymphoid organs, such as spleen or lymph nodes. As with many other primitive vertebrates, lymphoid cells of cyclostomes are produced in distinct organs that, as with the bone marrow of higher vertebrates, are able to house and differentiate hematopoietic stem cells under the influence of local cell microenvironments. In the Atlantic hagfish, the pronephros was thought to be the equivalent of

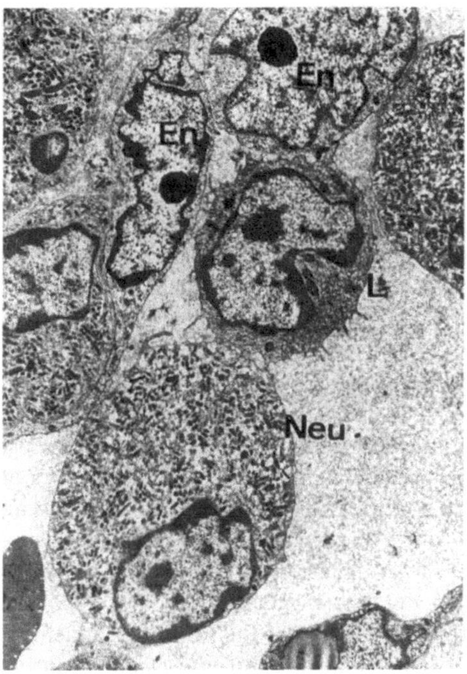

Fig. 5. Circulating lymphoidlike cells (*L*) and neutrophils (*Neu*) of the Atlantic hagfish, *Myxine glutinosa*. Endothelial cells (*En*). ×4000

the thymus of jawed vertebrates (FÄNGE 1966). It is a paired organ which consists of numerous ciliated tubules whose nephrostomes open into the pericardial coelomic cavity. Its other end enters the wall of the pronephric vein to reach the so-called central mass consisting of renal epithelial cells and foci of hematopoietic tissue consisting of erythroid cells, lymphoid elements and macrophages (ZAPATA et al. 1984). Additionally, RIVIERE et al. (1975) described small, lymphocyte-like cells able to uptake labeled thymidine in the muscle velar complex of the Pacific hagfish which was inferred as a phylogenetic precursor to the thymus. Presumably, however, these cells represent satellite cells of the skeletal muscle which, as with lymphocytes, are capable of active division.

The other hematopoietic organ of myxinoids is formed by hematopoietic masses in the intestinal lamina propria reported by some authors as a phylogenetic precursor of the spleen (TOMONAGA et al. 1973). This tissue structurally resembles the bone marrow of higher vertebrates, consisting of mature and developing granulocytes and a few macrophages scattered in cell clusters between large fat cells in a supporting meshwork of fibroblastic reticular cells (Fig. 6; TANAKA et al. 1981; ZAPATA et al. 1995).

In lampreys, lymphoid accumulations in the gills of ammocoetes (premetamorphosed individuals) were considered as possible homologues of the thymus (FINSTAD et al. 1964). Ultrastructural examination revealed that they represent in fact filtering areas in which phagocytic endothelia and free macrophages are

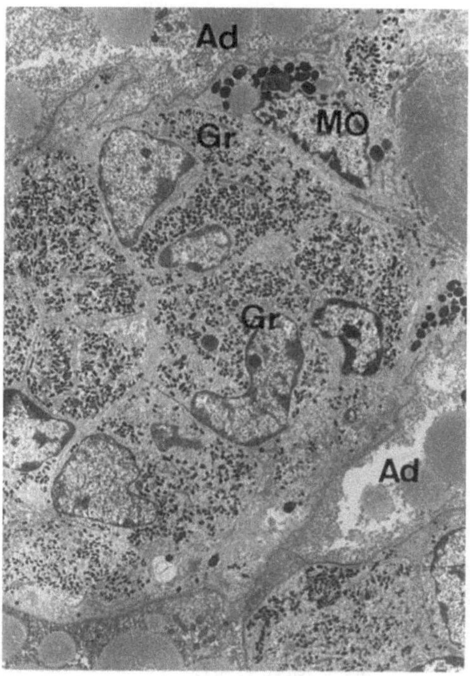

Fig. 6. Masses of hematopoietic tissue in the intestinal lamina propria of the Atlantic hagfish, *Myxine glutinosa*. Developing and mature granulocytes (*Gr*) and macrophages (*MO*) appear between big fat cells (*Ad*). ×3000

thought to clean the lumen of pharyngeal cavity (PAGE and ROWLEY 1982; ARDAVIN and ZAPATA 1988). Blood cells of lampreys are formed in distinct organs throughout their life cycle, including the typhlosole and the nephric fold in ammocoetes, and in the so-called supraneural body and adult opistonephros in the postmetamorphic animals (ARDAVIN et al. 1984; ARDAVIN and ZAPATA 1987). These organs have the same histological organization consisting of cell cords which house all the blood cell lineages, including lymphocytes and plasma cells, arranged between sinusoidal blood vessels, renal tubules in the case of kidney, and fat cells in the supraneural body (Fig. 7).

During metamorphosis, the ammocoete hematopoietic organs (largely the typhlosole, which completely regresses) disappear because their stromal components change and are unable to support blood cell differentiation. Subsequently the supraneural body, a fat column dorsally running along the central nervous system, which has no hematopoietic capacity in ammocoetes, becomes the major hematopoietic organ of adult lampreys.

3.2 The Primary, Central Lymphoid Organs of Jawed Vertebrates

The Chondrichthyes, including holocephalans and elasmobranchs (Fig. 2), are the most primitive vertebrates exhibiting true lymphoid organs, including thymus and

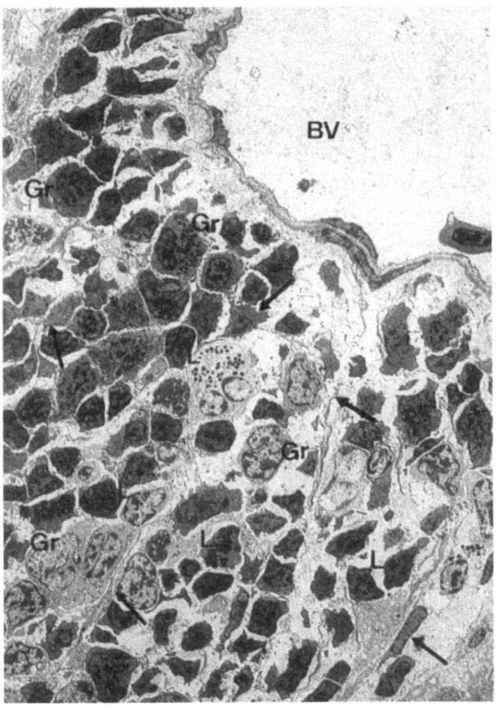

Fig. 7. Hematopoietic tissue in the typhlosole of an ammocoete of the anadromous sea lamprey, *Petromyzon marinus*. Developing and mature leukocytes, including granulocytes (*Gr*) and varying sized lymphocytes (*L*) occur in a meshwork of fibroblastic reticular cells (*arrows*). *BV*, Blood vessel. ×2400

morphofunctional equivalents of bone marrow. The lymphoid organs appear, therefore, in the first vertebrates known to elicit adaptive immune responses.

3.2.1 The Thymus

A thymus gland occurs in all jawed vertebrates. Apart from small anatomical differences, such as the occurrence of two thymus glands and the number of pharyngeal pouches that contribute to its formation, there are few histological variations in the thymus throughout the vertebrate phylogeny, with the possible exception of teleost fish.

In most vertebrates the thymus is a lobulated bilateral organ, completely surrounded by a connective tissue capsule, which contains two well-defined areas, the cortex and the medulla. The thymic parenchyma consists of a supporting meshwork of epithelial cells which houses, largely in the cortex, maturing thymocytes. In the medulla, in which the epithelial cells predominate, there additionally are secretory epithelial cells and epithelial cysts, but Hassall's bodies appear for the first time in birds; in some mammals (i.e., the rat) they are infrequent structures. Large epithelial cells, resembling the epidermal glands of skin (Clothier and Balls 1985; Barrutia et al. 1989) occur, however, in the thymic medulla of *Xenopus laevis*. Macrophages and interdigitating dendritic cells have been morphologically identified also in the thymus of ectotherms, although whether they play the same role as in the mammalian thymus requires further confirmation. Myoid cells in the thymus of both amphibians and reptiles are more frequent than in mammals (Raviola and Raviola 1967). In fish thymus, however, their number is low (reviewed in Zapata and Cooper 1990). These isolated muscle elements are round or oval, large cells, the cytoplasm of which is filled with myofilaments that form sarcomerelike units arranged around the nucleus. Experiments using chicken-quail chimeras suggest that myoid cells are derived from neural crest (Nakamura et al. 1986). In addition, the thymi of lower vertebrates include neuroendocrinelike cells and a low percentage of B lymphocytes and plasma cells.

The situation in teleosts deserves special attention. In the teleost thymus a clear cortex-medulla demarcation cannot be distinguished (Zapata et al. 1996b). Furthermore, the organ remains close to the pharyngeal epithelium and retains, in adult fish, its embryonic connection to the pharyngeal cavity (Fig. 8). As in other vertebrates, the organ is basally surrounded by a continuous connective tissue capsule which, when the gland is large, is projected in trabeculae into the organ. Supporting epithelial cells constitute a network from this basal zone to the outer region in which they connect with the epithelial cells delimiting the pharyngeal cavity (Fig. 9). The spaces of this network are occupied largely by lymphoblasts, medium and small lymphocytes (Fig. 9).

Tatner and Manning (1982) speculated that this histological organization allows antigens to enter directly into the embryonic thymus from the gill cavity. We recently demonstrated, however, that trout fry of different ages immersed in a ferritin solution trapped ferritin through gills, not through the thymic primordium. In very immature (4-day-old) fry exposed to ferritin for a long time, ferritin ap-

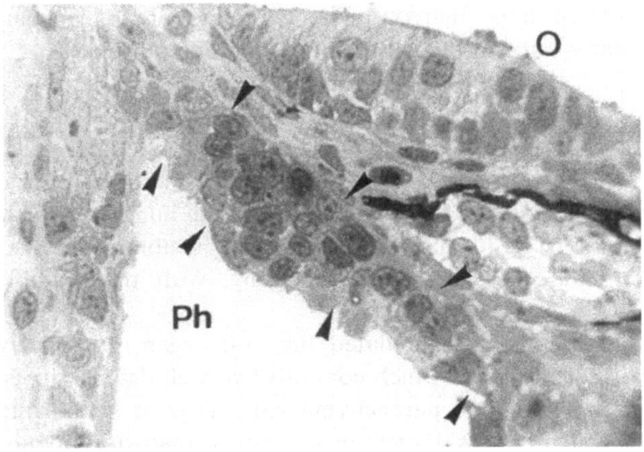

Fig. 8. Thymus (*arrowheads*) of a 7 day-old zebrafish, *Danio rerio* (teleost fish). Note the continuity of the gland with the pharyngeal cavity (*Ph*). *O*, Otic vesicle. ×400

peared in the thymic parenchyma but not in the pharyngeal epithelium which covers the organ. This suggests that the ferritin passively gains access to the thymus through either the basement membrane under the connective tissue capsule or the extracellular spaces adjacent to the limits between the thymic primordium and the pharyngeal epithelium (Castillo et al. 1998). These results suggest that under natural conditions gills would trap antigens and nonantigenic materials before they gain access into the thymus.

In summary, we conjecture that the teleost histological organization may reflect an incomplete migration of the thymic primordium from the gill buds (where it originates) to the underlying mesenchyme during ontogeny. Somewhat surprisingly, these special features do not seem to lead to a lack of functional capabilities of the teleost thymus. In fact several molecular markers known to be involved in T lymphocyte development are expressed in the thymi of lower vertebrates. In sharks and skates (Luer et al. 1995; A. Miracle, personal communication), trout, zebrafish, axolotl, and *Xenopus* (reviewed in Hansen and Zapata 1998), recombinase-activating gene (Rag) and terminal deoxynucleotidyl transferase (Tdt) are strongly expressed. In addition, the highest TcR β expression occurs in the thymi of sharks, rainbow trout, axolotl and *Xenopus* (Fellah et al. 1993; Partula et al. 1995; Chretien et al. 1997; McKinney et al. 1997), and expression of Lck (a src tyrosine kinase involved in T cell maturation and activation) has recently been demonstrated in the zebrafish thymus (Trede and Zon 1998). Interestingly, the vast majority of thymocytes in *Xenopus* express CD5 (Jurgens et al. 1995) and transcripts of a presumptive ancestral CD3 gene (Dzialo and Cooper 1997). Furthermore, *Xenopus* cortical thymocytes express CTX, a new molecule of Ig superfamily, as well as CD8, presumably representing DP ($CD4^+CD8^+$) cells, although a specific reagent for detecting *Xenopus* CD4 is not available (Chretien et al. 1996). In other lower vertebrates, mainly teleosts, mAb presumably reacting

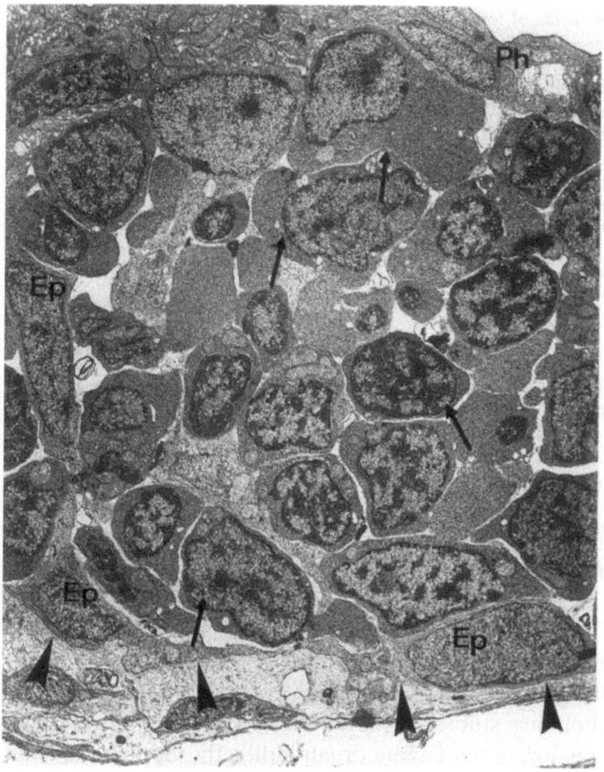

Fig. 9. Thymus of a 7 day-old zebrafish, *Danio rerio* (teleost fish). A network of thymic epithelial cells (*Ep*) connects the connective tissue capsule (the limits of which are marked by *arrowheads*) with the pharyngeal epithelium (*Ph*). The holes of supporting network are occupied by varying sized lymphocytes (*arrows*). ×4500

with T cell markers (Scapigliati et al. 1995; Passer et al. 1996; Rombout et al. 1997) require further molecular characterization.

3.2.2 Bone Marrow and Bone Marrow Equivalents

The bone marrow is the major locus for B lymphocyte production in adult higher vertebrates. The capability of this organ to house and differentiate hematopoietic progenitors is dependent on stroma consisting largely of fibroblastic reticular cells, adventitial cells, macrophages, fat cells and sinusoidal endothelial cells. A hematopoietic bone marrow appears for the first time in the most evolved urodelans of Plethodontidae family. In more primitive vertebrates there is no bone marrow but many distinct organs (Table 2) house hematopoietic stem cells in a stroma similar to that of mammalian bone marrow. Unfortunately, despite the fact that the histological organization of most of these organs has been examined, their immunological functions are unknown.

Table 2. Known bone marrow equivalents of lower vertebrates (*sans* reptilians)

Jawless fishes	
Myxinoids	Pronephros, hematopoietic tissue of the intestinal lamina propria
Lampreys	Typhlosole, nephric fold (ammocoetes); adult opistonephros, supraneural body (postmetamorphic and adult lampreys)
Chondrichthyes	
Holocephali	Orbital and subcranial hematopoietic tissue
Elasmobranchii	Epigonal and Leydig organs; meningeal tissue (*Dasyatis akajei, Triakis scyllia; Scyliorhinus torazame, Mustelus manazo*)
Osteichthyes	
Bichir	Kidney, meninges
Chondrostei	Cranium, heart, kidney, gonads
Holostei[a]	Cranium, heart, kidney
Teleostei	Kidney
Sarcopterygii	
Dipnoi	Spiral valve, kidney, gonads
Coelacanth	To be studied
Amphibians	
Apodans	Liver, kidney (?)
Urodelans	Liver, kidney – pronephros and mesonephros (Family Protidae); bone marrow (Family Pletodonthidae); meningeal tissue (*Megalobathrachus japonicus*)
Anurans	Liver (Family Pipidae); embryonic and larval kidney

[a] Holostei is a paraphyletic group that includes gars and bowfins.

Most elasmobranchs contain masses of hematopoietic tissue in the esophagus and gonad walls, the so-called Leydig and epigonal organs, respectively (reviewed in Zapata et al. 1996a). Preliminary studies in the clearnose skate, *Raja eglanteria*, which demonstrate Rag expression in the Leydig organ indirectly support the tenet that this organ, together with the spleen, may be involved in B lymphopoiesis in elasmobranchs (reviewed in Hansen and Zapata 1998). Lympho-hematopoietic tissue has been found in the "meninx primitiva" of some elasmobranchs (Chiba et al. 1988). In other primitive fish, including Chondrostei, gars, bowfins (formerly Holostei; Scharrer 1944), bichirs (unpublished data), the urodeles, *Ambystoma* (Dempster 1930) and *Megalobathrachus* (Sano and Imai 1961), and early human embryos (Kappers 1958), hematopoietic tissue occurs in both meninges and choroid plexuses. A similar tissue appears in the orbita and subcranium of the holocephalan, *Chimaera monstrosa* (Stahl 1967; Mattisson and Fänge 1986; Mattisson et al. 1990). In all these organs, cell cords arranged between sinusoidal blood vessels harbor granulopoiesis as well as lymphocytes, macrophages and plasma cells. Furthermore, in some elasmobranch species, the meningeal lymphoid tissue extensively expands and projects into the brain ventricles in remarkable macrophage-lymphocyte cell clusters presumably preventing the entrance of noxious agents into the brain parenchyma (Torroba et al. 1995).

In many lower vertebrates (Table 2) the kidney (pronephros and/or mesonephros) is a hematopoietic organ, the histological organization of which is remarkably similar to the bone marrow of higher vertebrates (Fig. 10). In teleosts, the kidney is comprised of two distinct, but histologically similar regions: (a) the anterior, cephalic, or head kidney; and (b) the middle and posterior, trunk kidney.

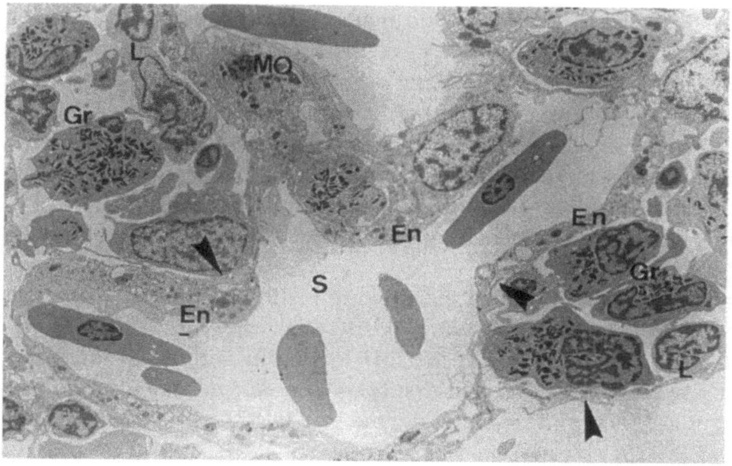

Fig. 10. Pronephros of *Rutilus rutilus* (teleost fish). Note the resemblance of its histological organization to that of mammalian bone marrow. Cell cords supported by fibroblastic reticular cells (*arrowheads*) contain different types of developing and mature granulocytes (*Gr*), lymphocytes (*L*) and macrophages (*MO*). Sinusoidal blood vessel (*S*), endothelial cells (*En*). ×2650

Both regions contain hematopoietic tissue but it is more developed in the head kidney in which the renal function has disappeared (reviewed in ZAPATA et al. 1996b). Enlarged sinusoids separate cell cords formed by heterogeneous fibroblastic reticular cells (QUESADA et al. 1990; PRESS et al. 1994). Sinusoidal endothelial cells, free macrophages and fibroblastic reticular cells are actively phagocytic elements of teleost kidney. On the other hand, both morphological and functional data support the kidney of teleosts as the major source of B lymphocytes. In this regard, recent reports which demonstrate the expression of nonrearranged and multiple forms of IgL transcripts in the pronephros of both Atlantic cod and rainbow trout (DAGGFELDT et al. 1993), as well as the strong expression of Rag and Tdt (reviewed in HANSEN and ZAPATA 1998), confirm the occurrence of B cell precursors in the teleost kidney. Furthermore, the occurrence of antigen-presenting cells, and T and B lymphocytes, in the teleost kidney, supports a relevant role of renal lymphoid tissue in the immune responsiveness of these bony fish.

In amphibians which do not have hematopoietic bone marrow, such as the primitive urodelans of the family Protidae, and, presumably, apodans (limbless newts), the kidney retains its hematopoietic capacity. Conversely, in the family Plethodontidae, in which an active hematopoietic bone marrow is present, the kidney has lost the capacity for producing blood cells (CURTIS et al. 1979). Correspondingly, the kidney of embryonic and larval anurans loses its hematopoietic capacity when the bone marrow begins to generate blood cells after metamorphosis. In fact, liver, kidney, spleen, and bone marrow produce lymphocytes at some time throughout the life cycle of anurans.

The perihepatic cell layer is an important lymphomyeloid organ of both apodans and urodelans, and retains its capabilities in the most primitive family of anurans, Pipidae, in which a hematopoietic bone marrow is already producing blood cells (reviewed in ZAPATA and COOPER 1990). Furthermore, before metamorphosis the liver is an active lymphopoietic organ in most anurans (reviewed in HANSEN and ZAPATA 1998). In the liver of both apodans and urodelans, groups of developing and mature granulocytes, as well as lymphocytes and numerous plasma cells, occur under the connective tissue capsule and among the hepatocytes closely associated with the hepatic sinusoids through which the mature cells migrate. The spleen, however, is thought to be the major site for B development in adult urodeles (HANSEN and ZAPATA 1998).

In anurans, liver, spleen and bone marrow are involved in B-cell formation. Rag expression is detected as early as 3 days postfertilization (dpf) in the *Xenopus* liver and at 5–6/dpf, rearrangements of the IgH loci occur within the organ (reviewed in HANSEN and ZAPATA 1998). The first IgL arrangements take place at 7–13 days (MUßMANN et al. 1998) and, accordingly, the first IgM^+ cells can be detected around 10–12/dpf (HADJI-AZIMI et al. 1990; MUßMANN et al. 1998). IgY positive cells appear late: about 24/dpf in either liver or spleen (HSU and DU PASQUIER 1984) and after metamorphosis, the number of B lymphocytes increases in the liver (CHEN and TURPEN 1995). Bone marrow and spleen are, however, the main loci for B lymphopoiesis in adult *Xenopus* as supported by the expression of both Rag (GREENHALGH et al. 1993) and Tdt (HANSEN and ZAPATA 1998). In contrast, in *Rana pipiens*, the first cytoplasmic Ig positive cells appear in the pronephros; and 2 days later, IgM positive cells occur in both pronephros and liver (ZETTERGREN et al. 1980). In the larval period pronephros and mesonephros are the main B cell-producing organs, whereas in adult frogs, pre-B cells were identified only in the bone marrow (EIPERT et al. 1979; ZETTERGREN et al. 1980).

3.3 The Spleen Is the Major Secondary Peripheral Lymphoid Organ That Can Be Traced Throughout Vertebrate Phylogeny

The spleen, a blood-filtering organ involved in the immune response, appears, as does the thymus, in Chondrichthyes. Its basic histological pattern remains largely unchanged in all vertebrates but there are important species-specific differences in the amount of splenic lymphoid tissue. Moreover, increasing architectural complexity occurs in more evolved anurans and reptiles (reviewed in ZAPATA and COOPER 1990). The organ is divided in two distinct areas, the white pulp and the red pulp, which harbor different cell populations as a reflection of their different functions. The white pulp of all studied fishes consists of lymphoid masses of varying sizes in which lymphocytes, antigen-presenting cells, and plasma cells form cell clusters (Fig. 11). This cell composition implicates fish spleen as a source of antibodies. Presumably, in elasmobranchs (as mentioned above for *Xenopus*) the spleen also contains pre-B cells, suggesting Rag expression in the organ (BERNSTEIN et al. 1994). In fish as well as in apodans and anurans the white pulp and the red pulp are intermingled without any demarcation reminiscent of the marginal zone of mammalian spleen. After immunization, the lymphoid masses of white pulp are,

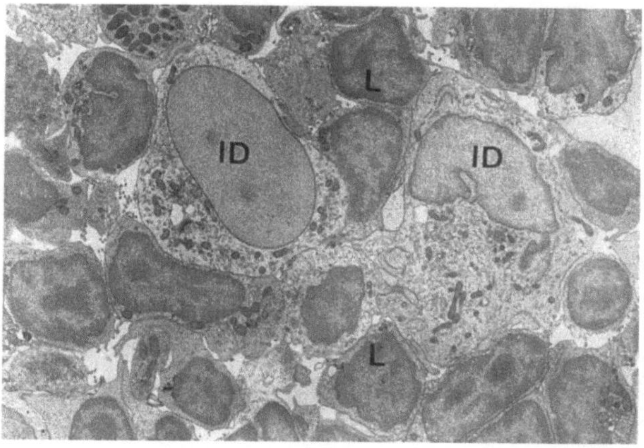

Fig. 11. Clusters of lymphocytes (*L*) and presumptive interdigitating dendritic cells (*ID*) in the spleen of a bichir (primitive bony fish). ×3500

however, more evident and are easily distinguishable from the surrounding red pulp. A "physical barrier" which truly provides evidence of morphofunctional regionalization of the organ appears in the spleen of anurans and is subsequently more defined in reptiles.

Despite the lack of regionalization and of a concrete marginal zone (characteristics of mammalian spleen for antigen trapping and cell trafficking in and out of splenic white pulp), the fish spleen contains ellipsoidal blood vessels presumably involved in antigen trapping. Remarkably, the histoenzymatical pattern of ellipsoidal macrophages from teleost spleen is similar to that of mammalian marginal zone macrophages (Press et al. 1994). Ellipsoids are terminal branches of splenic arteries equipped with a sheath of reticular cells, macrophages, and connective tissue fibers, which are prominent in some fish. This is especially true in elasmobranchs, in which lipid deposits accumulate (Dustin 1975; Zapata et al. 1996a). Ellipsoids trap various substances including antigens, nonantigenic materials, invading pathogens, degenerated blood cells, etc. (Fig. 13), and thus are imperative for initiating specific immune responses.

Peripheral lymphoid organs of lower vertebrates do not contain germinal centers even after repeated immunization. This phenomenon could be related to the occurrence of relatively few follicular dendritic cells in these animals and/or with their inability to form a network of immune complexes, thus retaining cell processes closely associated with B lymphocytes (reviewed in Zapata et al. 1995; Hansen and Zapata 1998). However, some authors have suggested that the so-called melanomacrophage centers (MMCs) which occur in the fish lymphoid organs could be phylogenetic precursors of the germinal centers of endothermic vertebrates (Lamers 1985). The MMCs appear as isolated melanomacrophages (Fig. 13) and/or large cell clusters (Figs. 12, 14) in all the fish studied, including Agnatha, but are also observed in amphibians and reptiles (reviewed in Zapata et al. 1996b). MMCs

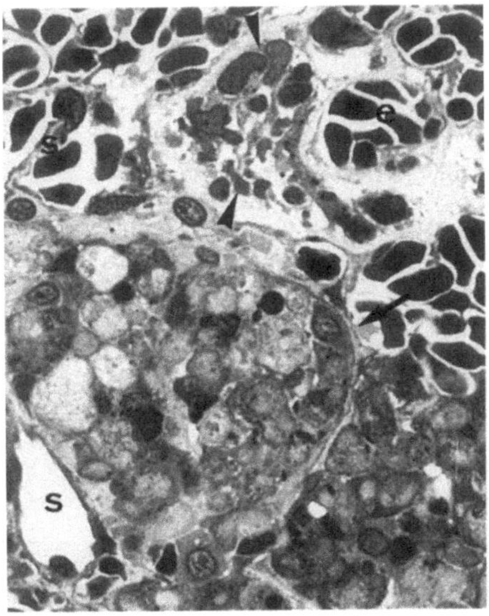

Fig. 12. Spleen of a sheep erythrocyte immunized goldfish, *Carassius auratus* (teleost fish). Note the enlarged walls (arrowhead) of the ellipsoidal blood vessels (*e*) and a large, partially encapsulated melanomacrophage center (*arrows*). Sinusoidal blood vessels (*S*) occur in the red pulp. ×500

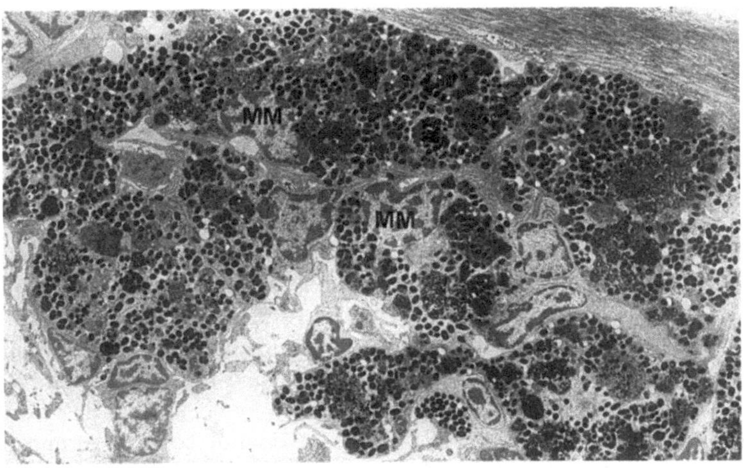

Fig. 13. Group of isolated melanin-containing macrophages (*MM*) in the olfatory sac of a bichir (primitive bony fish). ×4150

are aggregates of closely packed macrophages which contain diverse materials, principally lipofucsin, melanin, and hemosiderin. Teleost MMCs frequently appear in the axillary of the ellipsoidal branches of the spleen, totally or partially encapsulated (Fig. 12) and sometimes associated with a lymphocytic cuff and infiltrated with granulocytes and pyroninophilic cells.

MMCs are merely scavengers in which macrophages accumulate. Other authors have also pointed out that the function of MMCs is the centralization of

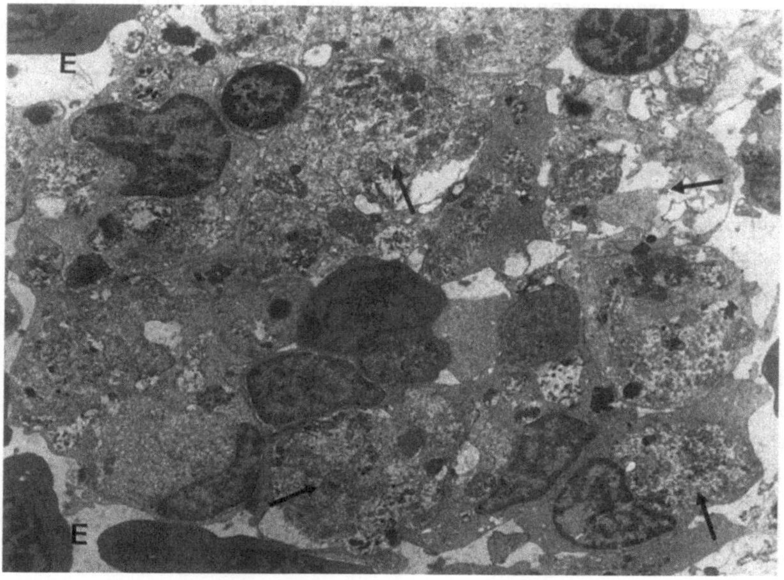

Fig. 14. Melanomacrophage center in the splenic red pulp of goldfish, *Carassius auratus* (teleost fish). The cell limits are difficult to see and numerous cell debris occur (*arrows*) inside the center. Erythrocytes (*E*). ×8560

endogenous and exogenous materials for further destruction, detoxification or recycling (VOGELBEIN et al. 1987). TSUJII and SENO (1990) claimed that the formation of large MMCs in the fish lymphoid organs could be related to a poor enzyme repertoire of fish macrophages. We have reported that phagocytic cells of the largest MMCs represent residual elements that show an exhausted lysosomal activity (HERRAEZ and ZAPATA 1986, 1991). Thus, the number and size of MMCs were shown to increase in the spleen and kidney of goldfish immunized with sheep red blood cells (SRBC), but not after administration of formalin-killed *Yersinia ruckeri* (HERRAEZ and ZAPATA 1986, 1987). In both cases the antigens were trapped by the ellipsoidal macrophages of spleen (Fig. 12) and the endothelial cells of renal sinusoids. Consequently, splenic macrophages and phagocytic cells of kidney transported the engulfed antigens to the MMCs. However, ultrastructural analysis demonstrated that bacteria and not erythrocytes, were completely digested in the cytoplasm of phagocytic melano-macrophages and that subsequent changes in the number or size of MMCs were not induced.

The spleen in both apodans and urodeles consists mainly of a red pulp with a few, small lymphoid aggregates. Electron microscopy shows these lymphoid aggregates to contain lymphoid cells, mature granulocytes and giant, pigment-containing phagocytic cells. In addition, in the cell cords of red pulp there are numerous lymphoid cells. After antigenic stimulation, the number and size of the lymphoid aggregations increase, becoming relatively distinct from the red pulp (ZAPATA and COOPER 1990).

The histological organization of anuran spleen has not been well-studied, with the exception of *Xenopus laevis* (TURNER and MANNING 1973; BALDWIN and COHEN 1981; BALDWIN and SMINIA 1982; BALDWIN 1983) and *Bufo calamita* (BARRUTIA et al. 1983, 1985a,b). Despite some phylogenetic distance between these two species in both cases the lymphoid tissue is concentrated in follicles scattered throughout the splenic parenchyma totally or partially isolated from the red pulp by a layer of flat reticular cells through which there is an intense cell traffic (Fig. 15). In the lymphoid follicles supporting reticular elements, macrophages, presumptive interdigitating dendritic cells, lymphocytes, and both mature and developing plasma cells have been ultrastructurally identified (BARRUTIA et al. 1983, 1985b). Remarkably, in both species, giant, nonlymphoid cells which resemble the antigen-trapping dendritic cells of mammalian peripheral lymphoid organs are seen (Fig. 16; BALDWIN and COHEN 1981; BALDWIN and SMINIA 1982; BARRUTIA et al. 1983, 1985a).

Direct and indirect evidence confirms the occurrence of both T- and B-lymphocytes in the *Xenopus* spleen. Flow cytometric studies demonstrate that about 65%–70% of splenocytes express the T-cell marker CD5 (no $CD5^+$ B cells occur in *Xenopus*; JURGENS et al. 1995), whereas IgM^+ B cells represent 30% of splenic lymphocytes and 33% are $CD8^+$ cells (reviewed in HORTON et al. 1998). In addition, all splenic lymphocytes express MHC class II molecules (ROLLINS-SMITH et al. 1996). Immunohistochemical studies identify the perifollicular zone of *Xenopus* splenic lymphoid follicles as a thymic-dependent area in which $CD5^+$ and $CD8^+$ lymphocytes occur (HORTON 1994; HORTON et al. 1998). After early thymectomy, $CD5^+$, $CD8^+$ splenocytes are almost undetectable in *Xenopus* spleen. Remarkably,

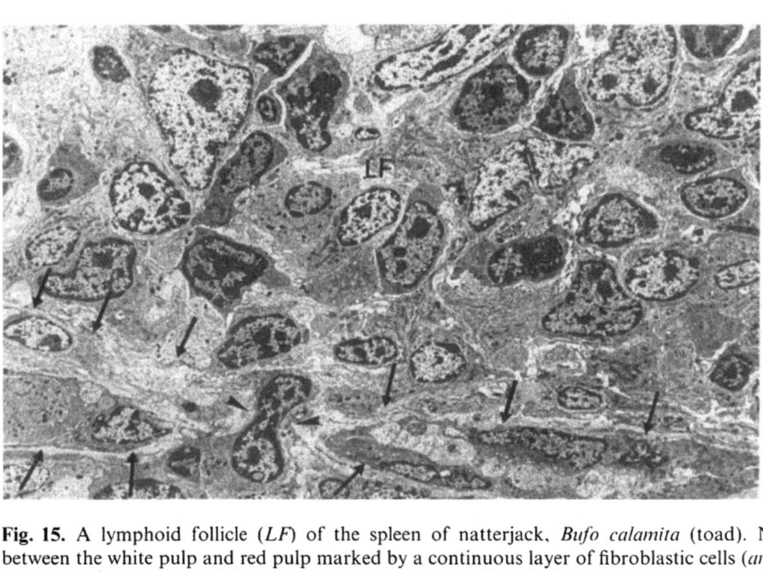

Fig. 15. A lymphoid follicle (*LF*) of the spleen of natterjack, *Bufo calamita* (toad). Note the limits between the white pulp and red pulp marked by a continuous layer of fibroblastic cells (*arrows*). This cell layer is frequently crossed by lymphoid cells (*arrowheads*). ×3350

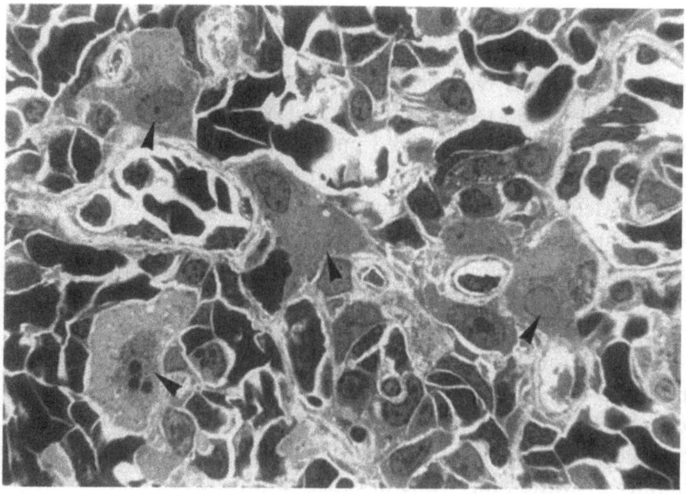

Fig. 16. Groups of giant cells (arrowheads) appear in the spleen of natterjack, *Bufo calamita* (toad), after immunization. ×500

presumptive interdigitating cells, a cell type that occurs in the T-dependent areas of mammalian peripheral lymphoid organs, also occupy this perifollicular area in the spleen of *Bufo calamita* (Barrutia et al. 1985b).

Changes in the splenic lymphoid tissue after immunization are also indicative of the cell content of splenic lymphoid follicles. Carbon particles can be detected in the red pulp shortly after administration (within 1–3h) but in most cases appear later in the lymphoid follicles (Cooper and Wright 1976; Barrutia et al. 1983). Carbon-engulfing cells correspond in all cases to macrophages. Alternatively, extracellular trapping of ^{125}I-labeled *Salmonella adelaide* on dendritic cells of splenic red pulp was found in *Bufo marinus* (Diener and Marchalonis 1970). In other species the trapping is more randomly associated with white pulp (Secombes and Manning 1980). Furthermore, in *X. laevis* and *Rana catesbeiana* larvae (Manning and Turner 1972; Moticka et al. 1973), the injection of human γ-globulin (in adjuvants) and sheep erythrocytes, respectively, induced the appearance of vacuoles in the splenic area that the authors considered to correspond to regions of phagocytic activity. Ultrastructural analysis of these areas after hyperimmunization using adjuvant demonstrated the existence of giant, dividing cells, so-called XL cells, that trap foreign material on the surfaces of their cytoplasmic extensions (Baldwin and Cohen 1981). These cells, which show capacity for antigen trapping and transporting in vivo and in vitro, could function as a bridge between T cells presumably in the marginal zone, and B cells in the lymphoid follicle around the central artery. Consequently, antigens extravasated into the red pulp would be processed by macrophages, trapped on the surface of XL cells, and then transported to the marginal zone of XL cells to be presented to lymphocytes (Baldwin and Sminia 1982; Baldwin 1983).

By comparison, giant cells in *Bufo calamita* occur in the red pulp near the marginal zone rather than in the white pulp (as the XL cells). Sometimes giant cells or their processes penetrate into the lymphoid follicles making contact with perifollicular presumptive T cells (BARRUTIA et al. 1983, 1985a). This suggests that giant cells could move freely throughout the red pulp as well as in and out of the white pulp. Their numbers increase enormously after intraperitoneal immunization with SRBC (Fig. 16) coincident with increased numbers of circulating monocytes in blood sinusoids and/or the red pulp. The presumably giant, dendritic cells of *B. calamita,* which are generated from circulating monocytes, are able to process antigens and cooperate with the lymphocytes in the lymphoid follicles scattered throughout the splenic parenchyma (BARRUTIA et al. 1985a).

A next step in the evolutionary trend of splenic histological organization of vertebrates occurs in reptiles. In some reptilian families the spleen shows increasing compartmentalization which suggests a higher specialization of their immune responses. In contrast to the condition of more primitive vertebrates, in some reptiles the spleen consists mainly of white pulp while in others both the red and the white pulp are well developed (reviewed in ZAPATA and COOPER 1990).

Although the spleen of some lizards does not show good demarcation of the white pulp from the red pulp, in the chelonians (turtles) studied, there is a definitive demarcation into a red and a white pulp (MURATA 1959; BORYSENKO and COOPER 1972; ZAPATA et al. 1981d). In addition, the lymphoid tissue of white pulp can be divided in the so-called periarteriolar lymphoid sheath (PALS) and the periellipsoidal lymphoid sheath (PELS). In *Mauremys caspica*, the PELS comprises an outer zone devoid of reticular fibers and an outer one containing reticular fibers circumferentially arranged (ZAPATA et al. 1981d). Ellipsoids are absent in the spleen of some lizards and snakes (MURATA 1959), although they have been described in the spleen of *Uromastyx aegyptia* (HUSSEIN et al. 1978b) and could be lacking in that of the turtles *Lessemys punctatus* and *Chelydra serpentina* (ZAPATA and COOPER 1990).

The ultrastructure of splenic lymphoid tissue has been principally studied in the chelonian *M. caspica* (ZAPATA et al. 1981d; LECETA and ZAPATA 1991). In the PALS, small lymphocytes, lymphoblasts, and moderately phagocytic interdigitating cells predominate. The inner zone of PELS contains small and medium lymphocytes associated with large dendritic cells. These dendritic cells trap immune complexes resembling the follicular dendritic cells of mammalian germinal centers (KROESE and VAN ROOIJEN 1983; KROESE et al. 1985). They have little phagocytic capacity (KROESE and VAN ROOIJEN 1983; KROESE et al. 1985; LECETA and ZAPATA 1991), are weakly positive for acid phosphatase, negative for nonspecific esterase and ATPase (KROESE et al. 1985), and extend their cell processes between the surrounding lymphocytes, forming cell clusters (ZAPATA et al. 1981d). By comparison, in the outer zone, there are less lymphoid cells and many more reticular cells and macrophages which engulf carbon particles (LECETA and ZAPATA 1991).

This compartmentalized histological organization of the spleen of some reptiles suggests the existence of a specialized immune function(s) in each splenic area. Experiments using thymectomy as well as characterization of the splenic lymphoid tissue with specific antisera, suggest that splenic PALS of reptiles consists predominantly of T-cells, whereas in the PELS, B lymphocytes are more abundant. In the lizard *Calotes versicolor*, adult thymectomy or treatment with rabbit anti-*Calotes* thymocyte serum results in severe depletion of the juxta-arteriolar zone of splenic white pulp (Pitchappan and Muthukkaruppan 1977). Heteroserum specific to IgM of *M. caspica* stains most cells of the inner zone of PELS and some of the outer zone but the PALS is totally negative (Leceta and Zapata 1991).

Furthermore during humoral immune responses lymphoblast and plasma cell maturation occurs basically in the outer lymphocyte sheath of the spleen of *Chelydra serpentina*. Subsequently, mature plasma cells migrate to both sinusoids and cell cords of the red pulp (Borysenko 1985). Also, in *M. caspica*, changes in the splenic lymphoid tissue after a second challenge with SRBC affect mostly PELS (Leceta and Zapata 1986). All of these results together strongly suggest a phylogenetic relationship between the inner zone of PELS of reptiles and the lymphoid follicles of mammalian spleen, although there are no germinal centers in reptilian spleen. Alternatively, the outer zone of PELS could be equivalent to the marginal zone of mammalian spleen, a site of cell migration and antigen trapping.

3.4 Lymphoid Aggregates Which Occur in the Gut Lamina Propria (GALT) of Lower Vertebrates Do Not Represent True, Isolated Lymphoid Organs

Intraepithelial lymphocytes have been morphologically identified in the gut of all vertebrates, including Agnatha, but lymphoid aggregates only appear in the lamina propria from Chondrichthyes. In addition to, these aggregates do not constitute true, isolated lymphoid organs such as the cecal tonsils of birds or the tonsils and Peyer's patches of mammals. They are not encapsulated. The gut-associated lymphoid aggregates (GALT) may attain a relatively large size and a mucosal immune system presumably occurs in all gnathostomata (Hart et al. 1988; Zapata and Cooper 1990; Zapata et al. 1996b).

Small lymphoid aggregates occur in the spiral valve and/or duodenum of several elasmobranchs (Zapata et al. 1996a) and massive lymphoid infiltrates have been reported in the central region of the spiral valve of various sharks, rays (Tomonaga et al. 1986) and in the chain dogfish, *Scyliorhinus canicula* (Hart et al. 1988). In these locations plasma cells seem to be restricted to the lamina propria, whereas granulocytes, macrophages, and lymphocytes also appear in the intestinal epithelium. Tomonaga et al. (1986) indicated, however, that some intraepithelial lymphoid cells of elasmobranch gut contain cytoplasmic Ig. Both IgM-forming cells and IgR-forming cells apparently constitute two distinct B cell subsets in the lamina propria of the skate *Raja kajonei* (Tomonaga et al. 1984).

Different classes of primitive fish, including paddlefish, gar, bowfin and the South American lungfish, *Lepidosiren paradoxa*, possess GALT (Fichtelius et al. 1968; Fänge 1984). Among lower vertebrates the GALT of teleost fish has been particularly studied. As described above for elasmobranchs, it consists principally of differently sized lymphocytes, plasma cells and macrophages. In addition, several types of granulocytes have been identified. Especially relevant is the occurrence of PAS positive cells (VALLEJO and ELLIS 1989) and eosinophil granular cells (BERGERON and WOODWARD 1982), which could mediate hypersensitive reactions in the teleost gut. Macrophages of the intestinal epithelium and of the lamina propria could be involved in scavenging and/or antigen presentation. In carp they show significant differences from macrophages isolated from other lymphoid organs, a condition also found in mammals. Intestinal macrophages adhere poorly to glass or plastic, form close cell clusters with lymphocytes, express antigen determinants on their outer surface and, presumably, bind Ig (ROMBOUT et al. 1986, 1989, 1993; ROMBOUT and VAN DER BERG 1989).

In amphibians, including both anurans and urodeles, lymphoid aggregates appear also in the gut (GOLDSTINE et al. 1975; ZAPATA and COOPER 1990). In *Pleurodeles waltlii,* they occur predominantly in the small intestine, but a notable development of lymphoid tissue in the large intestine and cloaca is lacking (ARDAVIN et al. 1982). These lymphoid accumulations infiltrate the intestinal epithelium producing decreased numbers of mucous cells and numerous discontinuities of basement membrane. In anurans, development of GALT is greater than in urodeles and lymphoid tissue occurs throughout the gut from the small intestine to cloaca, but it shows a similar histological organization.

"Classical" reports considered the lymphoid tissue associated with the reptilian gut as a phylogenetic precursor of the avian bursa of Fabricius. More recent studies combining ultrastructural and immunological analysis have, in contrast, confirmed its relationship to the mucosal immune system. Apart from some lizards (e.g., *Agama stellio* and *Chamaeleon chamaeleon*) in which no GALT has been reported (EL RIDI et al. 1981) other studied species contain numerous densely cell populated lymphoid aggregates throughout the gut (ZAPATA and COOPER 1990). In fact, the development of reptilian GALT is greater than in any other ectothermic vertebrate.

As in elasmobranchs, Ig positive cells in teleosts are largely restricted to the lamina propria whereas intraepithelial lymphocytes are Ig negative, probably representing T lymphocytes or natural killer cells (BIELEK 1988; ROMBOUT et al. 1993). A similar situation has been found in the gut of sea bass, *Dicentrarchus labrax*. In this species, most lymphoid cells identified in the GALT are recognized by a monoclonal antibody DLT 15 which reacts with antigenic determinants expressed on thymocytes and peripheral T cells (SCAPIGLIATI et al. 1995; ABELLI et al. 1997). Furthermore, reverse transcriptase polymerase chain reaction experiments on DLT 15 positive gut lymphocytes demonstrate that these cells express the TcRβ chain (SCAPIGLIATI et al. 1997) supporting their T cell nature. The sea bass intestine is particularly enriched with DLT 15 positive cells as compared with the upper gut levels (ABELLI et al. 1997).

In comparison, cells producing IgX (an analogous, but not homologous, Ig of mammalian IgA) are located chiefly in the gut epithelium of adult *Xenopus laevis* (MUßMANN et al. 1996) and TOCHINAI (1976) reported that *Xenopus* GALT is not affected by early thymectomy. In contrast, indirect evidence suggests the occurrence of thymus-derived cells in the lymphoid aggregates of small and large intestine, but not in those of esophagus of the lizard, *Chalcides ocellatus* (AKEF 1978). In addition, few plaque-forming cells have been found in the reptilian cloaca (ROTHE and AMBROSIUS 1968; KANAKAMBIKA and MUTHUKKARUPPAN 1972) or in the intestine (SIDKY and AUERBACH 1968), although we identified numerous plasma cells in both locations (ZAPATA and SOLAS 1979; SOLAS and ZAPATA 1980).

Although the existence of a common mucosal immune system has been claimed for all lower vertebrates, the issue has been analyzed only to a limited extent in teleosts. On the other hand, the evolutionary origins of IgA as well as the nature of secretory antibody activity in primitive vertebrates remains largely unknown. Antigen administration results in increased numbers of intraepithelial leukocytes (DAVINA et al. 1982), induces the production of specific antibodies in the mucosae and bile but not in the serum (HART et al. 1988; DAVIDSON et al. 1993), and serves as a "protective" immune response in skin mucus (HART et al. 1988). Also in both amphibians and reptiles increased numbers of lymphoid aggregates appear after intraperitoneal and intracloacal immunization (ARDAVIN 1981; SOLAS et al. 1981) and in general a direct correlation between the number of GALT and humoral immune responses has been established in some lizards (EL DEEB 1978; El RIDI et al. 1981). IgM has been identified in the gut mucous and bile of both elasmobranchs and teleosts (HART et al. 1988). However, except for the biliary Ig of sheepshead (teleost) which exhibits important differences in both structure and molecular weight to the serum H chain (LOBB and CLEM 1981), only slight differences have been found between mucosal and serum Igs of fish (HART et al. 1988; FUDA et al. 1992; ROMBOUT et al. 1993). Similar Ig levels have been reported in the bile and serum of dogfish (HART et al. 1988), whereas in sheepshead and carp there are lower values in the biliary secretions (LOBB and CLEM 1981). On the other hand, no secretory component in the intestine and bile has been reported in dogfish (HART et al. 1988) and teleosts (LOBB 1987; ROMBOUT et al. 1993), but it has been described in the skin mucus of sheepshead and the serum of the nurse shark, *Ginglymostoma cirratum*, where it contains molecules showing a high affinity for mammalian secretory component (UNDERDOWN and SOCKIN 1978).

As support for the occurrence of a common mucosal immune system in teleosts, bath immunization of catfish, *Ictalurus punctatus*, enhances secretory immunity but is unable to stimulate the systemic immune system (LOBB 1987). Other data suggest that fish mucosal and systemic immune compartments could both be active although exhibiting different kinetics. Following intraperitoneal immunization with *Aeromonas salmonicida* in the trout, the first antibody-secreting cells appear in the pronephros 2 weeks later, but only after 7 weeks is there a significant response in the intestine. Conversely, antibody-secreting cells appeared at the same time (3 weeks postimmunization) after intraoral antigen administration (DAVIDSON et al. 1994). In both elasmobranchs and teleosts there is evidence of antigen processing in the intestine although the relative relevance of enterocytes and macrophages in the process remains to be elucidated (HART et al. 1988; ROMBOUT et al. 1993). The route of administration is another important feature. In this regard anal immunization appears to provide a better protective response than oral or immersion delivery (HART et al. 1988).

IgY occurs as a secretory isotype in the gut of young axolotls (FELLAH et al. 1985); and in adult *Xenopus laevis*, bile and gut secretions contain both IgM and IgX (JURG 1977), but no secretory components have been found (MUßMANN et al. 1996). VAERMAN et al. (1975) were unable, however, to find IgA in bile and intestinal secretions of the tortoise, *Testudo hermanni*, although they reported the occurrence of IgM.

Apart from the GALT and the occurrence of lymphoid cells in skin, there is some evidence about the lymphoid tissue associated with other mucosae in lower vertebrates. In the snapping turtle, *Chelydra serpentina*, BORYSENKO and COOPER (1972) described lymphoid tissue in the tracheal mucosa and that a few small lymphoid accumulations occur in the lung of both reptiles (BORYSENKO and COOPER 1972) and anurans (MANNING and HORTON 1969). On the other hand, diffuse lymphoid tissues permanently appear in the avian pineal system and Harderian gland (reviewed in ZAPATA and COOPER 1990).

3.5 The Lymph Nodes Appear Late in Vertebrate Phylogeny

3.5.1 Lymphomyeloid Structures in Ectothermic Vertebrates

In contrast to the spleen, true lymph nodes occur only in endothermic vertebrates and reach their full development in eutherian mammals. Although a few lymph- and vein-associated lymphoid accumulations have been found in anurans and reptiles, these are histologically very different than mammalian lymph nodes and seem to filter blood rather than (exclusively) lymph. The lymphomyeloid organs of anurans are located in the branchial region of larvae and in the throat and axillary regions of adults (Table 3). They exhibit a similar histological organization consisting of cell cords, which largely contain lymphocytes, macrophages, and plasma cells, arranged in a meshwork of sinusoidal blood vessels and a few lymphatic vessels (COOPER 1976; ZAPATA et al. 1981b).

The lymph gland is a bilateral structure of the branchial chamber near the developing anterior limb (COOPER 1976) which occurs in most frogs (except *Rana temporaria*; PLYTYCZ and BIGAJ 1982) but not in other anuran genera (*Alytes*,

Table 3. Lymphomyeloid organs among three phylads of anuran amphibians

	Xenopus	*Rana*	*Bufo*
Larvae			
Lymph gland	–	+	–
Cavity bodies	+[a]	+	+
Adults			
Propericardial bodies	–	+	+
Procoracoidal bodies	–	+	+
Jugular bodies	–	+	+

[a] Four ventral pairs, two dorsal pairs.

Xenopus, *Bufo*, *Bombina*). They disappear after metamorphosis in some frogs, but persist in other ones (e.g., *Rana esculenta*). COOPER et al. (1975) emphasized but it has thus far not demonstrated the relevance of this organ as a source of stem cells for T- and B-lymphocytes. Because they are ideally located to trap antigens and because their macrophages actively phagocytose injected carbon, lymph glands seem to be rather involved in immune responses to antigen transported via blood and/or lymph. In fact, they respond to foreign erythrocytes by producing blasts and plaque-forming cells, and their removal results in impaired antibody synthesis (COOPER 1976).

Four pairs of ventral cavity bodies and two pairs of dorsal cavity bodies occur in the branchial region of *Xenopus laevis* (TOCHINAI 1975), and 30 pairs of ventral cavity bodies have been described in *Rana pipiens* (HORTON 1971a,b). By the end of metamorphosis, the cavity bodies disappear (HORTON 1971a,b).

The largest lymphomyeloid organ of some adult anurans is the jugular body, a bilateral, encapsulated organ consisting of cell cords, formed by reticular cells and fibers, arranged between sinusoidal blood vessels. Occasional lymphatic vessels, characterized by extremely thin, irregular walls (ZAPATA et al. 1981b) occur in the jugular body. In the cell cords variously sized lymphocytes, lymphoblasts, and plasma cells abound. In addition, granulocytes and macrophages have also been observed (ZAPATA et al. 1981b,c). Jugular bodies trap antigens (DIENER and MARCHALONIS 1970) and after immunization undergo marked pyroninophilia with the appearance of numerous antibody-producing plasma cells (reviewed in ZAPATA et al. 1981b). In *Rana bermeja* immunization with sheep erythrocytes via the dorsal lymph sac resulted in an important increase in both mature and developing eosinophils within the jugular body. Fourteen days after immunization the number of plasma cells begins to decrease and accumulations of degranulated eosinophils appear (ZAPATA et al. 1981c).

The condition in reptiles becomes slightly more complex. Lymphoid tissue has been found in the axillae of the gecko, *Gehyra variegata*, associated with both lymphatic and blood vessels (JOHNSTON 1973). In other reptiles, including the lizard *Trachydosaurus rugosus*, the turtle, *Chelydra serpentina*, and the snake, *Elaphe quadrivirgata*, similar aggregates of lymphoid tissue associated with the lymphatic network have been histologically detected (BIGGS 1957; KOTANI 1959). The axillary sinus of gecko could be continuous with the lateral lymphatics and, thus, the lymph flows into the sinus and bathes the vein and vascular lymphoid tissue. This anatomical organization represents an important difference between the gecko's organs and the described lymphomyeloid organs of anurans, in which all blood cell lineages may be found. In this regard the gecko lymph nodules appear to be closer to the mammalian lymph nodes. In fact the primitive architecture of the lymphatic nodules of gecko, i.e., a lymphoid tissue supported by a reticular meshwork surrounding a blood vessel and all enclosed within a lymphatic vessel resembles the lymph nodes described in the echidna. In this monotreme mammal several lymph nodules lie within the lumen of a lymphatic vessel supported by a vascular bundle. Neither primary follicles nor medulla occur in this primitive lymph node (DIENER et al. 1967).

3.5.2 Lymph Node Structures in Endothermic Vertebrates

Three types of histologically different lymph nodes have been described in birds although they could reflect distinct developmental stages and/or functional activities (BIGGS 1957; MCCORKLE et al. 1979; OLAH and GLICK 1983). The simplest ones represent nonencapsulated lymphoid infiltrates embedded in the fat tissue. A second group corresponds to slightly bigger lymphoid aggregates that may develop germinal centers after antigenic immunization. These two types of avian lymph nodules do not contain really lymphatic vessels and the germinal centers appear far from the lymphatic trunk, presumably receiving antigen stimulus from the blood circulation instead of the lymphatics. True lymph nodes possess well-developed lymphatic sinuses, lymphaticovenous anastomosis in the periphery and indirect evidence suggests that this could exhibit a certain regionalization with distinct T- and B-dependent lymphoid areas.

4 Concluding Remarks, Caveats, and Future Prospects

This chapter gives a broad overview of the phylogeny of vertebrates, with special emphasis on "primitive" species and fishes. Phylogenetic hypotheses are provided for chordates from Ascidians to the mammals, and systematic terminology concerning certain groups is clarified. This phylogenetic framework provides the basis for which evolutionary studies on the immune system are performed. The second part of this chapter uses this phylogenetic framework for surveying the immunological structures and organization of immune cells and tissues. These observations convincingly show that just as there is substantial variability in the form and bauplan of vertebrate taxa, there is notable variation in the appearance and organization of immune structures in these organisms. Generalizations can be made to the extent that immune structures from primitive vertebrates, despite their organizational differences, are readily recognizable, and its likely that they function in a similar manner across all species. Other evolutionary trends are noted, including the emergence of GALT, lymphoid organs, spleen and thymus, as well as phylad-specific structures.

Descriptive studies have been carried out at the cellular, molecular, and histological levels on a number of chordate/vertebrate species. On the one hand, we have only scraped the surface of our understanding due to the enormity of the problem. On the other hand, however, it is clear that most of the differences observed are of the "variations-on-a-theme" flavor. There are over 40,000 species of vertebrates (more than half of which are fishes) and in some cases these numbers clearly do not reflect evolutionary history due to differential extinction rates among taxa (ROMER 1966; CARROLL 1988, 1997; FOOTE and SEPKOSKI 1999). This is certainly true, for example, for lungfishes and coelacanths, and for many cartilaginous fishes. Thus, while we feel that future descriptive studies must continue, the choice of taxa needs to be judiciously made and should be steeped in both a phylogenetic and hypothetico-deductive framework in order to extract the most information possible. Practically speaking, it would be highly advantageous if the organism being examined is experimentally pliant.

Most recently, the availability of large-insert cloning vectors such as P1 artificial chromosomes (PACs; IOANNOU et al. 1994; AMEMIYA et al. 1996) and F-factor based bacterial artificial chromosomes (BACs; SHIZUYA et al. 1992) has allowed construction of clone banks from selected lower vertebrate species. These vectors can stably accommodate inserts of over 200 kb pairs as bacterial plasmids, and specific problems regarding immune system evolution can be addressed, in part, by using these reagents. The underlying tenet here is that while the proteins of the immune system are required for function, evolutionary changes occur at the level of the genome. These changes could involve the genes themselves or the manner in which the genes are organized and regulated. Specific problems for which the PAC/BAC system are useful include: (a) identification of putative MHC I- and II-like molecules in very primitive invertebrate species via chromosomal walking from class III type genes (see Rast et al., this volume); (b) identification and characterization of novel gene families that resemble Ig and TcR (see Yoder and Litman, this volume; LITMAN et al. 1999); (c) assessment of the evolutionary dynamics and genomic composition of MHC in primitive vertebrates (GRASER et al. 1998); (d) assessment of the inter-cluster distances in IgH genes of the shark, and subsequent analysis of putative regulatory regions (OTA et al. 1995; AMEMIYA et al. 1995); and (e) modification and subsequent transgenic expression to assay cis-regulatory elements of rag genes (JESSEN et al. 1999; YU et al. 1999). Numerous other permutations of the PAC/BAC system, including their tractability in "functional genomics" experiments, can be invoked to specifically address certain questions. Finally, a more holistic approach to problem-solving should be implemented (DU PASQUIER and FLAJNIK 1999). Indeed, the incorporation of PAC/BAC methodology for gene identification and functional genomics, as well as the availability of expressed sequence tags (EST) databases, should be utilized (when applicable) together with more conventional cellular, molecular, and histological approaches to examine the phylogeny and diversification of the immune system.

Acknowledgements. A.G.Z. is supported, in part, by grant PB98-0332 from the Spanish Ministry of Education and Science. C.T.A. is supported, in part, by the Center for Human Genetics (Boston University School of Medicine) and by a grant from the National Science Foundation (IBN-9905408). We thank Dr. Louis Du Pasquier for helpful suggestions on the manuscript.

References

Abelli L, Picchietti S, Romano N, Mastrolia L, Scapigliati G (1997) Immunohistochemistry of gut-associated lymphoid tissue of the sea bass *Dicentrarchus labrax* (L.). Fish and Shellfish Immunol 7:235–245

Akef M (1978) Immunological studies in some Egyptian reptiles. MSc Thesis. Faculty of Sciences, Cairo University, Egypt

Amemiya CT, Ohta Y, Litman RT, Haire RN, Rast JP, Litman GW (1993) Implications of the immunoglobulin VH gene organization in the coelacanth, *Latimeria chalumnae*. Proc Natl Acad Sci USA 90:6661–6665

Amemiya CT, Ota T, Rast J, Zon L, Litman GW (1996) Organization and evolution of immunoglobulin genes, and identification of genes involved in regulation and ontogeny of the immune system. Current Topics Molec Evol 1:159–168

Amemiya CT, Ota T, Litman GW (1996) Construction of P1 artificial chromosome (PAC) libraries from lower vertebrates. In: Birren B, Lai E (eds) Nonmammalian Genomic Analysis: A Practical Guide, Academic Press, NY, pp 223–256

Amores A, Force A, Yan Y-L, Joly L, Amemiya C, Fritz A, Ho RK, Langeland J, Prince V, Wang YL, Westerfield M, Ekker M, Postlethwait JH (1998) Zebrafish hox clusters and vertebrate genome evolution. Science 282:1711–1714

Ardavín C (1981) Tejido linfoide asociado al tubo digestivo en el anfibio urodelo *Pleurodeles waltlii*. Efecto de la inmunización primaria y secundaria. MSc Thesis, Faculty of Biology, Complutense Univ, Madrid, Spain

Ardavín C, Gomariz RP, Barrutia MG, Fonfría J, Zapata AG (1984)The lympho-hemopoietic organs of the anadromous sea lamprey *Petromyzon marinus*. A comparative study throughout its life cycle. Acta Zool Stockh 65:1–15

Ardavín C, Zapata AG (1987) Ultrastructure and changes during metamorphosis of the lympho-hemopoietic tissue of the larval anadromous sea lamprey, *Petromyzon marinus*. Dev Comp Immunol 11:79–93

Ardavín C, Zapata AG (1988) Lymphoid components of the branchial cavernous body of the ammocoete of *Petromyzon marinus*. Acta Zool Stockh 69:23–28

Ardavín C, Zapata AG, Villena A, Solas MT (1982) Gut-associated lymphoid tissue (GALT) in the amphibian urodele *Pleurodeles waltlii*. J Morph 173:35–41

Bailey WJ, Kim J, Wagner GP, Ruddle FH (1997) Phylogenetic reconstruction of vertebrate Hox cluster duplications. Mol Biol Evol 14:843–853

Baldwin III WM (1983) Antigen trapping cells in *Xenopus laevis*: Tissue distribution. Dev Comp Immunol 7:709–710

Baldwin III WM, Cohen N (1981) A giant cell with dendritic properties in spleens of the anuran amphibian *Xenopus laevis*. Dev Comp Immunol 5:461–474

Baldwin III WM, Sminia T (1982) Antigen trapping cells: advantages of phylogenetic studies. Dev Comp Immunol Suppl 2:59–67

Barrutia MG, Leceta J, Fonfría J, Garrido E, Zapata AG (1983) Non-lymphoid cells of the anuran spleen: An ultrastructural study in the natterjack *Bufo calamita*. Am J Anat 167:83–94

Barrutia MG, Torroba M, Fernández MJ, Vicente A, Zapata AG (1989) Macrophages and epithelial cells of the thymus gland. An ultrastructural study in the natterjack *Bufo calamita*. Tissue and Cell 21: 65–81

Barrutia MG, Villena A, Gomariz RP, Razquin B, Zapata AG (1985a) Ultrastructural changes in the spleen of the natterjack *Bufo calamita* after antigenic stimulation. Cell Tissue Res 239:435–441

Barrutia MG, Razquin B, Gomariz RP, Zapata AG (1985b) Presence of presumptive interdigitating cells in the spleen of the natterjack *Bufo calamita*: Experientia 41:1393–1394

Bergeron T, Woodward B (1982) Ultrastructure of the granule cells in the small intestine of the rainbow trout (*Salmo gairdneri*) before and after stratum granulosum formation. Can J Zool 61:133–138

Bernstein RM, Schluter SF, Lake DF, Marchalonis JJ (1994) Evolutionary conservation and molecular cloning of the recombinase activating gene 1. Biochem Biophys Res Commun 205:687–692

Bielek E (1988) Ultrastructural analysis of leucocyte interaction with tumour targets in a teleost *Cyprinus carpio*. Dev Comp Immunol 12:809–821

Biggs PM (1957) The association of lymphoid tissue with the lymph vessels in the domestic chicken *Gallus domesticus*. Acta Anat 29:36–47

Boeddrich A, Burgtorf C, Francis F, Hennig S, Panopoulou G, Steffens C, Borzym K, Lehrach H (1999) Sequence analysis of an amphioxus cosmid containing a gene homologous to members of the aldo-keto reductase gene superfamily. Gene 230:207–214

Borysenko M (1985) Ultrastructural analysis of normal and immunized spleen of the snapping turtle *Chelydra serpentina*. J Morph 149:243–264

Borysenko M, Cooper EL (1972) Lymphoid tissue in the snapping turtle *Chelydra serpentina*. J Morph 138:487–498

Carroll RL (1988) Vertebrate Paleontology and Evolution, WH Freeman and Co., New York

Carroll RL (1997) Patterns and Process of Vertebrate Evolution, Cambridge University Press, New York

Castillo A, Razquin B, Villena A, Zapata AG, López-Fierro P (1998) Thymic barriers to antigen entry during the posthatching development of the thymus of rainbow trout, *Oncorhynchus mykiss*. Fish and Shellfish Immunol 8:157–170
Chen X, Turpen J (1995) Intraembryonic origin of hepatic hematopoiesis in *Xenopus laevis*. J Immunol 154:2557–2567
Chibá A, Torroba M, Honma Y, Zapata AG (1988) Occurrence of lympho-haemopoietic tissue in the meninges of the stringray *Dasyatis akajei* (Elasmobranchii, Chondrichthyes). Am J Anat 183:268–276
Chretien I, Robert J, Marcuz A, García-Sanz JA, Courtet L Du Pasquier L (1996) CTX, a novel molecule specifically expressed on the surface of cortical thymocytes in *Xenopus*. Eur J Immunol 26:780–791
Chretien I, Marcuz A, Fellah J, Charlemagne J, Du Pasquier L (1997) The T cell receptor β genes of *Xenopus*. Eur J Immunol 27:763–771
Cloutier C, Ahlberg P (1996) Interrelationships of basal sarcopterygians. In: Stiassny, MLJ, Parenti LR, Johnson GD (eds) Interrelationships of Fishes, Academic Press, New York, pp 445–479
Clothier RH, Balls M (1985) Structural changes in the thymus glands of *Xenopus laevis* during development. In: Balls M, Bownes M (eds) Metamorphosis, Claredon Press, Oxford, pp 332–359
Cohen DM (1970) How many recent fishes are there? Calif Acad Sciences 28:341–346
Compagno LJV (1973) Interrelationships of living elasmobranchs. Zool J Linn Soc 53, Suppl 1,15–61
Compagno LJV (1977) Phyletic relationships of living sharks and rays. Amer Zool 17:303–322
Cooper EL (1976) Immunity mechanisms. In: Lofts B (ed) Physiology of the Amphibia. Vol III. Academic Press, New York, pp 163–272
Cooper EL, Brown BA, Wright RK (1975) New ideas on amphibian immunity: the lymph gland, a generator of both T and B cells. Am Zool 11:183–192
Cooper EL, Wright RK (1976) The anuran amphibian spleen. An evolutionary model for terrestrial vertebrates. In: Battisto JR, Streilen JW (eds) Immuno-Aspects of the Spleen. Elsevier/North-Holland, Amsterdam, pp 47–60
Cracraft J (1981) Toward a phylogenetic classification of birds of the world (class Aves). Auk 98:681–714
Curtis SK, Cowden RR, Knagel JW (1979) Ultrastructure of the bone marrow of the salamander *Plethodon glutinosus* (Caudata Plethodontidae). J Morph 123:405–440
Daggfeldt A, Bengten E, Pilstrom L (1993) A cluster type organization of the loci of the immunoglobulin light chain in Atlantic cod (*Gadus morhua* L) and rainbow trout (*Oncorhynchus mykiss* Walbaum) indicated by nucleotide sequences of cDNAs and hybridization analysis. Immunogenetics 38:199–209
Davidson GA, Ellis AE, Secombes CJ (1993) Route of immunization influences the generation of antibody secreting cells in the gut of rainbow trout, *Oncorynchus mykiss* (Walbaum 1792) Dev Comp Immunol 17:373–376
Davidson GA, Ellis AE, Secombes CJ (1994) A preliminary investigation into the phenomenon of oral tolerance in rainbow trout (*Oncorhynchus mykiss*, Walbaum 1792) Fish and Shellfish Immunol 4: 141–151
De Carvalho MR (1996) Phylogenetic interrelationships of Neoselachians. In: Stiassny MLJ, Parenti LR, Johnson GD (eds) Interrelationships of Fishes, Academic Press, New York, pp 35–62
Dempster WT (1930) The morphology of the amphibian endolymphatic organ. J Morph 50:71–126
De Pinna MCC (1996) Teleostean monophyly. In: Stiassny MLJ, Parenti LR, Johnson GD (eds) Interrelationships of Fishes, Academic Press, New York, pp 147–162
Denton TE, Howell WM (1973) Chromosomes of the African polypterid fishes, *Polypterus palmas* and *Calamoichthys calabaricus* (Pisces: Brachiopterygii). Experientia 29:122–124
Diener E, Marchalonis JJ (1970) Cellular and humoral aspects of the primary immune response of the toad *Bufo marinus*. Immunology 18:279–293
Diener E, Wistar R, Ealey EHM (1967) Phylogenetic studies on the immune response. II. The immune response of the Australian echidna, *Tachyglossus aculeatus*. Immunology 13:329–338
Du Pasquier L (1993) Evolution of the immune system. In: Paul WE (ed) Fundamental Immunology Third Edition, Raven Press Ltd, New York, pp 199–233
Du Pasquier L, Flajnik M (1999) Origin and evolution of the vertebrate immune system. In: Paul WJ (ed) Fundamental Immunology, Fourth Edition, Lippincott-Raven Publishers, Philadelphia, pp 605–650
Dustin P (1975) Ultrastructure and function of the ellipsoids of the spleen. Their relationship with fat metabolism and red blood cells. Haematologica 60:136–154
Dzialo RC, Cooper MD (1997) An amphibian CD3 homologue of the mammalian CD 3γ and δ genes. Eur J Immunol 27:1640–1647
Eipert EF, Klempau AE, Lallone RL, Cooper EL (1979) Bone marrow antibody synthesis in *Rana*. Cell Immunol 46:275–280

El Deeb SO (1978) Study of the immune system of some Egyptian lizards. MSc Thesis, Faculty of Sciences, Cairo University, Egypt

El Ridi R, El Deeb S, Zada S (1981) The gut-associated lymphoepithelioid tissue (GALT) of lizards and snakes. In: Solomon JB (ed) Aspects of Developmental and Comparative Immunology I. Pergamon Press, Oxford, pp 233–239

Erdmann MV, Caldwell RL, Moosa MK (1998) Indonesian 'king of the sea' discovered. Nature 395:335

Fänge R (1966) Comparative aspects of excretory and lymphoid tissue. In: Smith RT, Miescher PA, Good RA (eds) Phylogeny of Immunity. University of Florida Press, Gainsville, pp 140–145

Fänge R (1984) Lymphomyeloid tissues in fishes. Vidensk Meddr Dansk Naturh Foren 145:152–158

Fänge R, Zapata AG (1985) Lymphohemopoietic system and bodily defenses of hagfish. In: Good RA (ed) Ontogeny and Phylogeny of the Immune System. The Iwanami Immunology Series, Iwanami Shoten Publ, Japan, pp 59–76

Feduccia A (1996) The Origin and Evolution of Birds, Yale Univ Press, New Haven

Fellah JS, Iscaki S, Vaerman JP, Charlemagne J (1985) Transient developmental expression of IgY and secretory component like protein in the gut of the axolot (*Ambystoma mexicanum*) Dev Immunol 2:181–190

Fellah JS, Kerfourn F, Guillet F, Charlemagne J (1993) Conserved structure of amphibian T-cell antigen receptor β chain. Proc Nat Acad Sci USA 90:6811–6814

Fichtelius KE, Finstad J, Good RA (1968) Bursa equivalents of bursaless vertebrates. Lab Invest 19: 339–351

Fink SV, Fink WL (1996) Interrelationships of ostariophysan fishes. In: Stiassny MLJ, Parenti LR, Johnson GD (eds) Interrelationships of Fishes, Academic Press, New York, pp 209–249

Finstad J, Papermaster B, Good RA (1964) Evolution of the immune responses. II. Morphologic studies on the origin of the thymus and organized lymphoid tissue. Lab Invest 13:490–512

Foote M, Sepkoski JJ (1999) Absolute measures of the completeness of the fossil record. Nature 398: 415–417

Forey P, Janvier P (1993) Agnathans and the origin of jawed vertebrates. Nature 361:129–134

Fuda H, Hara A, Yamazaki F, Kobayashi K (1992) A peculiar immunoglobulin M (IgM) identified in eggs of chum salmon (*Oncorhynchus keta*). Dev Comp Immunol 16:415–423

Fujii T (1982) Electron microscopy of the leucocytes of the typhlosole in ammocoetes with special attention to the antibody producing cells. J Morph 173:87–100

Garcia-Fernandez J, Holland PW (1994) Archetypal organization of the amphioxus Hox gene. *Nature* 370:563–566

Gardiner BG, Maisey JG, Littlewood DT (1996) Interrelationships of basal neopterygians. In: Stiassny MLJ, Parenti LR, Johnson GD (eds) Interrelationships of Fishes, Academic Press, New York, pp 117–146

Goldstine SN, Manickavel V, Cohen N (1975) Phylogeny of gut-associated lymphoid tissue. Am Zool 15:107–118

Grande L, Bemis WE (1996) Interrelationships of Acipenseriformes, with comments on "chondrostei." In: Stiassny MLJ, Parenti LR, Johnson GD (eds) Interrelationships of Fishes, Academic Press, New York, pp 85–115

Graser R, Vincek V, Takami K, Klein J (1998) Analysis of zebrafish Mhc using BAC clones. Immunogenet 47:318–325

Greenhalgh P, Olesen CE, Stainer LA (1993) Characterization and expression of recombination activating genes (Rag-1 and Rag-2) in *Xenopus laevis*. J Immunol 151:3100–3110

Greenwood PH, Rosen DE, Weitzman SH, Myers GS (1966) Phyletic studies of teleostean fishes, with a provisional classification of living forms. Bull Am Mus Nat Hist 131:339–456

Hadji-Azimi I, Coosemans V, Canicatti C (1990) B-lymphocyte populations in *Xenopus laevis*. Dev Comp Immunol 14:69–84

Hadji-Azimi I, Schwager J, Thiebaud C (1982) B-lymphocyte differentiation in *Xenopus laevis* larvae. Dev Biol 90:253–258

Hagen M, Filosa MF, Youson JH (1983) Immunocytochemical localization of antibody-producing cells in adult lamprey. Immunol Letters 6:87–92

Hanley PJ, Hook JW, Raftos DA, Gooley AA, Trent R, Raison RL (1992) Hagfish humoral defense protein exhibits structural and functional homology with mammalian complement components. Proc Natl Acad Sci USA 89:7910–7914

Hansen JD, Zapata AG (1998) Lymphocyte development in fish and amphibians. Immunol Rev 166

Hart S, Wrathmell AB, Harris JE, Grayson TH (1988) Gut immunology in fish: A review. Dev Comp Immunol 12:453–480

Hedges SB, Poling LL (1999) A molecular phylogeny of reptiles. Science 283:998–1001
Helfman GS, Collette BB, Facey DE (1997) The diversity of fishes. Blackwell Science, Toronto
Hennig W (1966) Phylogenetic systematics. Univ Illinois Press, Urbana
Herráez P, Zapata AG (1986) Structure and function of the melano-macrophage centers of the goldfish *Carassius auratus*. Vet Immunol Immunopathol 12:117–126
Herráez P, Zapata AG (1987) Trapping of intraperitoneally-injected *Yersinia ruckeri* in the lymphoid organs of *Carassius auratus*: The role of melanomacrophage centers. J Fish Biol 31 Suppl A, 235–237
Herráez P, Zapata AG (1991) Structural characterization of the melanomacrophage centers (MMC) of goldfish *Carassius auratus*. Eur J Morphol 29:89–102
Holland PW, Garcia-Fernandez J, Williams NA, Sidow A (1994) Gene duplications and the origins of vertebrate development. Dev Suppl 1994:125–33
Holland PW, Garcia-Fernandez J (1996) Hox genes and chordate evolution. Dev Biol 173:382–395
Horton JD (1971 a) Histogenesis of the lymphomyeloid complex in the larval of leopard frog *Rana pipiens*. J Morph 134:1–19
Horton JD (1971 b) Ontogeny of the immune system in amphibians. Am Zool 11:219–228
Horton JD (1994) MPHIBIANS. In: Turner RJ (ed) Immunology: a comparative approach. Chichester, Wiley, pp 101–135
Horton JD, Horton TL, Dzialo R, Granevor I, Minter R, Ritchie, Gartland L, Watson MD, Cooper MD (1998) T-cell and natural killer cell development in thymectomized *Xenopus*. Immunol Rev 166: 245–258
Hsu E, Du Pasquier L (1984) Ontogeny of the immune system in *Xenopus*. II. Antibody repertoire differences between larvae and adults. Differentiation 28:116–122
Hussein MF, Badir N, El Ridi R, Akef M (1978) Effect of seasonal variations on lymphoid tissues of the lizards *Mabuya quinquetaeniata* and *Uromastyx aegyptia* Forsk. Dev Comp Immunol 2:469–478
Ioannou P, Amemiya CT, Garnes J, Kroisel PM, Shizuya H, Chen C, Batzer MA, de Jong PJ (1994) A new bacteriophage P1-derived vector for the propagation of large human DNA fragments. Nature Genet 6:84–89
Ishiguro A, Kobayashi K, Suzuki M, Titani K, Tomonaga S, Kurosawa Y (1992) Isolation of a hagfish gene encodes a complement component. EMBO J 11:829–837
Janvier P (1981) The phylogeny of the craniata, with particular reference to the significance of fossil "Agnathans". J Vert Paleont 1:121–159
Jeffery WR (1997) Evolution of ascidian development. BioScience 47:417–425
Jessen JR, Willet CE, Lin S (1999) Artificial chromosome transgenesis reveals long-distance negative regulation of *rag1* in zebrafish. Nature Genetics 23:15–16
Johnston MLR (1973) Perivascular lymphoid tissue associated with the auxiliar lymph sinus and lateral vein of *Gehyra variegata* (Reptilia: Gekkonidae). J Morph 139:431–438
Jurgens JB, Gartland LA, Du Pasquier L, Horton JD, Gobel TW, Cooper MD (1995) Identification of a candidate CD5 homologue in the amphibian *Xenopus laevis*. J Immunol 155:4218–4223
Kanakambika P, Muthukkaruppan VR (1972) The immune response to sheep erythrocytes in the lizard, *Calotes versicolor*. J Immunol 109:415–419
Kappers JA (1958) Structural and functional changes in the telencephalic choroid plexus during human ontogenesis. In: Wolstenholme GEW, O'Connor CM (eds) Churchill, London, pp 3–31
Kilarski W, Plytycz B (1981) The presence of plasma cells in the lamprey (Agnatha). Dev Comp Immunol 5:361–366
Kotani M (1959) Lymhgefasse, lymphatische apparate und extravasculare. Safbahnen der schlange (*Elaphe quadrivirgata*, Boie). Acta Sch Med Univ Kyoto, Japan 36:121–171
Kroese FG, Leceta J, Dopp EA, Herráez P, Nieuwenhuis P, Zapata AG (1985) Dendritic immune complex trapping cells in the spleen of the snake *Python reticulatus*. Dev Comp Immunol 9:641–652
Kroese FG, van Rooijen N (1983) Antigen trapping in the spleen of the turtle *Chrysemys scripta elegans*. Immunology 49:61–68
Kronenberg M, Brines R, Kaufman J (1994) MHC evolution: a long term investment in defense. Immunol Today 15:4–6
Kumar S, Hedges SB (1998) A molecular timescale for vertebrate evolution. Nature 392:917–919
Lamers CHJ (1985) The reaction of the immune system of fish to vaccination. PhD Thesis, University of Wageningen, Netherlands
Leceta J, Zapata AG (1986) Seasonal variations in the immune response of the tortoise *Mauremys caspica*. Immunology 57:483–487

Leceta J, Zapata AG (1991) White pulp compartments in the spleen of the turtle, *Mauremys caspica*. A light-microscopic, electron-microscopic, and immuno-histochemical study. Cell Tissue Res 266: 605–613

Litman GW, Rast JP, Hulst MA, Litman RT, Shamblott MJ, Haire RN, Hinds-Frey KR, Buell RD, Margittai M, Ohta Y, Zilch AC, Good RA, Amemiya CT (1992) Evolutionary origins of immunoglobulin gene diversity. In: Gergely J, Benczur M, Erdei A, Falus A, Füst GY, Medgyesi G, Petranyi GY, Rajnavölgyi (eds) Progress in Immunology VIII, Springer-Verlag, Budapest, pp 107–114

Litman GW, Anderson MK, Rast JP (1999) Evolution of antigen binding receptors. Annu Rev Immunol 17:109–147

Lobb CJ (1987) Secretory immunity induced in catfish, *Ictalurus punctatus*, following bath immunization. Dev Comp Immunol 11:727–738

Lobb CJ, Clem LW (1981) Phylogeny of immunoglobulin structure and function. XI.I Secretory immunoglobulins in the bile of the marine teleost *Archosargus probatocephalus*. Mol Immunol 18: 615–619

Luer C, Walsh CJ, Bodine AB, Wyffels JT, Scott TR (1995) The elasmobranch thymus: anatomical, histological and preliminary functional characterization. J Exp Zool 173:342–354

Maisey JG (1984) Chondrichthyan phylogeny: A look at the evidence. J Vert Paleontol 4:359–371

Maisey JG (1986) Heads and tails: a chordate phylogeny. Cladistics 2:201–256

Manning MJ, Horton JD (1969) Histogenesis of lymphoid organs in larvae of the South African clawed toad *Xenopus laevis* (Daudin). J Embryol Exp Morphol 22:265–277

Manning MJ, Turner RJ (1972) Some responses of the clawed toad *Xenopus laevis* to soluble antigens administered in adjuvant. Comp Biochem Physiol 42 a, 737–747

Mattisson A, Fänge R (1986) The cellular structure of lymphomyeloid tissues in *Chimaera monstrosa* (Pisces, Holocephali). Biol Bull 171:660–671

Mattisson A, Fänge R, Zapata AG (1990) Histology and ultrastructure of the cranial lympho-haemopoietic tissue in *Chimaera monstrosa* (Pisces, Holocephali). Acta Zool Stockh 71:97–106

McCorkle FM, Stinson RS, Olah I, Glick B (1979) The chicken's femoral lymph nodules: T and B cells and the immune response. J Immunol 123:667–669

McEachran JD, Dunn KA, Miyake T (1996) Interrelationships of the batoid fishes (Chondrichthyes: Batoidea). In: Stiassny MLJ, Parenti LR, Johnson GD (eds) Interrelationships of Fishes, Academic Press, New York, pp 63–84

McKinney EC, Henderson R, Ota Y, Flajnik MF (1997) Cloning and expression of the T cell receptor B chain from the nurse shark. Dev Comp Immunol 21:238

Miyake T, McEachran JD (1991) The morphology and evolution of the ventral gill arch skeleton in batoid fishes (Chondrichthyes: Batoidea). Zool J Linn Soc 102:75–100

Moticka EJ, Brwon BA, Cooper EL (1973) Immunoglobulin synthesis in bullfrog larvae. J Immunol 110:855–861

Mußmann R, Du Pasquier L, Hsu E (1996) Is *Xenopus* IgX an analog of IgA?. Eur J Immunol 26: 2823–2830

Mußmann R, Courtet M, Du Pasquier L (1998) Development of the early B cell population in *Xenopus*. Eur J Immunol 28:2947–2959

Murata H (1959) Comparative studies of the spleen in submammalian vertebrates. II. Minute structure of the spleen, with special reference to the periarterial lymphoid sheath. Bull Yamaguchi Med School 6:83–105

Nakamura H, Nakayo KE, Yasuda M (1986) The ontogeny of thymic myoid cells in the chicken. Dev Growth Differ 8:185–190

Ohno S (1970) Evolution by gene duplication. Springer-Verlag, New York

Ohno S (1973) Fish and nature's extensive experiments with gene duplication. In: Schröder JH (ed) Genetics and Mutagenesis of Fish, Springer-Verlag, New York, pp 213–219

Olah I, Glick B (1983) Avian lymph node: light and electron microscopic study. Anat Rec 205:287–299

Ota T, Rast JP, Margittai M, Litman GW, Amemiya CT (1996) Studies on the early evolution of immunoglobulin genes. Curr Topics Molec Evol 1:151–158

Page M, Rowley AF (1982) A morphological study of pharyngeal lymphoid accumulations in larval lampreys. Dev Comp Immunol Suppl 2:35–40

Partula S, de Guerra A, Fellah JS, Charlemagne J (1995) Structure and diversity of the T cell antigen receptor β-chain in a teleost fish. J Immunol 155:699–706

Passer BJ, Chen CHC, Miller NW, Cooper MD (1996) Identification of a T lineage in the catfish. Dev Comp Immunol 20:441–450

Pebusque MJ, Coulier F, Birnbaum D, Pontarotti P (1998) Ancient large-scale genome duplications: phylogenetic and linkage analyses shed light on chordate genome evolution. Mol Biol Evol 15: 1145–1159

Pendleton JW, Nagai BK, Murtha MT, Ruddle FH (1993) Expansion of the Hox gene family and the evolution of chordates. Proc Natl Acad Sci USA 90:6300–6304

Pitchappan RM, Muthukkaruppan VR (1977) Thymus-dependent lymphoid regions in the spleen of the lizard *Calotes versicolor*. J Exp Zool 199:177–188

Plytycz B, Bigaj J (1982) Tadpoles of *Rana temporaria* do not possess the lymph gland. Dev Comp Immunol 6:781–784

Press CML, Dannevig BH, Landsverk T (1994) Immune and enzyme histochemical phenotypes of lymphoid and non-lymphoid cells within the spleen and head kidney of Atlantic salmon (*Salmo salar* L). Fish and Shellfish Immunol 4:79–93

Quesada J, Villena ML, Agulleiro B (1990) Structure of the spleen of the sea bass (*Dicentrarchus labrax*): A light and electron microscopy study. J Morph 206:273–281

Raviola E, Raviola G (1967) Striated muscle cells in the thymus of reptiles and birds: An electron microscopic study. Am J Anat 121:623–645

Riviere HB, Cooper EL, Reddy AL, Hildemann WH (1975) In search of the hagfish thymus. Am Zool 15:39–49

Rollins-Smith LA, Needham DAP, Davis AT, Blair PJ (1996) Late thymectomy in *Xenopus* tadpoles reveals a population of T cells that persists through metamorphosis. Dev Comp Immunol 20:165–174

Rombout JW, Block LJ, Lamers CH, Egberts E (1986) Immunization of carp (*Cyprinus carpio*) with *Vibrio angillarum* bacteria: Indications for a common mucosal immune system. Dev Comp Immunol 10:341–351

Rombout JW, Taverna-Thiele AJ, Villena M (1993) The gut-associated lymphoid tissue (GALT) of carp (*Cyprinus carpio* L): An immunocytochemical analysis. Dev Comp Immunol 17:55–66

Rombout JW, van den Berg AA (1989) Immunological importance of the second gut segment of carp. I. Uptake and processing of antigens by epithelial cells and macrophages. J Fish Biol 35:13–22

Rombout JW, van den Berg AA, Berg CTGA, Witte P, Egberts E (1989) Immunological importance of the second gut segment of the carp. III. Systemic and/or mucosal immune responses after immunization with soluble and particulate antigens. J Fish Biol 35:179–186

Rombout JW, van de Wal JW, Chompanjen A, Taverne N, Taverne-Thiele JJ (1997) Characterization of a T cell lineage marker in carp (*Cyprinus carpio* L.). Dev Comp Immunol 21:35–46

Rothe F, Ambrosius H (1968) Beitrage zur Immunobiologie poikolothermer Wirbeltiere Zellen bei Schildkröten. Acta Biol Med Ger 21:525–536

Sano Y, Imai C (1961) Über ein blutbildendes Orag in der sella Turcica des Riesensalamanders (*Megalobathrachus japonicus*). Z Zellforsch 53:471–480

Satoh N, Jeffery W (1995) Chasing tails in ascidians: developmental insights into the origin and evolution of chordates. Trends Genetics 11:354–359

Scapigliati G, Mazzini M, Mastrolia L, Romano N, Abelli L (1995) Production and characterization of a monoclonal antibody against thymocytes of sea bass *Dicentrarchus labrax* (L.) (Teleostei, Percicthydae). Fish and Shellfish Immunol 5:393–405

Scapigliati G, Romano N, Meloni S, Baldassini MR, Mazzini M, Abelli L (1997) Preliminary molecular and functional characterization of T-lymphocytes immunopurified from the teleost *Dicentrarchus labrax* (L.). Dev Comp Immunol 21:243 (Abstract)

Schaeffer B, Williams M (1977) Relationships of fossil and living elasmobranchs. Amer Zool 17:293–302

Scharrer E (1944) The histology of the meningeal myeloid tissue in the ganoids *Amia* and *Lepisosteus*. Anat Rec 88:291–310

Schlake T, Schorpp M, Nehls M, Boehm T (1997) The nude gene encodes a sequence-specific DNA binding protein with homologs in organisms that lack an anticipatory immune system. Proc Natl Acad Sci USA 94:3842–3847

Secombes CJ, Manning MJ (1980) Comparative studies on the immune system of fishes and amphibians. I. Antigen localization in the carp *Cyprinus carpio*. J Fish Dis 3:399–412

Shirai S (1996) Phylogenetic interrelationships of Neoselachians (Chondrichthyes, Euselachii). In: Interrelationships of Fishes, In: Stiassny MLJ, Parenti LR, Johnson GD (eds) Interrelationships of Fishes, Academic Press, New York, pp 9–34

Shizuya H, Birren B, Kim UJ, Mancino V, Slepak T, Tachiiri Y, Simon M (1992) Cloning and stable maintenance of 300-kilobase-pair fragments of human DNA in *Escherichia coli* using an F-factor-based vector. Proc Natl Acad Sci USA 89:8794–8797

Sidky YA, Auerbach R (1968) Tissue culture analysis of immunological capacity of snapping turtles. J Exp Zool 167:187–196
Solas MT, Leceta J, Zapata AG (1981) Structure of the cloacal lymphoid complex of *Mauremys caspica*. Dev Comp Immunol Suppl 1:151–156
Solas MT, Zapata AG (1980) Gut-associated lymphoid tissue (GALT) in reptiles: intraepithelial cells. Dev Comp Immunol 4:87–99
Stahl BJ (1967) Morphology and relationships of the Holocephali with special reference to the venous system. Bull Mus Comp Zool 135:141–213
Tanaka Y, Saito Y, Gotoh (1981) Vascular architecture and intestinal hematopoietic nests of two cyclostomes, *Eptatretus burgeri* and ammocoetes of *Entosphenus reissneri*: A comparative morphological study. J Morph 170:71–93
Tatner MF, Manning MJ (1982) The morphology of the trout *Salmo gairdneri* Richardson thymus: some practical and theoretical considerations. J Fish Biol 21:27–32
Tischendorf F (1969) Die Milz. In: Bargmann W (ed) Handbuch der Mikroskopischen Anatomie des Menschen vol VI/6, Springer-Verlag, Berlin
Tochinai S (1975) Distribution of lympho-epithelial tissues in the larval South African clawed toad *Xenopus laevis* Daudin. J Fac Sci Hokkaido Univ Ser VI Zool 19:803–811
Tochinai S (1976) Lymphoid changes in *Xenopus laevis* following thymectomy at the initial stage of its histogenesis. J Fac Sci Hokkaido Univ Ser 6:20–175
Tomonaga S, Hirokane H, Awaya K (1973) The primitive spleen of the hagfish. Zool Mag 82:215–217
Tomonaga S, Kobayashi K, Hagiwara K, Yamaguchi K, Awaya K (1986) Gut-associated lymphoid tissue in the elasmobranchs. Zool Sci 3:453–458
Tomonaga S, Kobayashi K, Kajii T (1984) Two populations of immunoglobulin-forming cells in the skate *Raja kanojei*: their distribution and characterization. Dev Comp Immunol 8:803–812
Torroba M, Chibá A, Vicente A, Varas A, Sacedón R, Jiménez E, Honma Y, Moral M, Zapata AG (1995) Macrophage-lymphocyte cell clusters in the hypothalamic ventricle of some elasmobranch fish. Ultrastructural analysis and possible functional significance. Anat Rec 242:400–410
Trede NS, Zon LI (1998) Development of T-cells during fish embryogenesis. Dev Comp Immunol 22: 253–263
Tsujii T, Seno S (1990) Melanomacrophage centers in the aglomerular kidney of the sea horse (teleost): Morphologic studies on their formation and possible function. Anat Rec 226:460–470
Turner RJ, Manning MJ (1973) Response of the toad *Xenopus laevis* to circulating antigens. II. Responses after splenectomy. J Exp Zool 183:35–46
Underdown BJ, Socken DJ (1978) A comparison of secretory component-immunoglobulin interactions amongst different species. Adv Exp Med Biol 107:503–511
Vaerman JP, Picard J, Heremans JF (1975) Structural data on chicken IgA and failure to identify the IgA of the tortoise. In: Hildemann WH, Benedict AA (eds) Immunologic Phylogeny. Adv Exp Med Biol 64:185–195
Varner J, Neame P, Litman GW (1991) A serum heterodimer from hagfish (*Eptatretus stoutii*) exhibits structural similarity and partial sequence identity with immunoglobulin. Proc Natl Acad Sci USA 88:1746–1750
Vallejo AN, Ellis AE (1989) Ultrastructural study of the response of eosinophil granule cells to *Aeromonas salmonicida* extracellular products and histamine liberators in rainbow trout, *Salmo gairdneri* Richrdson. Dev Comp Immunol 13:133–148
Vogelbein WK, Fournier JW, Overstreet RM (1987) Sequential development and morphology of experimentally induced hepatic melanomacrophage centers in *Rivulus marmoratus*. J Fish Biol 31: 145–153
Yu W, Nagaoka H, Jankovic M, Misulovin Z, Suh H, Rolink A, Melchers F, Meffre E, Nussenzweig MC (1999) Continued RAG expression in late stages of B cell development and no apparent re-induction after immunization. Nature 400:682–687
Zapata AG, Ardavín C, Gomariz RP, Leceta J (1981a) Plasma cells in the ammocoetes of *Petromyzon marinus*. Cell Tissue Res 221:203–208
Zapata AG, Cooper EL (1990) The Immune System: Comparative Histophysiology. John Willey and Sons. Chichester
Zapata AG, Chibá A, Varas A (1996b) Cells and tissues of the immune system of fish. In: Iwama GK, Nakanishi T (eds) The Fish Immune System. Organism, Pathogen, and Environment, Academic Press, Vancouver, pp 1–62
Zapata AG, Fänge R, Mattisson A, Villena A (1984) Plasma cells in adult Atlantic hagfish *Myxina glutinosa*. Cell Tissue Res 235:691–693

Zapata AG, Leceta J, Barrutia MG (1981d) Ultrastructure of splenic white pulp of the turtle *Mauremys caspica*. Cell Tissue Res 220:845–855

Zapata AG, Solas MT (1979) Gut-associated lymphoid tissue (GALT) in reptilia: Structure of mucosal accumulations. Dev Comp Immunol 3:477–487

Zapata AG, Torroba M, Sacedón R, Varas A, Vicente A (1996a) Structure of the lymphoid organs of elasmobranchs. J Exp Zool 275:125–143

Zapata AG, Torroba M, Vicente A, Varas A, Sacedón R, Jiménez E (1995) The relevance of cell microenvironments for the appearance of lympho-haemopoietic tissues in primitive vertebrates. Histol and Histopathol 10:761–778

Zapata AG, Villena A, Cooper EL (1981b) Ultrastructure of the jugular body of *Rana pipiens*. Cell Tissue Res 221:193–202

Zapata AG, Villena A, Razquin B, Cooper EL (1981c) The jugular body in anuran amphibians: role in immunity. Dev Comp Immunol 5 Suppl 1:129–135

Zettergren LD, Kubagawa H, Cooper MD (1980) Development of B cells in *Rana pipiens*. In: Manning MJ (ed) Phylogeny of Immunological Memory, Elsevier/North-Holland, pp 177–185

Zug GR (1993) Herpetology: an introductory biology of amphibians and reptiles, Academic Press, San Diego

Origin of Lymphocyte Lineages

Recombination-Activating Genes, Transposition, and the Lymphoid-Specific Combinatorial Immune System: A Common Evolutionary Connection

J.D. HANSEN and J.F. MCBLANE

1 Introduction

The hallmarks that distinguish the gnathostome (jawed vertebrates) adaptive immune system from the immune system of agnathans (jawless fish), which both diverged from a common ancestor over 500 million years ago (MYA), include the presence of somatically rearranging antigen receptors (immunoglobulins, Ig; T cell receptors, TCR; new antigen receptor, NAR) via recombination activating molecules (Ragl and 2) and specific antigen-presentation molecules (MHC class I and II: DU PASQUIER and FLAJNIK 1999). The process of somatic V(D)J recombination of antigen receptor gene segments, mediated by the Rag gene products, is the most significant characteristic of adaptive immunity in the jawed vertebrates. During this process a complex of Ragl and Rag2 proteins initiates recombination by introducing DNA double strand breaks (DSBs) at evolutionarily conserved recombination signal sequences (RSS). These RSS flank the Ig and TCR (and NAR) germline variable region (VDJ) segments which are found in all gnathostomes (DU PASQUIER and FLAJNIK 1999; SCHATZ et al. 1992). Expression of the Rag proteins is

Basel Institute for Immunology, Grenzacherstrasse 487, 4005 Basel, Switzerland
e-mail: hansen@bii.ch

regulated at both the transcriptional and post-transcriptional levels, and co-expression of Ragl and Rag2 is limited to lymphoid committed cells. In order to guarantee that the V(D)J recombinase acts only within the B and T lymphocyte lineage, the reaction is tightly regulated at multiple stages, including expression of the Rag genes themselves as well as controlled access of the recombination machinery to its substrates within chromatin. To this end, specific *cis*- and *trans*-acting regulatory elements have been maintained during evolution to safeguard this process (Hansen and Zapata 1998).

From an evolutionary perspective, the sudden and simultaneous appearance of Ig, TCR, Rag, and specific antigen presenting molecules in the elasmobranchs is striking. Evidence from the fossil records indicates that the elasmobranch and agnathan fish are separated by only some 100 million years, a relatively short evolutionary window. However, the exact time at which the adaptive immune system emerged remains uncertain. Part of the problem is that there are only two known representatives of agnathans (hagfish and lampreys) among the living, and unfortunately the placoderms, the key evolutionary step between the agnathans and elasmobranchs (sharks/rays/skates), are extinct. These were the first jawed vertebrates and undoubtedly harbor crucial clues as to when, where, and how the adaptive immune response evolved. Several theories have been advanced to explain this rapid emergence, including whole genome duplication events (polyploidization: reviewed in Kasahara, this volume). Recently dissection of the V(D)J recombination mechanism and genetic analyses of the genes and molecules mediating this process have shed light on the possible origin of the antigen receptor loci, lending weight to models proposed when the Ig and TCR genes were first cloned. Two decades ago Sakano and colleagues (1979) first highlighted the similarities between the germline structure of the antigen receptor gene segments and the structure of mobile genetic elements in prokaryotes, the transposons. In both cases, recombination occurs between repeated DNA sequences at the ends of a mobilized DNA fragment. Ten years later, using a recombination-complementation assay, Schatz and colleagues (1989) in Boston identified a single murine genomic DNA library fragment which activated V(D)J recombination in nonlymphoid cells. Amazingly, this 18-kb fragment contained two tightly linked genes, the recombination activating genes Ragl and Rag2 (Oettinger et al. 1990), both of which are essential for V(D)J recombination (Mombaerts et al. 1992; Shinkai et al. 1992). Now, some 20 years after Sakano's original suggestion, formal demonstration of the putative transposition hypothesis seems to be at hand. Recent biochemical data strongly suggest that the Rag genes were introduced into the vertebrate genome via a putative transposable element which subsequently led to the creation of our current antigen receptor recombinatorial system (Agrawal et al. 1998; Hiom et al. 1998; Lewis 1994). Since the Rag genes have been isolated from a variety of vertebrate classes, and the mechanistic details of the recombinatorial process are known, we can utilize this information to gain further insight into the origin of the combinatorial immune system. Therefore this review focuses on aspects related to the mechanistic and genomic similarities of antigen receptor genesis meditated by the Rag locus to those of certain transposition systems, and how the Rag-mediated processes were restricted to the B and T lineages of all gnathostomes.

2 General and Detailed Feature of V(D)J Recombination

2.1 General

All lymphocytes are descendants of a common pluripotent hematopoietic stem cell, and upon differentiation every developing lymphocyte is destined to express a unique antigen-binding specificity. This is not achieved by differential expression of one gene out of an array of millions but rather each cell constructs a unique antigen receptor gene by assembling a number of gene segments into a whole. This site-specific somatic recombination mechanism, responsible for building all Ig and TCR genes, is termed V(D)J recombination. Due to the large number of gene segments available at each of the seven antigen receptor loci (Table 1), most of the diversity in the mammalian Ig and TCR repertoires results from the random assortment of various combinations of gene segments into functional genes. For example, at the Igκ locus in developing mouse B cells, a functional κ light-chain gene results from the joining of two gene segments, by deletion (Fig. 1) or inversion of the intervening DNA; one of approximately 140 variable (Vκ) gene segments is joined to one of four functional joining (Jκ) segments to generate the VJκ coding unit. At most loci additional diversity results from insertion and deletion of bases at the junction between two gene segments. Insertions may be nontemplated base additions (N nucleotides) catalyzed by the enzyme terminal deoxynucleotidyl transferase (Bollum 1974; Lewis 1994) or short stretches of nucleotides which are palindromic to the end of the adjacent coding segment (P nucleotides; Lafaille et al. 1989; McCormack et al. 1989). The same mechanism assembles gene segments into functional coding units at the other antigen receptor loci. A second murine immunoglobulin light chain locus (Igλ) also joins a V to a J segment. At the IgH locus three types of gene segments are joined: variable (V_H), diversity (D_H), and joining (J_H). Following transcription and translation of the rearranged genes, successful pairing of heavy and light-chain (κ or λ) proteins results in a B cell expressing a unique antigen-binding specificity. In T lymphocytes, functional TCR genes are

Table 1. Estimated number of functional gene segments in mice

Locus	IgH	Igκ	Igλ	TCRα	TCRβ	TCRγ	TCRδ
#V	~200	~140	2	75	23	6	>16
#D	12	0	0	0	2	0	2
#J	4	4	4	~40	12	4	2

For more details on the human and murine repertoires, please refer to the following reviews: Ardern et al. 1995; Davis and Chien 1998; Max 1998.

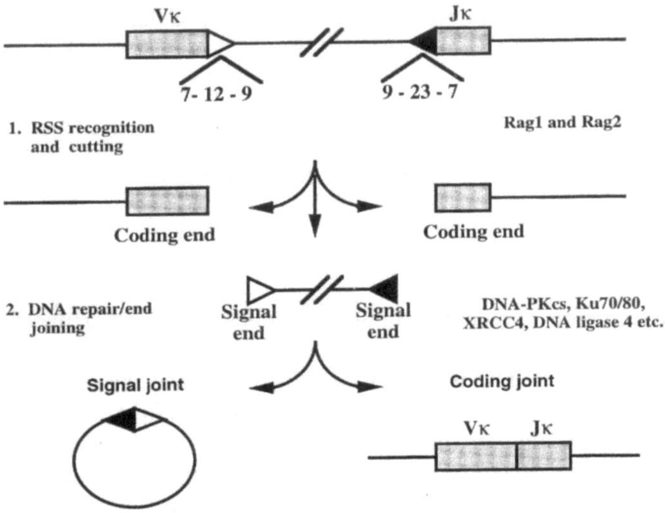

Fig. 1. Assembly of antigen receptor genes by somatic recombination. The pathway of the V(D)J recombination reaction at the Igκ locus via deletion is outlined. *Shaded boxes*, Vκ and Jκ coding segments; *triangles*, RSS. The RSS is composed of a conserved heptamer (7) and nonamer (9) separated by a spacer of 12bp (*open triangle*) or 23bp (*filled triangle*). The reaction occurs in two stages. In the first step, (1 RSS recognition and cutting), the Ragl and Rag2 proteins together bind the RSS sequences and introduce DNA DSB at the border between each RSS and its adjacent coding segment. The products of this DNA cleavage (*middle*) are a pair of signal ends and a pair of coding ends. In the second stage of the reaction (2 DNA repair/end joining), the coding ends are joined by a collection of DSB repair factors to form a coding junction, while the signal ends are joined to form a signal junction

assembled either by joining V and J segments (TCRα, TCRγ), or V, D and J segments (TCRβ, TCRδ; Table 1).

Recombination activity is targeted by the evolutionarily conserved RSS located immediately adjacent to each germline coding segment (Fig. 2). Each RSS is composed of a palindromic heptamer (consensus 5′ CACAGTG 3′) and an A/T rich nonamer (consensus 5′ ACAAAAACC 3′), separated by a spacer sequence of 12 or 23 base pairs (± -1bp; Schatz et al. 1992). Recombination only occurs between two gene segments flanked by RSS of varying spacer length: a 12-spacer RSS recombines with a 23-spacer RSS (the 12/23 rule; Lewis 1994). These criteria ensure assembly of antigen receptor gene segments in the correct configuration.

2.2 Detailed Mechanism

The recombination reaction can be functionally separated into two stages (Fig. 1). In the initial stage a DNA double-strand break (DSB) is introduced at the border between a coding segment and its flanking RSS, generating a signal and a coding end at each segment. This DNA cleavage step is performed by the products encoded within the Rag locus (Rag1 and Rag2). In the second stage a collection of

DSB repair proteins including DNA PKcs, Ku70, Ku80, XRCC4, and DNA ligase 4 mediate imprecise joining of the two coding ends into a coding joint, and precise joining of the two signal ends into a signal joint. This DNA repair step has recently been reviewed elsewhere (CRITCHLOW and JACKSON 1998; GRAWUNDER et al. 1998; KANAAR et al. 1998). The activities and possible origins of the Rag proteins are discussed below.

2.3 DNA Cleavage by Rag1 and Rag2

In vivo both a 12-spacer RSS and a 23-spacer RSS are required for V(D)J recombination to take place. This restriction is also imposed in vitro with purified Rag proteins under reaction conditions containing Mg^{2+} as the divalent cation (VAN GENT et al. 1996b). However, cutting at an isolated signal can also occur in vitro by purified Rags when Mn^{2+} is used instead (MCBLANE et al. 1995; VAN GENT et al. 1995). For simplicity, cutting at a single RSS is described in Fig. 2a. The reaction is initiated by recognition and binding of the RSS by Rag1 and Rag2 (AGRAWAL and SCHATZ 1997; HIOM and GELLERT 1997; SWANSON and DESIDERIO 1998). Once a stable binding complex has formed (containing Rag1, Rag2, and the RSS), DNA cleavage at the border of the RSS proceeds in two steps. In the first step Rag1 and Rag2 introduce a nick immediately 5′ of the heptamer sequence, generating a 3′OH group on the coding side. In the second step this hydroxyl group is used as a nucleophile by the Rag proteins to break the phosphodiester bond in the opposite strand, also precisely at the coding end/signal end border. Both proteins are required for each step of the cleavage reaction, and the efficiency can be improved by the addition of nuclear factors high-mobility group (HMG) 1 or 2, which probably bend DNA and improve binding of the Rag proteins to the RSS (SAWCHUK et al. 1997; VAN GENT et al. 1997). This direct transesterification reaction requires no additional energy source (VAN GENT et al. 1996a) and results in the upper and lower strands of the coding segment being covalently linked into a hairpin structure; the liberated signal end is blunt and 5′ phosphorylated. Hairpin coding ends and blunt signal ends are seen at antigen receptor loci in lymphocytes (ROTH et al. 1992a,b; SCHLISSEL et al. 1993) and are also generated in vitro by purified Rag1 and Rag2 proteins (MCBLANE et al. 1995). Following formation of the DSB, the Rag proteins remain associated in a stable postsynaptic complex with both signal ends (AGRAWAL and SCHATZ 1997; HIOM and GELLERT 1998). In coupled reactions involving both a 12-spacer RSS and a 23-spacer RSS all four broken ends are held in the postsynaptic complex (Fig. 2b; HIOM and GELLERT 1998); this complex may be involved in coordinating the DNA cleavage and rejoining halves of the reaction (RAMSDEN et al. 1997). Before modification of the coding ends and coding joint formation can occur, the hairpins at the coding ends must first be opened. It has been demonstrated that the Rag proteins themselves can perform this function via a hydrolysis reaction analogous to the initial nicking step (BESMER et al. 1998). Whether they perform hairpin opening in vivo remains unclear. Alternatively, a hairpin nuclease may be recruited to the postcleavage

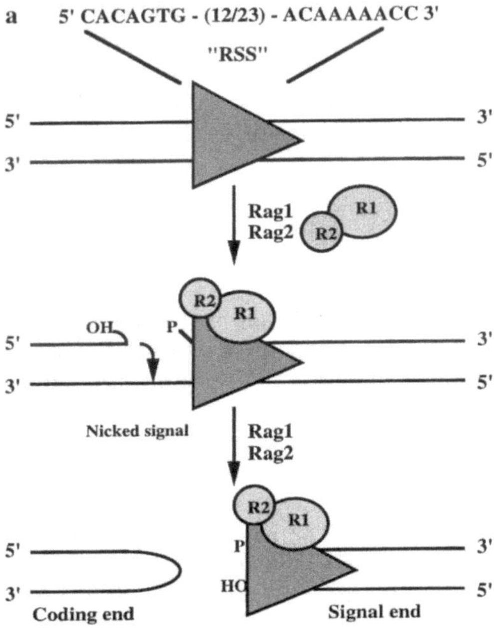

Fig. 2. a The mechanism of DNA cleavage by Ragl and Rag2 (McBlane et al. 1995). The RSS substrate (*top*; *shaded triangle*) is cut in a two step reaction. In the first step Ragl and Rag2 introduce a nick immediately 5′ of the RSS heptamer (shown here as the *top strand*), liberating a 3′OH group on the coding side (*middle*). In the second step this 3′OH attacks the phosphodiester bond in the lower strand (*curved arrow*, *middle*), and is coupled to the phosphate group in the process. The result is a DSB (*bottom*) with a hairpin coding end and a blunt signal end. **b** Signal cleavage proceeds through a series of specific protein/DNA complexes (Agrawal and Schatz 1997; Hiom and Gellert 1997; 1998; Swanson and Desiderio 1998). A pair of RSS sequences (*triangles*) are bound by Ragl and Rag2 (*shaded elipses*) and a synaptic complex is formed. Following signal cleavage, the Rag proteins remain associated with all four cleavage products in a postcleavage synaptic complex (Hiom and Gellert 1998). Hairpin opening and ligation of broken ends results in the coding junction and signal junction. **c** Generation of P and N nucleotide additions at coding junctions. Asymmetrical opening of a coding end hairpin (*top left*) can generate a 5′ flap (*middle left*). Fill-in synthesis, by extension of the opposite strand, would generate an insertion of palindromic (*P*) nucleotides (*blue* in the final junction). Nontemplated base additions (N nucleotides, *red*) may also occur due to the action of the enzyme TdT at open coding ends

synaptic complex. A candidate mammalian hairpin nuclease has recently been identified, composed of a complex containing the Mre11, Rad50, and NBS1 gene products (T.T. Paull and M. Gellert, personal communication). The Mre11 protein, which also has potent DNA exonuclease activity (Paull and Gellert 1998), could also be involved in processing coding ends prior to ligation. Addition and loss of bases at the open coding end results in increased diversity in the coding junction. If the hairpin is opened away from the tip, fill-in synthesis opposite the resulting flap generates insertion of P nucleotides at the coding junction which are palindromic to the end of the coding segment (Fig. 2c). The opened hairpin is also a substrate for the addition of nontemplated N nucleotides by the evolutionarily conserved enzyme TdT (Hansen 1997a) or loss of nucleotides by an exonuclease(s).

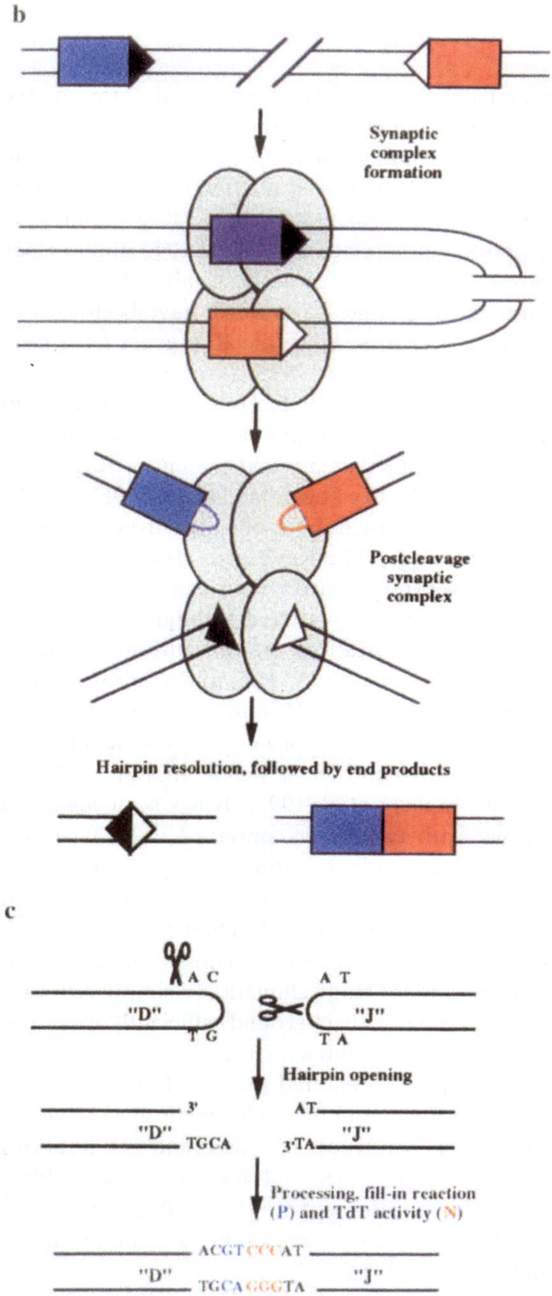
b
Synaptic
complex
formation
Postcleavage
synaptic
complex
Hairpin resolution, followed by end products
c
A C
A T
"D"
"J"
T G
T A
Hairpin opening
3'
AT
"D"
TGCA
3'TA
"J"
Processing, fill-in reaction
(P) and TdT activity (N)
ACGTCCCAT
"D"
TGCAGGGTA
"J"

Fig. 2b,c.

3 Rag Genomics

Using site directed mutagenesis and deletional analysis, truncated forms of the RAG proteins were characterized which define the functionally active "core" regions of Ragl and Rag2 (Cuomo and Oettinger 1994; Kirch et al. 1996; Sadofsky et al. 1994; Sadofsky et al. 1993; Silver et al. 1993). For the murine Rag proteins the active core regions are approximately 60% (Ragl) and 72% (Rag2) of the total protein (Ragl core: amino acids 384–1008, 1040 amino acids total and Rag2 core: amino acids 1–383, 540 amino acids total). These studies enabled purification of soluble forms of Ragl and Rag2 which have been used to dissect their RSS binding and cleavage activities. Since the Rag genes have been cloned from a variety of species (Bernstein et al. 1996; Carlson et al. 1991; Greenhalgh et al. 1993; Hansen and Kaattari 1995; Hansen and Kaattari 1996; Oettinger et al. 1990; Schatz et al. 1989; Willett et al. 1997), we can now examine the level of conservation within both the core and the entire Rag locus. The Ragl core displays an average of 80% identity (including blocks of 95%–100% identity) from the elasmobranchs through mammals, both of which diverged from a common ancestor some 450 million years ago. The Rag2 core also displays high identity (ca. 55%) among the vertebrates (not cloned yet in elasmobranchs). In addition, the nonconserved sites within Rag2 are scattered throughout the entire open reading frame, unlike Ragl in which substitutions are concentrated in the N-terminal region. Within Ragl a unique, well conserved C_3HC_4 dimerization domain is located at the border of the relatively nonconserved N-termini and the conserved core region. This motif is also present in several other proteins, including members of the yeast excision repair machinery, Rad16 and 18, as well as BCRA-1, TNFR, and PML (Hughes and Yeager 1997; Schatz et al. 1992). It has been suggested that this dimerization domain, along with two other conserved C_2H_2-type zinc fingers found within Ragl, is involved in homodimerization of Ragl (Rodgers et al. 1996). Just downstream of this region a homeodomain has been identified which is possibly involved in the association of Ragl with the RSS (Difilippantonio et al. 1996; Spanopoulou et al. 1996). Finally, interspecies domain-swapping experiments have demonstrated that zebrafish-mouse Ragl chimeras or simply zebrafish Ragl transfected into murine cell lines (Ragl$^{-/-}$ background) efficiently carry out signal and coding joint formation of mammalian antigen receptors in vivo (Roman et al. 1997), demonstrating that critical features required for V(D)J recombination have been conserved in all gnathostomes. Furthermore, phylogenetic analysis of the core or entire Ragl and Rag2 amino acid sequences reveals branching positions which are in line with the fossil records (Hansen and Kaattari 1995, 1996; Hughes and Yeager 1997). The relative positions of Ragl and Rag2 within the phylogenetic trees support the notion that these two genes share a common source that predates the emergence of the cartilagenous fish.

One striking feature of the Rag locus is the strong evolutionary conservation of the basic genomic architecture (Fig. 3). In all vertebrates for which the entire locus has been characterized, Ragl and Rag2 are tightly linked and are convergently

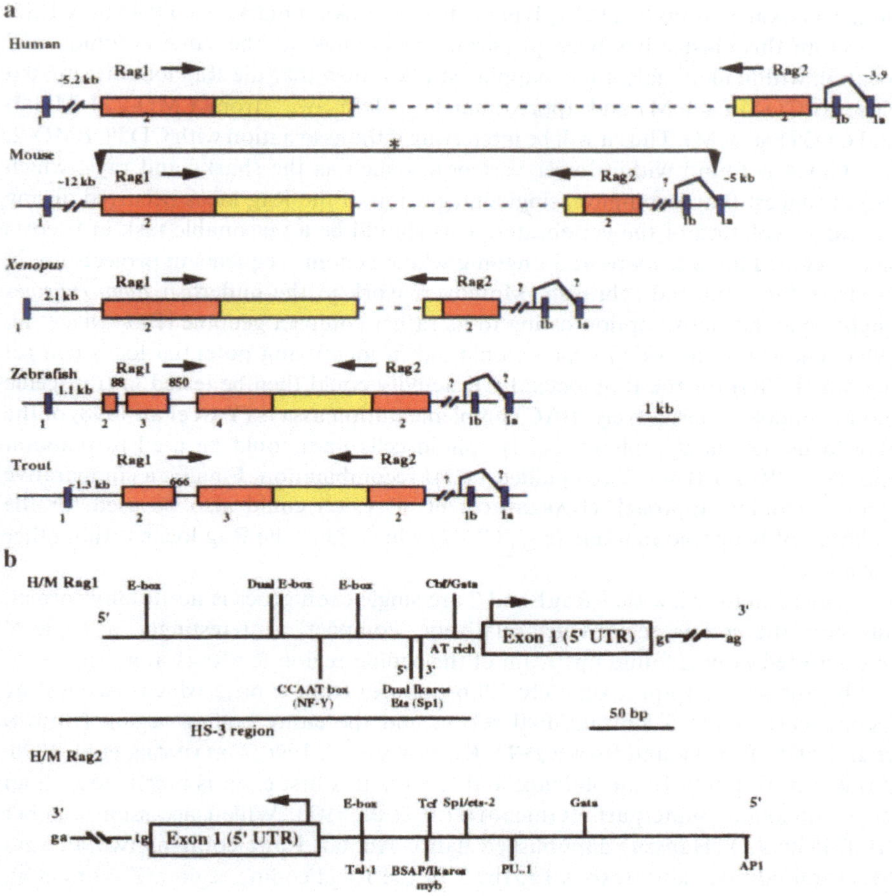

Fig. 3. a Genomic organization of the Rag locus among vertebrates. In all species examined Rag1 and Rag2 are convergently transcribed and are found on a single genomic fragment (GREENHALGH et al. 1993; HANSEN and KAATTARI 1996; OETTINGER et al. 1990; SCHATZ et al. 1992; WILLETT et al. 1997). For Rag1 a single untranslated exon (exon 1) is found upstream of the major coding region (exon 2) and either one or two small untranslated first exons lie upstream of the Rag2 coding exon (exon 2). In humans, alternative splicing dictates the choice for Rag2 exons 1a (3.9kb from coding region) and 1b (0.7kb from coding region; ZARRIN et al. 1997). In contrast to the amphibian and mammalian arrangement, the teleost Rag1 coding exon is split by an internal exon(s). Introns have not been reported within the Rag 2 coding exon. Colors indicate: 5′ untranslated exons (*blue*), coding regions (*orange*) and 3′ UTR (*yellow*). *Asterisk*, the murine genomic fragment that confers basal expression in lymphoid and nonlymphoid expression (DOBBELING et al. 1996; OETTINGER et al. 1990). **b** Schematic of the Rag core promoter regions in mammals (FULLER and STORB 1997; KURIOKA et al. 1996; LAURING and SCHLISSEL 1999; ZARRIN et al. 1997) which contain transcription factor motifs potentially involved in Rag transcription. For Rag2 only Exon 1a and the 5′ flanking region is depicted

transcribed (WILLETT et al. 1997). The Rag locus resides within a gene dense region on chromosome 11p13 and 2 (56cM) in humans and mice, respectively (ICHIHARA et al. 1993; NCBI website). The Rag locus is part of a large syntenic (mouse vs.

human) linkage group (ca. 9cM) lying between flanking markers CD59 and CD82. Therefore this cluster has been physically maintained at the same chromosomal location within mammals. Fine mapping studies show that the Rag locus is situated close to CD44 (< 1cM) and approximately 3–4cM away from LMO-2, HMG-1, and CD59 (ca. 3cM). Thus it will be interesting if the association with CD59, LMO-2, and CD44 is found within lower vertebrates such as the sharks and rays, which would suggest that a common, single integration of the Rag locus occurred during the early evolution of the vertebrates. This should be a reasonable task in teleosts since detailed linkage maps and ongoing whole genome sequencing projects are in progress for trout and zebrafish. Moreover, work in the pufferfish *Fugu rubripes* might be an attractive option owing to its rather compact genome (Brenner et al. 1993; Trower et al. 1996) which would aid in identifying potential locus control regions (LCRs) for the Rag locus. LCR activity could then be tested in transgenic mouse models. Alternatively, BAC complementation assays (Yan et al. 1998) of the Rag locus into nonlymphoid and lymphoid cells lines could be used to pinpoint specific LCRs for B and T cell guided V(D)J recombination. Finally, a comparative synteny cloning approach (Karlstrom et al. 1999) could also be used for the isolation of lymphoid markers (e.g., CD44) which flank the Rag locus within other vertebrates.

The common view that Rag1 and 2 are single exon genes is not totally correct, although the coding regions are relatively "compact". Interestingly, a single 5′ untranslated exon is found upstream of the coding region for Rag1 at a distance of 5.2kb (human) and approximately 12kb (mouse) from exon 2 which encodes the second part of the 5′ untranslated region and the entire coding region (Brown et al. 1997b, Fuller and Storb 1997; Kurioka et al. 1996; Oettinger et al. 1990; Zarrin et al. 1997). In amphibians and teleosts this first exon is much closer than its mammalian counterparts (Greenhalgh et al. 1993; Willett accession number AF074330; J.D. Hansen, unpublished data). Human Rag2 contains two alternatively spliced first exons (exons 1a, 1b) 5′ of the Rag2 coding region (Zarrin et al. 1997). Exon 1a appears to be the predominant choice in pre-B cell lines and thymoctyes. Differing from all other vertebrates, the teleost (trout and zebrafish) Rag1 genomic coding region was found to contain introns other than the split 5′ UTR exon (Hansen and Kaattari 1995; Willett et al. 1997). These two species, which diverged over 120 million years ago, share a common intron of roughly the same size which splits the N-terminal portion of the core. Whether this is specific for teleost fish is not known as the elasmobranch locus has yet to be fully defined. Looking at Fig. 3, it appears that the locus has expanded during the course of evolution. Indeed, the two teleost loci share a rather compact organization in which the 3′ UTRs of Rag1 and 2 contain overlapping polyadenylation sites. This may provide means of regulating Rag activity through a posttranscriptional antisense interference mechanism reminiscent of that used by some classes of transposases to regulate their levels of transposition. Perhaps further evidence for expansion is that only a small portion of the intergenic region is conserved between the two teleost Rag intergenic regions, and little homology is found between the teleost and murine intergenic regions. It should be noted that the zebrafish intergenic sequence contains a DANA-M1 retrotransposon insertion which, interestingly enough, is found in other developmentally regulated genes including LIM-3 and BMP-4.

The promoter region for Ragl resides immediately 5′ of exon 1 (BROWN et al. 1997b; FULLER and STORB 1997; KITAGAWA et al. 1996; Kurıoka et al. 1996; ZARRIN et al. 1997). Similar to other immunologically relevant genes including TdT, lck, V-preB, mb-1,λ5, CD19, and c-kit, Ragl contains a TATA-less promoter. Closer examination of the Ragl promoter region via sequence analysis, DNAse I hypersensitvity and gene reporter constructs defined a minimal core promoter. This core region (Fig. 3b) contains putative lymphoid-related transcription factor binding sites such as AP-1, Ikaros, E-boxes, and Sp-1. However, transfection assays with a variety of reporter constructs showed that this region is not lymphoid specific, indicating that other elements such as enhancers and other modifiers of chromatin structure are probably responsible for lymphoid/lineage-specific expression of Ragl. One intriguing finding is that the 3′ site of the dual Ikaros motif along with the CCAAT box is required for basal expression. Finally, DNAse I hypersensitivity studies reveal an altered chromatin structure coincident with Ragl transcription; four hypersensitive sites were detected within a 24kbp 5′ Ragl flanking fragment, one of which (HS-1) resides within the core promoter element. The other three hypersensitive sites have yet to be fully characterized. Taken together, these data suggest that *cis*-acting regulatory elements and modified chomatin structure are likely key elements in the lymphoid restricted expression of Ragl.

Due to the close proximity of Ragl and Rag2, it was possible that motifs responsible for directing lymphoid-specific expression of both genes could be found in the Rag2 promoter (LAURING and SCHLISSEL 1999; ZARRIN et al. 1997). As for Ragl, B and T lineage specific sequence motifs were also found in the mammalian Rag2 TATA-less promoter including PU.1, E2 a, PAX-5 (BSAP), Ikaros, Lcf-1/lef-1, Cbf, myb, Gata-3, and Tal-1 (Fig. 3b). Again, as with the Ragl promoter, the Rag2 core promoter is not totally sufficient to elicit lymphoid specificity. Interestingly, BSAP (Pax-5) activates transcription of murine Rag2 in transfected B and T cell lines and it binds the Rag2 promoter in vivo suggesting that it may be involved in regulating Rag expression (LAURING and SCHLISSEL 1999). Moreover a genomic fragment encompassing the 5′ end of the Ragl coding exon, the intergenic region and the 5′ flanking region of Rag2 (see Fig. 3), conferred basal transcriptional activity in both lymphoid and nonlymphoid cell lines (DOBBELING et al. 1996; OETTINGER et al. 1990). Thus, *cis*-elements flanking the promoter regions are probably responsible for the lymphoid and stage-specific expression of the Rag genes. Finally, as enhancers by definition can be located 5′, 3′, or within a gene several groups have looked at the Rag intergenic region to identify potential regulatory motifs or significant sequence homologies (BERTRAND et al. 1998a,b; HANSEN 1997b). An array of putative lymphoid specific transcription factor binding sites are found within the vertebrate intergenic region but this region appears not to function as an enhancer (BERTRAND et al. 1998a).

4 Origins of Rag1 and Rag2

Transposons are mobile genetic elements which either move or are copied into new sites in the genome (Berg and Howe 1989; Saedler and Gierl 1996). The transposase enzyme responsible for performing this DNA recombination reaction is encoded within the transposon element and directs a series of characteristic DNA cleavage and rejoining reactions to specific sequences repeated at the transposon termini.

The genomic organization of the Ig and TCR loci themselves first hinted at a possible link between the V(D)J recombination mechanism and DNA transposition (Sakano et al. 1979). Detailed knowledge of the Rag genes and the biochemical activities of their products strengthened this comparison (Lewis and Wu 1997; Thompson 1995). Recent biochemical data sealed this functional link, strongly suggesting that the Rag genes appeared in the vertebrate genome about 450MYA as passengers in a Rag transposon. The evidence accumulated to date in favor of the Rag transposon comes from 4 main sources:

1. Similarities between DNA targets for transposition and V(D)J recombination.

The target within the transposon for binding and cutting by the transposase is a pair of direct or inverted repeats at the transposon ends. The targets for the Rag proteins, a pair of RSS motifs, could also be viewed as a pair of DNA repeats at opposite ends of a DNA fragment undergoing recombination.

2. The genomic architecture of the Rag locus (detailed above).

The key feature here is the compact organization of the Rag locus, with Rag1 and Rag2 being immediately adjacent to each other as they would be if they were encoded within a single transposon. Also, the absence of introns within the coding unit of each gene in most species is suggestive of a bacterial origin.

3. Mechanism of DNA cutting and joining

The Rag-mediated cleavage pathway described above is mechanistically very similar to the chemical steps involved in DNA transposition reactions (Mizuuchi 1992, 1997) and retroviral integration (Engelman et al. 1991; van Gent et al. 1996a; reviewed in Craig 1996). All of these reactions expose a 3′ hydroxyl group as a first step in the reaction, then utilize this hydroxyl to attack a target site phosphodiester bond by transesterification. In V(D)J recombination the target is the opposite strand of DNA and leads to hairpin formation at the coding end. Transposition and integration reactions attack a phosphodiester bond at an unrelated target site, leading to insertion of the mobile element into a new site. Additional mechanistic parallels can be demonstrated by the activity of the Rag proteins in vitro. For example, Rags and HIV integrase are unusual in that they can both use alcohol instead of water as the nucleophile in the DNA cleavage reaction (Engelman et al. 1991; van Gent et al. 1996a; Vink et al. 1991). It has been suggested that this may reflect a similar flexibility in the active site of these nucleases (Besmer et al. 1998). A second mechanistic similarity with HIV integrase is the reversibility of the cutting reaction. HIV integrase catalyzes a reversal of HIV DNA insertion into target DNA in a reaction termed disintegration (Chow et al. 1992). Rag proteins can perform a functionally analogous reaction, reversing the V(D)J cleavage step by joining a hairpin coding end to a signal end to form a "hybrid joint" (Melek et al. 1998). Hybrid joints are also seen in vivo, even in cells deficient in DNA repair activities required for V(D)J recombination (Bogue et al. 1997).

The hairpin coding ends formed during V(D)J cleavage are unusual DNA recombination intermediates, but a number of DNA transposons also apparently move via a hairpin intermediate, including plant transposons Ac/Ds and Tam3 (Coen et al. 1989), *Drosophila* transposon hobo (Atkinson et al. 1993), and transposon Ascot-1 from the fungus *Ascobolus immersus* (Colot et al. 1998). In each case, transposition generates short, palindromic insertions at the vacated excision site. These palindromic insertions are probably formed by hairpin opening and end joining, analogous to the insertion of P nucleotides at V(D)J coding junctions (Fig. 2c). Hairpin intermediates have also been physically demonstrated during excision of the bacterial transposon Tn10 (Kennedy et al. 1998). Therefore, the similarity of the chemical steps during the cutting and joining reactions performed by Rag proteins and transposases/integrases is very suggestive of a common evolutionary origin for these enzymes.

4. DNA transposition by the Rag proteins

The most dramatic parallel between DNA transposition and V(D)J recombination was shown recently by two groups (Agrawal et al. 1998; Hiom et al. 1998; reviewed in Plasterk 1998; Roth and Craig 1998). In these studies, the Rag proteins performed not only DNA cleavage at the RSS, but also transpositional insertion of the RSS-containing cleavage products into unrelated target DNA (Fig. 4). When both 12-RSS and 23-RSS ends were inserted in a concerted transposition reaction, sequencing of the transposition products revealed the presence of 5-bp repeats flanking the RSS insertion sites. These repeats resulted from Rag-mediated joining of the 3′OH groups on the signal ends to opposite strands of target DNA with a 5-bp stagger. Following transpositional insertion and filling-in of the gaps adjacent to each RSS, a 5-bp pair duplication was generated at the target site (Fig. 4). Direct repeats of 2–12bp are generated at the majority of all transposon insertion sites (Saedler and Gierl 1996).

The structure of the antigen receptor loci is itself a compelling argument in favor of a transpositional origin for the immune system (Sakano et al. 1979), with both V(D)J recombination and DNA transposition involving a pair of inverted repeats at the ends of the recombining element. Accumulating lines of evidence including the genomic organization of the Rag genes and the mechanistic similarities with DNA transposases and integrases lends weight to the hypothesis that the Rags were part of a primordial transposon. In the simplest interpretation of this model, the original Rag mobile element contained the Rag transposase, encoded by

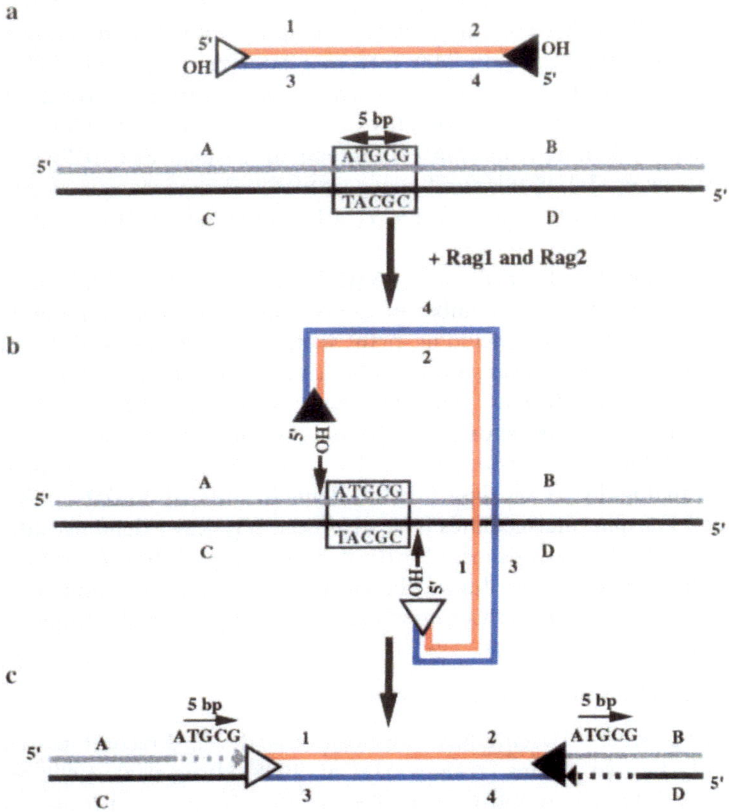

Fig. 4a–c. Model of how Rag-mediated transposition generates a duplication of the 5-bp target site (Hiom et al. 1998; Agrawal et al. 1998). **a** The "transposable element" comprises a linear DNA fragment with RSS motifs (*triangles*) at the ends. The upper and lower strands are indicated in *red* and *blue*; *numbers 1–4* are placed for orientation. The target DNA is shown below the transposable element. It contains a 5-bp sequence (*boxed*) which will be duplicated as a result of the transposition reaction. *A–D* are used to orient the target DNA. **b** In the first step Rag1 and Rag2 use 3′OH groups at the RSS ends to cut 5′ of the target sequence on each strand (*arrows*). By joining the RSS-3′OHs to the target site 5′ phosphates with a 5-bp stagger, a five-nucleotide gap is generated adjacent to each RSS. **c** Filling in these gaps by repair synthesis (*broken arrows*) generates a duplication of the target site

the Rag1 and Rag2 genes, flanked by inverted repeats which resemble the RSS now found at all rearranging loci (Fig. 5). The Rag genes and the RSS elements may have become separated following integration into the vertebrate genome, or they may have been separated during the transpositional event.

These hypothetical transpositional events, catalyzed by the Rag transposase, probably involved the stable integration of the element into the germline of a primitive vertebrate that predates the elasmobranchs, possibly within the now extinct placoderm fish or agnathans. Such an ancestral insertion event could have involved a simple antigen receptor molecule such as CRTAM, which is composed

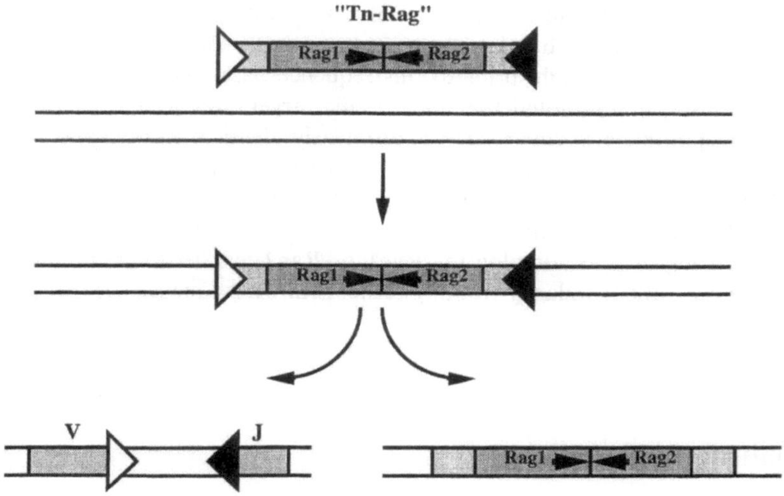

Fig. 5. Rag transposon model (based on AGRAWAL et al. 1998; HIOM et al. 1998). *Top*, the putative ancestral transposon (*Tn-Rag*). It encodes the Rag transposase (Rag1 and Rag2) and terminal repeats (*triangles*) which resemble the RSS motifs which now flank all germline antigen receptor gene segments. Integration into the genome of a primitive vertebrate split a target gene into V and J gene segments. Physical separation of the Rag genes and RSS-like sequences is shown here after integration of the Rag transposon but may also have occurred during the transposition event itself. Expansion of the repertoire of gene segments flanked by RSS elements and generation of rearranging loci probably occurred by genome duplication and translocations

of a single exon V domain, followed by C1-type constant region, a transmembrane domain and cytoplasmic domain (see Du Pasquier, this volume). Thus a transposition event mediated by the Rag genes most likely resulted in the invasion of an RSS module which split a V domain into V and J domains. Furthermore, the short D segments found at some antigen receptor loci may have resulted from ongoing Rag-mediated transpositional events or from further modification of the genome by duplication, inversion, and/or translocation. Other primordial candidate sites for transposition of the RSS cassette include members of the ever growing Tapasin gene family (see Du Pasquier, this volume). Quite possibly precursors of molecules such as NAR (which possesses both Ig and TCR-like features) might harbor remnants of an ancestral incursion of a V domain (ROUX et al. 1998). Expansion of the repertoire of gene segments (i.e., multicluster or translocon) would then occur either by continued transpositional recombination of signal sequence cassettes or, more likely, by genome-wide duplication and translocation events consisting of blocks of segments (AGRAWAL et al. 1998; LEWIS and WU 1997; MARCHALONIS and SCHLUTER 1998; THOMPSON 1995).

It has been suggested that Rag genes are related to prokaryotic invertases and integration host factors such as those found in *Escherichia coli* and *Salmonella* (BERNSTEIN et al. 1996; DIFILIPPANTONIO et al. 1996; SPANOPOULOU et al. 1996). Obviously it will be difficult to verify such a hypothesis without evidence for the existence of a candidate transposon which was responsible for this event. Perhaps the strongest link with a known transposon is to the Hin invertase of *Salmonella*. Interestingly enough, prior to this sequence based association of the Rags and invertases, it was noted that the specific target sites which invertases and certain transposable elements recognize are nearly identical to the conserved elements within the RSS (Difilippantonio et al. 1996; Dreyfus 1992; Hughes et al. 1992; Spanopoulou et al. 1996). Specifically, the invertases recognize a nonamerlike element and the transposases interact with a heptamerlike motif. Although tantalizing, a true homolog for Rag1 or Rag2 has yet to be found as these genes show little similarity to any other known sequence or to one another.

5 Determining Lineage Fate

In all gnathostomes the hematopoietic tissue is derived from the third germ layer, the mesoderm, in response to specific inducing stimuli including BMP-4 and Mix-1. From within the developing mesoderm a common progenitor emerges, the hemangioblast, which gives rise to stem cells responsible for both the vascular (vascular stem cells and blood lineages (hematopoietic stem cells, HSC) in response to factors such as Flk-1 and SCL (Hansen and Zapata 1998; see Fig. 6). The HSC is capable of self-renewal and has the ability to generate a diverse range of blood cell lineages based on several factors. These include cell to cell contact, growth factors, and cascades of transcription factors, all of which dictate determination and later differentiation of the primitive and definitive blood lineages. Gene inactivation studies in mice have provided several candidates for factors which have evolved to dictate blood development including Scl, Gata members, Ets family members, basic helix-loop-helix factors and Ikaros family members (for a review see Anderson and Rothenberg, this volume). In zebrafish, Zon and colleagues have investigated the early development of the lymphoid system by making use of specific mutations affecting blood development and by characterizing key factors involved in the determination and differentiation of HSCs (reviewed in Hansen and Zapata 1998). These critical factors include SCL, GATA-1, GATA-2, Flk-1, Flk-4 and Fli-1. Moreover, PU.1 has also been isolated from two teleost species (zebrafish and trout) and was found to be expressed within the early embryonic ICM (zebrafish) and adult primary lymphoid tissues (trout), suggesting that PU.1 plays a similar role for all vertebrates (Lieschke and Zon, personal communication; J.D.H., unpublished observation).

Ikaros, a key factor for HSC differentiation and lymphoid commitment within all vertebrates, may be critical for restricting Rag expression or antigen receptor accessibility to the lymphoid lineages (Georgopoulos et al. 1997; Hansen and Zapata 1998). By means of alternative splicing Ikaros can generate distinct isoforms which have varying affinity for specific promoter sites

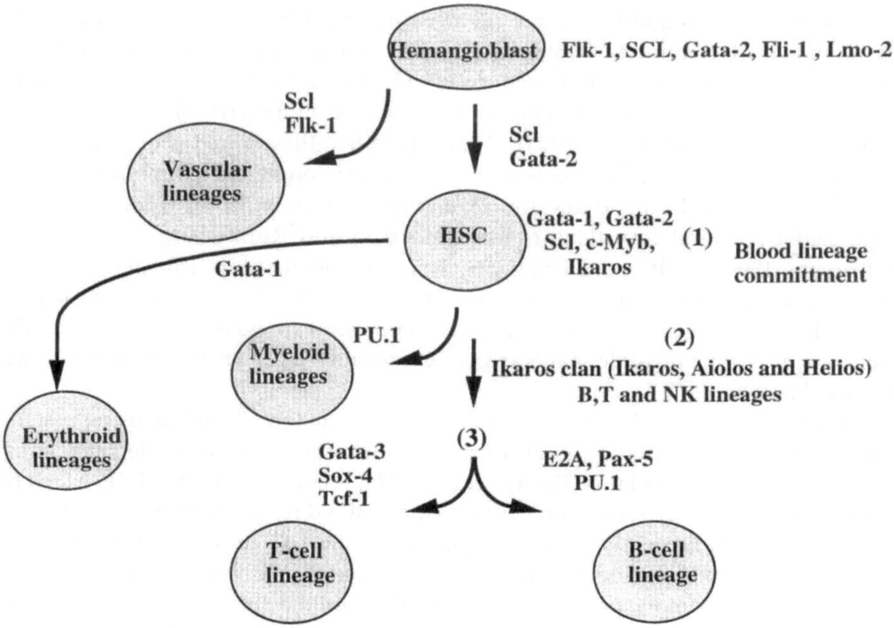

Fig. 6. Hierarchy of transcription factors which dictate blood lineage commitment in vertebrates. Gene targeting has determined several factors which are responsible for specific stages of HSC determination and differentiation leading to the lymphoid lineages as indicated (CLEVERS and GROSSCHEDL 1996; SINGH 1996). Stages at which Rag expression and antigen receptor loci accessibility might be regulated are indicated by numerical designations (*1–3*)

(GEORGOPOULOS et al. 1997; HAHM et al. 1994). Overall Ikaros is believed to regulate genes such as TdT, IL-2r, CD4/CD8, Ragl and 2, B29, TCRα, and CD3 γδε (CLEVERS and GROSSCHEDL 1996; SINGH 1996). Ikaros dominant negative knockout ($DN^{-/-}$) mice show a complete lack of development for the B, T, and natural killer (NK) pathways, demonstrating that this factor is a key determinant for lymphoid commitment and, that a common progenitor is likely to be involved (GEORGOPOULOS et al. 1994).

A second mutation (Ikaros-null) was later introduced in which fetal lymphoid development was lacking (B, T, and NK cells), but postnatally T cell progenitors were found which could colonize the thymus and give rise to oligoclonal T cells of both α/β and γ/δ lineages (WANG et al. 1996). This led investigators to postulate that other factor(s) were involved in the Ikaros pathway for lymphocyte development. The true role of the $DN^{-/-}$ mutation was made clear by the recent finding that Ikaros is actually a member of a small but closely related gene family including two other hematopoietically restricted genes, Aiolos and Helios (HAHM et al. 1998; KELLEY et al. 1998; MORGAN et al. 1997). It was later found that the DN–/– muation generates Ikaros isoforms that are capable of forming inactive heterodimers with Helios and Aiolos. Closer examination of these three genes suggests that Ikaros and Helios are involved in both HSC differentiation and the development of committed lymphoid progenitors (KELLEY et al. 1998). Helios is expressed predominantly within stem cells and in progenitors of the T cell lineage whereas recent data suggest that Aiolos may be more critical for B lineage commitment (KELLEY et al. 1998). Taken together these three genes appear to play key roles in the differentiation of HSC to the committed lymphoid lineages and thus are good candidates for regulating antigen receptor generation. As suggested by several investigators (see Anderson and Rothenburg, this volume; THOMPSON 1995) the B and T cell lineages presumably developed from a common ancestral progenitor, such as a primitive NK lineage progenitor. Owing to the fact that NK development is also dependent upon the expression of Ikaros family members, it is quite possible that Rag gene expression fell under the influence of the Ikaros clan.

Recently a novel role for Ikaros has been suggested in which Ikaros sequesters stage-specific genes into the heterochromatin for transcriptional silencing, thus acting as a negative regulator (BROWN et al. 1997a; KLUG et al. 1998). TdT, which is responsible for N insertions, is expressed in a lymphoid restricted manner similar to the Rag genes, and recently negative regulation of the TdT promoter by Ikaros has been demonstrated by ERNST et al. (1996) and HAHM et al. (1994). As mentioned above, the Rag promoters contain Ikaros motifs; could a similar scenario exist for the Rag locus during lymphoid lineage commitment in which Ikaros acts as a regulator of Rag expression? Overall, it appears that multiple transcription factors are responsible for regulating the development of B and T lineages. Several windows are apparent in which transcription factors can influence the pathways for lineage restriction and differentiation, expression of Rag genes and associated accessory molecules as well as influencing the accessibility of the recombining gene segments within chromatin (Fig. 6).

Finally, it is still difficult to fully explain the rapid appearance of genes involved in the adaptive immune system of vertebrates (Fig. 7). As many groups have failed so far in identifying homologs of Rag, Ig, or TCR in the two living representatives of agnathans, it is conceivable that other factors which regulate the recombinatorial process were already in place before the introduction of the Rag genes. A different approach for analyzing the evolutionary origins of the adaptive immune system would be to characterize those factors that are essential for lymphoidlike commitment in agnathans and to assess their expression patterns. Several hematopoietic transcription factors have already been isolated from elasmobranch and teleost fish (HANSEN and ZAPATA 1998; LITMAN et al. 1999), but little is known about the agnathans in this regard. Recently a unique Ikaros-family member was isolated from a hagfish blood cell cDNA library (Hansen, Cunningham, Flajnik, and Raison, manuscript in preparation) which opens new avenues for examining the immune systems within these elusive beasts. Therefore transcription factors could provide useful tools for addressing questions about the evolution of lymphopoiesis in all vertebrates.

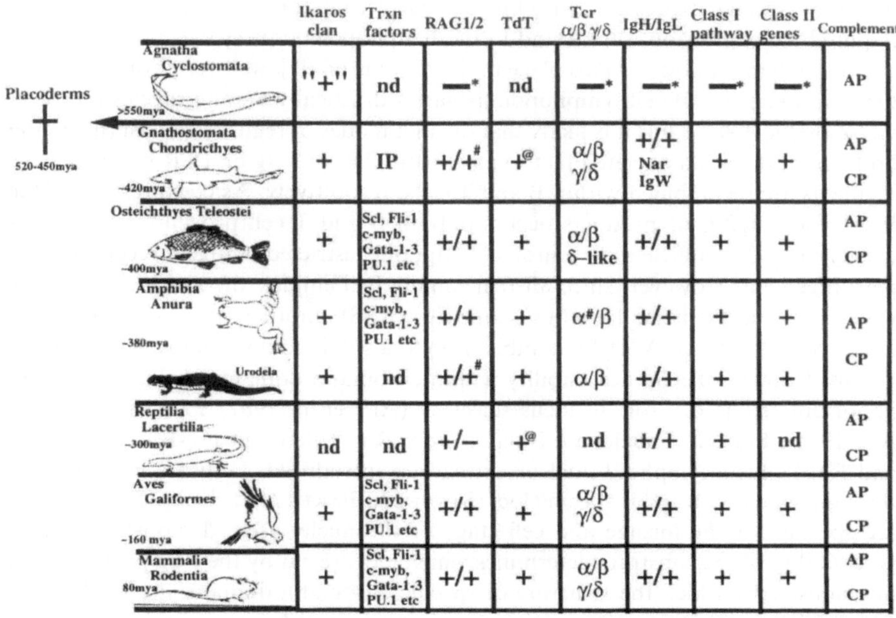

	Ikaros clan	Trxn factors	RAG1/2	TdT	Tcr α/β γ/δ	IgH/IgL	Class I pathway	Class II genes	Complement
Agnatha Cyclostomata >550mya	"+"	nd	—*	nd	—*	—*	—*	—*	AP
Gnathostomata Chondricthyes ~420mya	+	IP	+/+$^{\#}$	+$^{@}$	α/β γ/δ	+/+ Nar IgW	+	+	AP CP
Osteichthyes Teleostei ~400mya	+	Scl, Fli-1 c-myb, Gata-1-3 PU.1 etc	+/+	+	α/β δ-like	+/+	+	+	AP CP
Amphibia Anura ~380mya	+	Scl, Fli-1 c-myb, Gata-1-3 PU.1 etc	+/+	+	α$^{\#}$/β	+/+	+	+	AP CP
Urodela	+	nd	+/+$^{\#}$	+	α/β	+/+	+	+	
Reptilia Lacertilia ~300mya	nd	nd	+/–	+$^{@}$	nd	+/+	+	nd	AP CP
Aves Galiformes ~160 mya	+	Scl, Fli-1 c-myb, Gata-1-3 PU.1 etc	+/+	+	α/β γ/δ	+/+	+	+	AP CP
Mammalia Rodentia 80mya	+	Scl, Fli-1 c-myb, Gata-1-3 PU.1 etc	+/+	+	α/β γ/δ	+/+	+	+	AP CP

Fig. 7. Markers of the adaptive immune system across vertebrate phylogeny. In addition to IgH and IgL, the elasmobranchs possess two unique members, NAR and IgW. *nd*, Not determined; *, tentative; *IP*, in progress; #, assumed presence; @, based upon the presence of N insertions in these species. Class I pathway refers to the presence of class I heavy chains, LMP and TAP. Complement: *AP*, alternative pathway and classical pathway. No information is available for the placoderms, which were extinct prior to the emergence of the cartilaginous and bony fish. Time scales (millions of years ago, *mya*) are based upon the fossil records (CARROL 1988). For more details, please refer to the following reviews (DU PASQUIER and FLAJNIK 1999; HANSEN and ZAPATA 1998; LITMAN et al. 1999; MARCHALONIS et al. 1998)

6 Restricting Antigen Receptor Assembly to the B and T Cell Lineages

If the Rag genes were randomly introduced into the germline of a primitive vertebrate, how did their expression become restricted to cells of the lymphoid lineage(s)? At least four possible scenarios exist: (a) Integration into a chromosomal domain predestined to be expressed within a definitive lymphoid progenitor. This seems unlikely, although the Rag locus is located within a syntenic group that includes a number of genes expressed in lymphoid cells (see Sect. 3), several of these genes are also expressed outside of the lymphoid system. (b) Rag expression was initially more widespread and later selective pressures may have ensured that expression would only be permitted in those cell lineages which could tolerate and benefit from the DNA cuts Rag1 and Rag2 introduce. (c) The Rags were inactive until factors specific for the Rag LCR and/or promoters coevolved to induce lymphoid-specific expression. (d) Only the correct mixture of transcription factors required for expression were found in the lymphopoietic program.

T cell lymphopoiesis takes place in the thymus of all jawed vertebrates whereas the major sites for B cell lymphopoiesis varies drastically among species (Hansen and Zapata 1998). Thus it is likely that distinct modes of regulation exist (including both *cis*- and *trans*-associated factors) to ensure that only Ig or TCR germline gene segments are recombined within B and T cells, respectively. As coexpression of the Rag1 and Rag2 gene products occurs only in B and T cell progenitors, this specificity may be sufficient to enforce lymphoid-restricted antigen receptor gene rearrangement. However, since all rearranging loci employ the same recombinase machinery and the same targeting sequences (RSS), another level of control must have evolved to target V(D)J recombination to a specific subset of loci within B and T cells. Control of RSS accessibility within chromatin domains probably plays an important regulatory role in locus selection (Alt et al. 1987; Yancopoulos and Alt 1985). Indeed, if nuclei are isolated from Rag-1 deficient pre-B or pre-T cells and Rag1 plus a lymphoid nuclear extract are introduced, DSB are generated by the supplemented Rags at specific loci (Stanhope-Baker et al. 1996). These breaks are specific for the lineage and cell stage of the nuclei used, demonstrating that accessibility of the substrate determines which loci are cut by the Rag proteins. The mechanisms by which the structure of specific chromatin domains are modified in preparation for V(D)J recombination remain unclear, although even the most basic unit of chromatin structure, the mononucleosome, inhibits RSS cleavage by Rags (Kwon et al. 1998; McBlane and Boyes, manuscript submitted). The stimulatory influence of Ig and TCR enhancer sequences on the efficiency of V(D)J recombination also suggests that RSS sequences are more accessible within open chromatin domains (Cedar and Bergman 1999; Hempel et al. 1998; Schlissel and Stanhope-Baker 1997; Sleckman et al. 1996). Recently members of the Ikaros family have been found to be involved in chromatin remodeling and deacetylation processes within developing T lymphocytes, suggesting that this family is critical for antigen receptor accessibility (Kim et al. 1999). Thus transcription factors which bind and activate specific enhancers may contribute to chromatin opening and the induction of locus-specific V(D)J recombination. Evolution of a developmentally regulated gene rearrangement process therefore required the acquisition of multiple layers of control. These include lineage-specific expression of the Rag genes and controlled access of the Rag substrates within chromatin. The molecular details of these regulatory processes are still being investigated in several vertebrate models.

7 Concluding Remarks

We describe multiple lines of evidence above which suggest that the Rag genes were introduced into the vertebrate genome over half a billion years ago as part of an ancestral transposon. Two outcomes of this fortuitous "jumping" of a primordial gene cassette into a primitive vertebrate genome could ultimately have led to the creation of our combinatorial immune system. First, the transposon integration event itself may have generated the original gene segments from which the current array of rearranging gene segments at Ig and TCR loci have evolved. Second, the introduction of the Rag transposase would allow somatic recombination of these gene segments into a diverse array of surface receptors. Additionally, multiple layers of regulation have evolved to limit activity of the rearrangement process to specific cell types and stages of cell development, and to target V(D)J recombination to a subset of potential targets within a single cell.

A growing number of transposable elements have been identified within eukaryotic genomes; thus it is possible that V(D)J rearrangement is not the only somatic recombination mechanism to have evolved from the products of an ancient transposition event. Based on studies in mammals and fish, Magor and colleagues (1999) recently postulated an enchancer related transposition event which gave rise to the process of class switching at IgH loci. In conclusion, further clues to the origin of Rags and the adaptive immune system will likely come from studies in more distant vertebrate ancestors such as the agnathan fish or possibly even a preserved specimen of the placoderms.

Acknowledgements. The authors thank Susan Gilfillan and Ulf Grawunder for their comments in regard to the manuscript. In addition, we extend our gratitude to Martin Gellert for helpful suggestions and to Louis Du Pasquier for the animal sketches used in Fig. 7. The Basel Institute for Immunology was founded and is supported by Hoffman-LaRoche AG, Basel.

References

Agrawal A, Eastman QM, Schatz DG (1998) Transposition mediated by RAG1 and RAG2 and its implications for the evolution of the immune system. Nature 394:744–751

Agrawal A, Schatz DG (1997) RAG1 and RAG2 form a stable postcleavage synaptic complex with DNA containing signal ends in V(D)J recombination. Cell 89:43–53

Alt FW, Blackwell TK, Yancopoulos GD (1987) Development of the primary antibody repertoire. Science 238:1079–1087

Ardern B, Clark SP, Kabelitz D, Mak TW (1995) Mouse T-cell receptor variable gene segment families. Immunogenetics 42:501–530

Atkinson PW, Warren WD, O'Brochta DA (1993) The hobo transposable element of Drosophila can be cross-mobilized in houseflies and excises like the Ac element of maize. Proc Natl Acad Sci USA 90:9693–9697

Berg DE, Howe MM (1989) Mobile DNA (Washington DC: Am Soc Microbiol)

Bernstein RM, Schluter SF, Bernstein H, Marchalonis JJ (1996) Primordial emergence of the recombination activating gene 1 (RAG1): sequence of the complete shark gene indicates homology to microbial integrases. Proc Natl Acad Sci USA 93:9454–9459

Bertrand FE, 3rd, Olson SL, Martin DA, Wu GE (1998a) Sequence analysis of the mouse RAG locus intergenic region. Dev Immunol 5:215–222

Bertrand FE, 3rd, Olson SL, Willett CE, Wu GE (1998b) Sequence of the rag1 and rag2 intergenic region in zebrafish (Danio rerio). Dev Immunol 5:211–214

Besmer E, Mansilla-Soto J, Cassard S, Sawchuk DJ, Brown G, Sadofsky M, Lewis SM, Nussenzweig MC, Cortes P (1998) Hairpin coding end opening is mediated by RAG1 and RAG2 proteins. Mol Cell 2:817–828

Bogue MA, Wang C, Zhu C, Roth DB (1997) V(D)J recombination in Ku86-deficient mice: distinct effects on coding, signal, and hybrid joint formation. Immunity 7:37–47
Bollum FJ (1974) Terminal deoxynucleotidyl transferase. In: Boyer PD (ed) The Enzymes. Academic Press, NY
Brenner S, Elgar G, Sandford R, Macrae A, Venkatesh B, Aparicio S (1993) Characterization of the pufferfish (Fugu) genome as a compact model vertebrate genome. Nature 366:265–268
Brown KE, Guest SS, Smale ST, Hahm K, Merkenschlager M, Fisher AG (1997a) Association of transcriptionally silent genes with Ikaros complexes at centromeric heterochromatin. Cell 91:845–854
Brown ST, Miranda GA, Galic Z, Hartman IZ, Lyon CJ, Aguilera RJ (1997b) Regulation of the RAG-1 promoter by the NF-Y transcription factor. J Immunol 158:5071–5074
Carlson LM, Oettinger MA, Schatz DG, Masteller EL, Hurley EA, McCormack WT, Baltimore D, Thompson CB (1991) Selective expression of RAG-2 in chicken B cells undergoing immunoglobulin gene conversion. Cell 64:201–208
Carrol RL (1988) Vertebrate paleontology and evolution (New York: Freeman)
Cedar H, Bergman Y (1999) Developmental regulation of immune system gene rearrangement. Curr Opin Immunol 11:64–69
Chow SA, Vincent KA, Ellison V, Brown PO (1992) Reversal of integration and DNA splicing mediated by integrase of human immunodeficiency virus. Science 255:723–726
Clevers HC, Grosschedl R (1996) Transcriptional control of lymphoid development: lessons from gene targeting. Immunol Today 17:336–343
Coen ES, Robbins TP, Almeida J, Hudson A, Carpenter R (1989) Consequences and mechanisms of transposition in *Antirrhinim majus*. In: Mobile DNA, Berg DE, Howe MM, (ed) Am Soc Microbiol (Washington, DC), pp 413–436
Colot V, Haedens V, Rossignol JL (1998) Extensive, nonrandom diversity of excision footprints generated by Ds- like transposon Ascot-1 suggests new parallels with V(D)J recombination. Mol Cell Biol 18:4337–4346
Craig NL (1996) V(D)J recombination and transposition: closer than expected [comment]. Science 271:1512
Critchlow SE, Jackson SP (1998) DNA end-joining: from yeast to man. Trends Biochem Sci 23: 394–398
Cuomo CA, Oettinger MA (1994) Analysis of regions of RAG-2 important for V(D)J recombination. Nucleic Acids Res 22:1810–1814
Davis MM, Chien Y-H (1998) T-cell antigen receptors. In: Paul WE (ed) Fundamental Immunology, Lippincott-Raven (Philadelphia), pp 341–366
Difilippantonio MJ, McMahan CJ, Eastman QM, Spanopoulou E, Schatz DG (1996) RAG1 mediates signal sequence recognition and recruitment of RAG2 in V(D)J recombination. Cell 87:253–262
Dobbeling U, Hobi R, Berchtold MW, Kuenzle CC (1996) V(D)J recombination is regulated similarly in RAG-transfected fibroblasts and pre-B cells. J Mol Biol 261:309–314
Dreyfus DH (1992) Evidence suggesting an evolutionary relationship between transposable elements and immune system recombination sequences. Mol Immunol 29:807–810
Du Pasquier L, Flajnik M (1999) Origin and evolution of the vertebrate immune system. In: Paul WE (ed) Fundamental Immunology, Lippincott-Raven (Philidelphia), pp 605–659
Engelman A, Mizuuchi K, Craigie R (1991) HIV-1 DNA integration: mechanism of viral DNA cleavage and DNA strand transfer. Cell 67:1211–1221
Ernst P, Hahm K, Trinh L, Davis JN, Roussel MF, Turck CW, Smale ST (1996) A potential role for Elf-1 in terminal transferase gene regulation. Mol Cell Biol 16:6121–6131
Fuller K, Storb U (1997) Identification and characterization of the murine Rag1 promoter. Mol Immunol 34:939–954
Georgopoulos K, Bigby M, Wang JH, Molnar A, Wu P, Winandy S, Sharpe A (1994) The Ikaros gene is required for the development of all lymphoid lineages. Cell 79:143–156
Georgopoulos K, Winandy S, Avitahl N (1997) The role of the Ikaros gene in lymphocyte development and homeostasis. Annu Rev Immunol 15:155–176
Grawunder U, West RB, Lieber MR (1998) Antigen receptor gene rearrangement. Curr Opin Immunol 10:172–180
Greenhalgh P, Olesen CE, Steiner LA (1993) Characterization and expression of recombination activating genes (RAG-1 and RAG-2) in Xenopus laevis. J Immunol 151:3100–3110
Hahm K, Cobb BS, McCarty AS, Brown KE, Klug CA, Lee R, Akashi K, Weissman IL, Fisher AG, Smale ST (1998) Helios, a T cell-restricted Ikaros family member that quantitatively associates with Ikaros at centromeric heterochromatin. Genes Dev 12:782–796

Hahm K, Ernst P, Lo K, Kim GS, Turck C, Smale ST (1994) The lymphoid transcription factor LyF-1 is encoded by specific, alternatively spliced mRNAs derived from the Ikaros gene. Mol Cell Biol 14:7111–7123

Hansen JD (1997a) Characterization of rainbow trout terminal deoxynucleotidyl transferase structure and expression. TdT and RAG1 co-expression define the trout primary lymphoid tissues. Immunogenetics 46:367–375

Hansen JD (1997b) Inspection of the 3′ UTR genomic region for RAG1 and RAG2 in rainbow trout (*Oncorhynchus mykiss*) reveals potential regulatory motifs. Dev Immunol 5:129–131

Hansen JD, Kaattari SL (1995) The recombination activation gene 1 (RAG1) of rainbow trout (*Oncorhynchus mykiss*): cloning, expression, and phylogenetic analysis. Immunogenetics 42:188–195

Hansen JD, Kaattari SL (1996) The recombination activating gene 2 (RAG2) of the rainbow trout *Oncorhynchus mykiss*. Immunogenetics 44:203–211

Hansen JD, Zapata AG (1998) Lymphocyte development in fish and amphibians. Immunol Rev 166: 199–220

Hempel WM, Leduc I, Mathieu N, Tripathi RK, Ferrier P (1998) Accessibility control of V(D)J recombination: lessons from gene targeting. Adv Immunol 69:309–352

Hiom K, Gellert M (1997) A stable RAG1-RAG2-DNA complex that is active in V(D)J cleavage. Cell 88:65–72

Hiom K, Gellert M (1998) Assembly of a 12/23 paired signal complex: a critical control point in V(D)J recombination. Mol Cell 1:1011–1019

Hiom K, Melek M, Gellert M (1998) DNA transposition by the RAG1 and RAG2 proteins: a possible source of oncogenic translocations. Cell 94:463–470

Hughes AL, Yeager M (1997) Molecular evolution of the vertebrate immune system. Bioessays 19: 777–786

Hughes KT, Gaines PC, Karlinsey JE, Vinayak R, Simon MI (1992) Sequence-specific interaction of the Salmonella Hin recombinase in both major and minor grooves of DNA. Embo J 11:2695–2705

Kanaar R, Hoeijmakers JH, van Gent DC (1998) Molecular mechanisms of DNA double strand break repair. Trends Cell Biol 8:483–489

Karlstrom RO, Talbot WS, Schier AF (1999) Comparative synteny cloning of zebrafish you-too: mutations in the Hedgehog target gli2 affect ventral forebrain patterning. Genes and Development 13:388–393

Kelley CM, Ikeda T, Koipally J, Avitahl N, Wu L, Georgopoulos K, Morgan BA (1998) Helios, a novel dimerization partner of Ikaros expressed in the earliest hematopoietic progenitors. Curr Biol 8:508–515

Kennedy AK, Guhathakurta A, Kleckner N, Haniford DB (1998) Tn10 transposition via a DNA hairpin intermediate. Cell 95:125–134

Kim J, Sif S, Jones B, Jackson A, Koipally J, Heller E, Winandy S, Viel A, Sawyer A, Ikeda T, Kingston R, Georgopoulos K (1999) Ikaros DNA-binding proteins direct formation of chromatin remodeling complexes in lymphocytes. Immunity 10:345–355

Kirch SA, Sudarsanam P, Oettinger MA (1996) Regions of RAG1 protein critical for V(D)J recombination. Eur J Immunol 26:886–891

Kitagawa T, Mori K, Kishi H, Tagoh H, Nagata T, Kurioka H, Muraguchi A (1996) Chromatin structure and transcriptional regulation of human RAG-1 gene. Blood 88:3785–3791

Klug CA, Morrison SJ, Masek M, Hahm K, Smale ST, Weissman IL (1998) Hematopoietic stem cells and lymphoid progenitors express different Ikaros isoforms, and Ikaros is localized to heterochromatin in immature lymphocytes. Proc Natl Acad Sci USA 95:657–662

Kurioka H, Kishi H, Isshiki H, Tagoh H, Mori K, Kitagawa T, Nagata T, Dohi K, Muraguchi A (1996) Isolation and characterization of a TATA-less promoter for the human RAG-1 gene. Mol Immunol 33:1059–1066

Kwon J, Imbalzano AN, Matthews A, Oettinger MA (1998) Accessibility of nucleosomal DNA to V(D)J cleavage is modulated by RSS positioning and HMG1. Mol Cell 2:829–839

Lafaille JJ, DeCloux A, Bonneville M, Takagaki Y, Tonegawa S (1989) Junctional sequences of T cell receptor gamma delta genes: implications for gamma delta T cell lineages and for a novel intermediate of V-(D)-J joining. Cell 59:859–870

Lauring J, Schlissel MS (1999) Distinct factors regulate the murine RAG-2 promoter in B- and T-cell lines. Mol Cell Biol 19:2601–2612

Lewis SM (1994) The mechanism of V(D)J joining: lessons from molecular, immunological, and comparative analyses. Adv Immunol 56:27–150

Lewis SM, Wu GE (1997) The origins of V(D)J recombination. Cell 88:159–162

Litman G, Anderson MK, Rast JP (1999) Evolution of antigen binding receptors. Annual Reviews in Immunology in press
Marchalonis JJ, Schluter SF (1998) A stochastic model for the rapid emergence of specific vertebrate immunity incorporating horizontal transfer of systems enabling duplication and combinational diversification. J Theor Biol 193:429–444
Marchalonis JJ, Schluter SF, Bernstein RM, Hohman VS (1998) Antibodies of sharks: revolution and evolution. Immunol Rev 166:103–122
Max EE (1998) Immunoglobulins: molecular genetics. In: Paul WE (ed) Fundamental Immunology. Lippincott-Raven (Philadelphia), pp 111–182
McBlane JF, van Gent DC, Ramsden DA, Romeo C, Cuomo CA, Gellert M, Oettinger MA (1995) Cleavage at a V(D)J recombination signal requires only RAG1 and RAG2 proteins and occurs in two steps. Cell 83:387–395
McCormack WT, Tjoelker LW, Carlson LM, Petryniak B, Barth CF, Humphries EH, Thompson CB (1989) Chicken IgL gene rearrangement involves deletion of a circular episome and addition of single nonrandom nucleotides to both coding segments. Cell 56:785–791
Melek M, Gellert M, van Gent DC (1998) Rejoining of DNA by the RAG1 and RAG2 proteins. Science 280:301–303
Mizuuchi K (1992) Transpositional recombination: mechanistic insights from studies of Mu and other elements. Annu Rev Biochem 61:1011–1051
Mizuuchi K (1997) Polynucleotidyl transfer reactions in site-specific DNA recombination. Genes Cells 2:1–12
Mombaerts P, Iacomini J, Johnson RS, Herrup K, Tonegawa S, Papaioannou VE (1992) RAG-1-deficient mice have no mature B and T lymphocytes. Cell 68:869–877
Morgan B, Sun L, Avitahl N, Andrikopoulos K, Ikeda T, Gonzales E, Wu P, Neben S, Georgopoulos K (1997) Aiolos, a lymphoid restricted transcription factor that interacts with Ikaros to regulate lymphocyte differentiation. EMBO J 16:2004–2013
Oettinger MA, Schatz DG, Gorka C, Baltimore D (1990) RAG-1 and RAG-2, adjacent genes that synergistically activate V(D)J recombination. Science 248:1517–1523
Paull TT, Gellert M (1998) The 3′ to 5′ exonuclease activity of Mre 11 facilitates repair of DNA double-strand breaks. Mol Cell 1:969–979
Plasterk R (1998) V(D)J recombination. Ragtime jumping [news; comment]. Nature 394:718–719
Ramsden DA, Paull TT, Gellert M (1997) Cell-free V(D)J recombination [see comments]. Nature 388:488–491
Rodgers KK, Bu Z, Fleming KG, Schatz DG, Engelman DM, Coleman JE (1996) A zinc-binding domain involved in the dimerization of RAG1. J Mol Biol 260:70–84
Roman CA, Cherry SR, Baltimore D (1997) Complementation of V(D)J recombination deficiency in RAG-1(–/–) B cells reveals a requirement for novel elements in the N-terminus of RAG-1. Immunity 7:13–24
Roth DB, Craig NL (1998) VDJ recombination: a transposase goes to work. Cell 94:411–414
Roth DB, Menetski JP, Nakajima PB, Bosma MJ, Gellert M (1992a) V(D)J recombination: broken DNA molecules with covalently sealed (hairpin) coding ends in scid mouse thymocytes. Cell 70: 983–991
Roth DB, Nakajima PB, Menetski JP, Bosma MJ, Gellert M (1992b) V(D)J recombination in mouse thymocytes: double-strand breaks near T cell receptor delta rearrangement signals. Cell 69:41–53
Roux KH, Greenberg AS, Greene L, Strelets L, Avila D, McKinney EC, Flajnik MF (1998) Structural analysis of the nurse shark (new) antigen receptor (NAR): molecular convergence of NAR and unusual mammalian immunoglobulins. Proc Natl Acad Sci USA 95:11804–11809
Sadofsky MJ, Hesse JE, Gellert M (1994) Definition of a core region of RAG-2 that is functional in V(D)J recombination. Nucleic Acids Res 22:1805–1809
Sadofsky MJ, Hesse JE, McBlane JF, Gellert M (1993) Expression and V(D)J recombination activity of mutated RAG-1 proteins [published erratum appears in Nucleic Acids Res 1994 Feb 11;22(3):550]. Nucleic Acids Res 21:5644–5650
Saedler H, Gierl A (1996) Transposable elements. In Current Topics in Microbiology and Immunology (Berlin: Springer)
Sakano H, Huppi K, Heinrich G, Tonegawa S (1979) Sequences at the somatic recombination sites of immunoglobulin light-chain genes. Nature 280:288–294
Sawchuk DJ, Weis-Garcia F, Malik S, Besmer E, Bustin M, Nussenzweig MC, Cortes P (1997) V(D)J recombination: modulation of RAG1 and RAG2 cleavage activity on 12/23 substrates by whole cell extract and DNA-bending proteins. J Exp Med 185:2025–2032

Schatz DG, Oettinger MA, Baltimore D (1989) The V(D)J recombination activating gene, RAG-1. Cell 59:1035–1048
Schatz DG, Oettinger MA, Schlissel MS (1992) V(D)J recombination: molecular biology and regulation. Annu Rev Immunol 10:359–383
Schlissel M, Constantinescu A, Morrow T, Baxter M, Peng A (1993) Double-strand signal sequence breaks in V(D)J recombination are blunt, 5′-phosphorylated, RAG-dependent, and cell cycle regulated. Genes Dev 7:2520–2532
Schlissel MS, Stanhope-Baker P (1997) Accessibility and the developmental regulation of V(D)J recombination. Semin Immunol 9:161–170
Shinkai Y, Rathbun G, Lam KP, Oltz EM, Stewart V, Mendelsohn M, Charron J, Datta M, Young F, Stall AM, et al (1992) RAG-2-deficient mice lack mature lymphocytes owing to inability to initiate V(D)J rearrangement. Cell 68:855–867
Silver DP, Spanopoulou E, Mulligan RC, Baltimore D (1993) Dispensable sequence motifs in the RAG-1 and RAG-2 genes for plasmid V(D)J recombination. Proc Natl Acad Sci USA 90:6100–6104
Singh H (1996) Gene targeting reveals a hierarchy of transcription factors regulating specification of lymphoid cell fates. Curr Opin Immunol 8:160–165
Sleckman BP, Gorman JR, Alt FW (1996) Accessibility control of antigen-receptor variable-region gene assembly: role of cis-acting elements. Annu Rev Immunol 14:459–481
Spanopoulou E, Zaitseva F, Wang FH, Santagata S, Baltimore D, Panayotou G (1996) The homeodomain region of Rag-1 reveals the parallel mechanisms of bacterial and V(D)J recombination. Cell 87:263–276
Stanhope-Baker P, Hudson KM, Shaffer AL, Constantinescu A, Schlissel MS (1996) Cell type-specific chromatin structure determines the targeting of V(D)J recombinase activity in vitro. Cell 85:887–897
Swanson PC, Desiderio S (1998) V(D)J recombination signal recognition: distinct, overlapping DNA-protein contacts in complexes containing RAG1 with and without RAG2. Immunity 9:115–125
Thompson CB (1995) New insights into V(D)J recombination and its role in the evolution of the immune system. Immunity 3:531–539
Trower MK, Orton SM, Purvis IJ, Sanseau P, Riley J, Christodoulou C, Burt D, See CG, Elgar G, Sherrington R, Rogaev EI, St. George-Hyslop P, Brenner S, Dykes CW (1996) Conservation of synteny between the genome of the pufferfish (Fugu rubripes) and the region on human chromosome 14 (14q24.3) associated with familial Alzheimer disease (AD3 locus). Proc Natl Acad Sci USA 93:1366–1369
van Gent DC, Hiom K, Paull TT, Gellert M (1997) Stimulation of V(D)J cleavage by high mobility group proteins. EMBO J 16:2665–2670
van Gent DC, McBlane JF, Ramsden DA, Sadofsky MJ, Hesse JE, Gellert M (1995) Initiation of V(D)J recombination in a cell-free system. Cell 81:925–934
van Gent DC, Mizuuchi K, Gellert M (1996a) Similarities between initiation of V(D)J recombination and retroviral integration. Science 271:1592–1594
van Gent DC, Ramsden DA, Gellert M (1996b) The RAG1 and RAG2 proteins establish the 12/23 rule in V(D)J recombination. Cell 85:107–113
Vink C, Yeheskiely E, van der Marel GA, van Boom JH, Plasterk RH (1991) Site-specific hydrolysis and alcoholysis of human immunodeficiency virus DNA termini mediated by the viral integrase protein. Nucleic Acids Res 19:6691–6698
Wang JH, Nichogiannopoulou A, Wu L, Sun L, Sharpe AH, Bigby M, Georgopoulos K (1996) Selective defects in the development of the fetal and adult lymphoid system in mice with an Ikaros null mutation. Immunity 5:537–549
Willett CE, Cherry JJ, Steiner LA (1997) Characterization and expression of the recombination activating genes (rag1 and rag2) of zebrafish. Immunogenetics 45:394–404
Yan YL, Talbot WS, Egan ES, Postlethwait JH (1998) Mutant rescue by BAC clone injection in zebrafish. Genomics 50:287–289
Yancopoulos GD, Alt FW (1985) Developmentally controlled and tissue-specific expression of unrearranged VH gene segments. Cell 40:271–281
Zarrin AA, Fong I, Malkin L, Marsden PA, Berinstein NL (1997) Cloning and characterization of the human recombination activating gene 1 (RAG1) and RAG2 promoter regions. J Immunol 159: 4382–4394

Transcription Factor Expression in Lymphocyte Development: Clues to the Evolutionary Origins of Lymphoid Cell Lineages?

M.K. Anderson and E.V. Rothenberg

1 Introduction

Lymphocyte development provides an excellent model system for studying the evolutionary divergence of cell lineages. Based on their appearance in vertebrate phylogeny, the origins of lymphoid cell lineages are likely to lie in events which occurred during a defined range of time at least 450 million years ago. The dramatic emergence of both B cells and T cells in the cartilaginous fish (Litman et al. 1999) indicates the occurrence of at least one major event pivotal for the development of lymphocytes as we know them, after the divergence of the vertebrates from the other chordates but before the diversification of the jawed vertebrates. One such event may have been the acquisition, perhaps by horizontal transfer, of the recombination-activating genes (RAGs; Agrawal et al. 1998). The large-scale gene duplication events which are believed to have occurred at approximately this same

Department of Biology, California Institute of Technology, Pasadena, CA 91125, USA time (Holland et al. 1994; Pebusque et al. 1998) may have provided another powerful mechanism for rapid evolutionary change. Lymphoid development is dependent upon networks of transcription factors, which serve not only to activate a series of temporally controlled gene batteries during differentiation but also to stabilize the mature phenotype. These transcription factors are generally members of multigene families whose origins are far more ancient than the lymphoid lineages in which they operate and provide a bridge across phylogenetic distances which have been thus far inaccessible to the study of rearranging antigen receptors. Furthermore, it is likely that duplication and/or divergence of both the *cis*-regulatory regions and the structural portions of transcription factor family members has contributed to the diversification of hematopoietic cell types in vertebrates. Therefore studies of the evolutionary acquisition of the transcription factor networks that regulate lymphoid lineage commitment should provide valuable insights into the origins of the lymphoid cell types and to the evolutionary diversification of cell types in general.

2 Transcription Factors in Mammalian Lymphoid Development

Transcription factors operate in lineage commitment by activating lineage-appropriate gene expression, repressing expression of lineage-inappropriate genes, and stabilizing the resulting gene expression patterns. Mammalian hematopoiesis involves a stepwise narrowing of lineage potential and commitment to at least ten different cell types. One of the earliest developmental choices in hematopoiesis appears to be lymphoid versus myeloid, although it is clear that alternative pathways exist which can generate certain hematopoietic cell types such as B cells, macrophages, and dendritic cells across this divide (Ardavin et al. 1993; Borrello and Phipps 1996). The primary pathway leading to lymphocyte lineage cells, however, probably begins with the divergence of a common lymphoid precursor from a pluripotent hematopoietic stem cell (Kondo et al. 1997). The lymphoid progenitors which become T cells in the thymus undergo a progressive loss of developmental potential, first losing potential for the B cell lineage, then for the natural killer (NK) cell lineage, and then for the thymic DC lineage, before committing to the T lineage (Wu et al. 1996; Carlyle et al. 1997). A schematic diagram of the choices available to a hypothetical common lymphoid precursor is shown in Fig. 1. Further developmental choices are subsequently available to T-lineage committed precursors, such as the αβ versus the γδ T cell lineages, and within the αβ lineage, the CD4 (helper) versus the CD8 (killer) T cell lineages.

The molecular mechanisms which underlie lymphoid development in the mammalian system involve an interplay between transcriptional networks and signaling through growth factor receptors (Cross et al. 1997). Gene knockout studies in mice have highlighted the role of certain transcription factors, such as

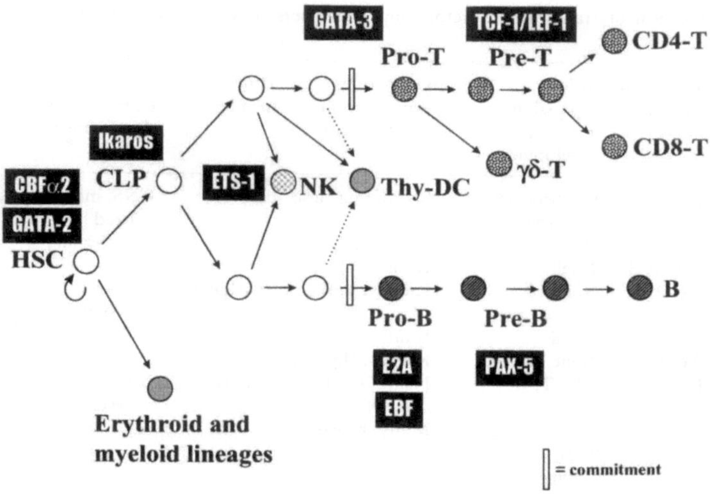

Fig. 1. Transcription factors in mammalian lymphoid development. The hematopoietic stem cell (*HSC*) is capable of either self-renewal (*curled arrow*) or differentiation (*straight arrow*) into any of the hematopoietic cell types. The common lymphoid precursor (*CLP*) can differentiate into B, T, NK, or thymic dendritic cells (*DC*), in an ordered progression of steps involving decrease and loss of lineage potential as well as commitment. *White bars*, the timing of commitment to the T or B cell lineage; *arrows*, the developmental potential of each stage; *dashed arrows*, a lessening of potential. Transcription factors necessary for progression beyond certain stages are depicted above, next to, or below the stage for which they are required

GATA-3 and TCF-1, in T cell development and others, such as E2A, EBF, and Pax-5, in B cell development. There are also transcription factors which affect multiple lymphoid lineages, such as Ikaros (Georgopoulos et al. 1994; Wang et al. 1996), PU.1 (Scott et al. 1994; McKercher et al. 1996), and AML-1/CBFα2 (Okuda et al. 1996), which may act at the level of a common precursor. Most of these transcription factors are members of multigene families with origins dating back before the divergence of the protostome and deuterostome lineages. Furthermore, most of the transcription factors which are more or less specific to hematopoietic development in the adult are used elsewhere during embryonic development (Table 1). This makes sense in a model in which previously existing transcription factors are recruited into new networks by alterations in their *cis*-regulatory sequences, possibly but not necessarily followed by alterations in their coding sequence. Presumably, alterations in coding sequence leading to specialization of a factor to a new role would require duplication first so as not to disrupt the original task of the protein. PU.1 is a notable exception to this general situation, however, as it has not been found outside the hematopoietic lineages, even during embryonic development, in either the mouse or the chicken. PU.1 may be the only known factor which has arisen specifically for use within the hematopoietic system without prior use elsewhere in the developing organism.

Table 1. Roles and expression of transcription factor family members in vertebrate and *Drosophila* development

Family	Subfamily	Vertebrate member	Role, expression in blood development	Other roles, expression in development	*Drosophila* member	Role, expression in development
GATA	GATA-1/2/3	GATA-1	Erythrocytes, megakaryocytes	Spermatogenesis	dGATAc	CNS, midgut, head patterning
		GATA-2	HSC	Urogenital, CNS		
		GATA-3	T cell	Kidney		
PAX	PAX-2/5/8	Pax-2	None	Kidney, brain	Sparkling	Nervous system
		Pax-5	B cell	Brain		
		Pax-8	None	Kidney, thyroid		
ETS	ETS/ERG	Ets-1	NK, T, B	Neural crest, heart	Pointed[a]	Nervous system
		Ets-2	T, myeloid cells	Many, broad		
		Fli-1	T, B, megakaryocytes	Neural crest, endothelium		
		Erg	B, myeloid cells	Neural crest, cartilage		
BHLH	E box	E2A	B, T cell	Many, broad	Daughterless	Sex determination oogenesis, neurogenesis
		HEB	T cell	CNS, olfactory		
	Scl/Tal	Scl	HSC, erythrocytes	Endothelium, brain	dScl	CNS
EBF	EBF/Olf	EBF/Olf-1	B cell	Olfactory neurons	Collier	Head patterning
		Olf-2	None	Olfactory neurons		
		Olf-3	None	Olfactory neurons		

This list is not comprehensive, but is rather a sampling to illustrate the multiple roles each family member can play in development, and the diversification of those roles and patterns of expression after duplication.
[a] Pointed is not the only Ets/Erg subfamily member in *Drosophila*. Other *Drosophila* genes listed here are the sole known members of their respective subfamilies and are not orthologous with any of the vertebrate subfamily members.

3 Gene Duplication and Lymphocyte Evolution

The appearance of lymphocyte lineages and adaptive immunity as a whole coincides with two proposed genome duplications in a chordate ancestor (Holland et al. 1994; Fig. 2). At least in modern mammals the duplicated genes are dispersed throughout the genome, potentially segregated into distinct regulatory environments (Table 2). These duplications provided opportunities for the diversification of gene function on a large scale, which may have facilitated in the emergence of several new features which are limited to the vertebrates (Davidson 1994). The developmental pathways of mammalian T and B lymphocytes exhibit several

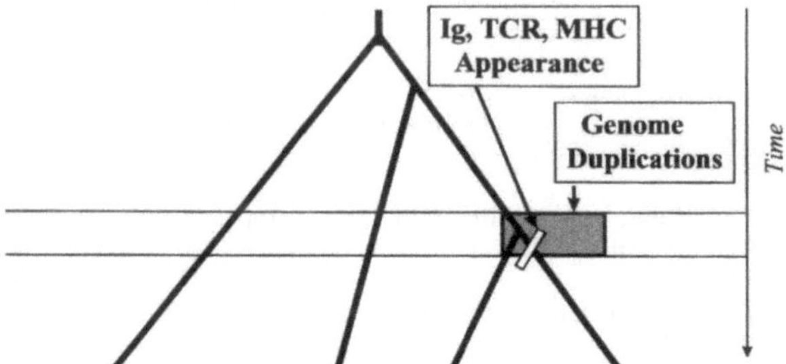

	Drosophila	Sea urchin	Lamprey	Mouse
Pax-2/5/8	1	1	?	3 (Z)
GATA-1/2/3	1	1	?	3 (Z)
Ets/Erg	1	2	?	4 (X)
E2A	1	1	?	3 (X)

Fig. 2. Evolutionary diversification of transcription factor subfamilies involved in lymphoid development. This figure, which is not drawn to scale, illustrates the overlap between the putative genome duplication events that occurred sometime between the divergence of the echinoderms and the divergence of the jawed vertebrates from the other deuterostome lineages (*hatched gray box*), and the emergence of the lymphoid lineages in the jawed vertebrates (*slanted white bar*). *Drosophila* is used as a representative of the protostome lineage for which there an abundance of molecular data available. Lamprey is included despite the paucity of molecular data available thus far because it is situated at a critical junction as the most closely related animal to the jawed vertebrates which lacks the molecules of adaptive immunity (Ig, TCR, and MHC). Furthermore, it remains unclear whether the two putative genome duplication events (or partial duplication events) occurred before the divergence of the jawed vertebrates from the jawless vertebrates, or whether one of these events may have occurred exclusively within the jawed vertebrate lineage (*horizontal lines, hatched box*). The mouse is used as a representative of jawed vertebrates due to the large amount of information available. *X*, a full complement of corresponding orthologs have been identified in an amphibian (*Xenopus*); *Z*, they have been identified in a teleost fish (zebrafish)

parallel characteristics, such as antigen receptor-dependent checkpoints, shifts in dependence on growth factor receptors, temporary use of surrogate light chains, and use of lymphocyte-specific molecules involved in rearrangement and diversification of the antigen receptors (i.e., Rag-1, Rag-2, TdT). These similarities, combined with the derivation of rearranging antigen receptors from a common ancestral molecule, suggest that at some point in evolutionary history there existed a common precursor to these two lymphocyte lineages. We postulate that a series of duplication events of certain key molecules (including Ig and TCR), followed by alterations in their *cis*-regulatory sequences, allowed the segregation of the cells bearing the either Ig or TCR molecules into divergent differentiation pathways. Once segregated, these cells would be free to develop alternative responses to signaling through these antigen receptors (and other receptors) which would define their different roles in the immune system.

We furthermore propose that the developmental pathways of blood cell types in mammals might retain traces of the evolutionary history of these cell types. It should be made clear at this point that we recognize the dangers inherent in this type of analysis. It is well-established that ontogeny does not always recapitulate

Table 2. Chromosomal locations of genes encoding transcription factor family members used in lympohid development

Factor	Alternative names	Mouse chromosome[a]	Human chromosome[b]	Comments[c]
bHLH family				
E2A	ITF1, TCF-3, TcfE2A, Alf-2, ME2	10, 43cM	19p13.3	Alone in mouse, linked to Lyl1, JunB in human
HEB	HTF4, Alf-1, ME1, Tcf12	9, 42cM	15q21	On same chromosome in mouse as Ets-1, Fli-1
E2-2	MITF2, ITF-2, ME2, TCF-4, SEF-2	2, 94cM	18q21.1	
Scl-1	Tal1	4, 49.5cM	1p32	Linked to Id3, CBFα3/AML2 and Pax7; closely linked to c-Jun
Lyl1		8, 38.5cM	19p13.2	Linked to JunD1 in mouse and to JunD and JunB in human; also closely linked to E2A in human
Id1	Idb1	2, unmapped	20q11	
Id2	Idb2	12, 7.0cM	2p25	Linked to Igκ in human
Id3	Idb3	4, 66cM	1p36.13-p36.12	Closely linked to CBFα3/ AML2 and Pax7; linked to Scl-1/Tal1
Id4	Idb4	13, unmapped	6p22-p21.3	
Olf/EBF				
EBF	Olf-1	11, 20cM	5q34	Linked to Hoxb (35cM)
Ebf2	O/E-3	14, 29cM	?	Linked to TCRα in mouse
Ebf3	O/E-2	7, unmapped	?	
Pax 2/5/8 family				
Pax2		19, 43.0cM	10q24.3-q25.1	Closely linked to Hox11/Tlx
Pax5		4, 20.7cM	9p13	In mouse, linked to c-Jun
Pax8		2, 10.5cM	2q12-q14	In mouse, linked to JunD2
Ets family				
Ets-1	Tpl1, Ets-1	9, 15cM	11q23.3	Linked to CD3γ, δ, ε, Thy-1, and Fli-1
Fli-1		9, 16cM	11q24.1-24.3	Linked to CD3γ, δ, ε, Thy-1, and Ets-1
Ets-2		16, 69.6cM	21q22.3	Closely linked to Erg and CBFα2
Erg		16, 69.1cM	21q22.2-q22.3	Closely linked to Ets-2, CBFα2

Table 2. (*Contd.*)

PU.1	Sfpi, Spil	2, 53cM	11p12-p11.22	Closely linked to Ragl/2; also linked to Hoxd, Pax6, and IL-1. On same chrom in mouse as E2–2 and Pax8, but far from both
Spi-B	SPIB	7, 23cM	19q13.3-q13.4	Linked closely to Bax, Flt3L and MyoD, and to FosB, IL-11, CD33
GATA				
GATA-1	Gatal	X, 1.9cM	Xp11.23	
GATA-2	Gata2	6, 38.5cM	3q21	Linked to Wnt7A and *mi*, linked in mouse to Hoxa, TCRβ and Igκ
GATA-3	Gata3	2, 7.0cM	10p15	M/H: closely linked to IL2R and to Bmil; in mouse, linked to Pax8

[a] From the Mouse Genome Database, Jackson Laboratories, Bar Harbor, Maine, USA (http://www.informatics.jax.org/).
[b] From OMIM (Online Mendelian Inheritance in Man, http://www.ncbi.nlm.nih.gov/Omim/) and GDB (The Genome Database, http://gdbwww.gdb.org/gdb/). Online Mendelian Inheritance in Man, OMIM (TM). Center for Medical Genetics, Johns Hopkins University (Baltimore, MD) and National Center for Biotechnology Information, National Library of Medicine (Bethesda, MD). The Genome Database, The Johns Hopkins University School of Medicine, Baltimore, Maryland USA.
[c] Linkages observed in both mouse and human unless otherwise indicated.

phylogeny. However, the analyses presented here do provide a model for lymphocyte evolution which can be tested and used as the foundation of further phylogenetic study. These types of studies could illuminate not only the evolutionary pathways of lymphocyte development but also provide insights into the mechanisms of transcription factor network evolution as well. Intensive study of mouse lymphocytes due to medical interest has led to an enormous amount of information about the molecular events involved in the development, signaling, and effector functions of these cells. We hope to exploit this system as a base with which to compare homologous systems in other species, as is being done with considerable success in other developmental systems (Holland et al. 1994; Fernandez et al. 1998; Mead et al. 1998). Furthermore, we hope to uncover elements of similarity which might give us a greater understanding of the nature of the origins of these unusual cell types and the networks that establish and maintain their unique phenotypes.

4 Recruitment of TF Family Members to the B Cell Lineage: Pax-5 and EBF

Two ancient transcription factor families contain members which are crucial for B cell development but not T cell development. The Pax (paired box) transcription factors are characterized by the presence of a paired domain, and often contain a homeodomain as well. The Pax factors can be divided into four main subgroups based on both the relatedness of their paired domains and on the inclusion/exclusion of other features (DAHL et al. 1997). Pax transcription factors are important in embryonic patterning in both invertebrates and vertebrates (DAHL et al. 1997). The Pax-2/5/8 subfamily contains one representative each in *Drosophila* and sea urchin (CZERNY et al. 1997), and three or more members in vertebrates (PFEFFER et al. 1998). In both zebrafish and mouse, Pax-2, Pax-5, and Pax-8 are all involved in anterior nervous system development. In addition, these factors have other specialized roles: Pax-2 in kidney development, Pax-8 in thyroid development, and Pax-5 in B cell development (URBANEK et al. 1994). The timing of the Pax-2/5/8 family duplication (CZERNY et al. 1997; PFEFFER et al. 1998) thus corresponds to the emergence of the lymphocyte phenotype, between the divergence of vertebrate and nonvertebrate deuterostomes and the emergence of the jawed vertebrates. This supports a model of recruitment of newly duplicated members to the lymphoid lineage. It is of interest that Pax-5 retains an overlapping, partially redundant role with Pax-2 in the formation of the midbrain-hindbrain junction in both zebrafish and mice. The interchangeability of the factors in the tissue in which their expression overlaps indicates that it may be the regulation of the Pax-5 factor, as much as its divergence in the coding region, that has diversified its usage.

Less information is available about the phylogenetic distribution of the EBF/Olf (E/O) family members, but it is clear that this family has an ancient origin as well, and is represented by one family member in *Drosophila* (CROZATIER et al. 1996) and one family member in *Caenorhabditis elegans* (PRASAD et al. 1998). There appear to be at least three family members in the mouse (WANG et al. 1997), all of which play overlapping roles in specification of olfactory receptors and neural tissue. Two family members, Xcoe2 and Xoe-3, have been described in *Xenopus* (DUBOIS et al. 1998). The use of the E/O family members in neurogenesis is conserved throughout the metazoans examined thus far (PRASAD et al. 1998). In mammals, however, the EBF/Olf-1 factor has been recruited for an additional, essential role in B cell development, and is required for progression past the pro-B cell stage (LIN and GROSSCHEDL 1995). There is also strong evidence that EBF/Olf-1 directly regulates B-lineage genes (HAGMAN et al. 1993; MARTENSSON and MARTENSSON 1997; SIGVARDSSON et al. 1997; KEE and MURRE 1998). It has not yet been determined whether the role of EBF-1/Olf-1 in B cell development is conserved in other vertebrates, probably due to the very recent cloning of these genes. The phenotype of EBF-$1^{-/-}$ mice indicates that the Olf-2 and Olf-3 factors can compensate for the loss of EBF-1/Olf-1 for olfactory neuron development but not for B cell development in these animals (LIN and GROSSCHEDL 1995). These three factors are very highly conserved in their amino acid sequences both within and outside the DNA binding domain (WANG et al. 1997). Thus, the divergence of the *cis*-regulatory sequences of these factors has probably been just as important as changes in the coding regions for specifying the different functions of each family member.

5 Recruitment of TF Family Members to the T Cell Lineage: GATA-3

The GATA family of transcription factors is also conserved throughout the metazoans, and is even found in yeast (MINEHART and MAGASANIK 1991). The GATA factors can be subdivided into two subfamilies based on DNA binding domain sequence: GATA-1/2/3 and GATA-4/5/6. The GATA-4/5/6 family members are involved in endoderm and cardiac development, while the GATA-1/2/3 factors are used primarily in hematopoietic development in mammals. GATA-2 is essential to hematopoietic stem cell development (TSAI et al. 1994), and GATA-3 is critical in T cell development but not B cell development (TING et al. 1996). GATA-1, which is the most derived member of this subfamily, is essential in the development of both erythrocytes and megakaryocytes (TAKAHASHI et al. 1998).

These factors are engaged in complex networks of regulation in mammalian hematopoiesis. For instance, as a cell differentiates towards the erythroid lineage, GATA-1 expression is turned on, which initiates a feedback loop resulting in the upregulation of GATA-1 and the downregulation of GATA-2. This stabilizes the emerging erythroid phenotype, in addition to turning on specific downstream erythoid-specific genes such as the erythropoietin receptor and the globin genes (WEISS et al. 1994; SESHASAYEE et al. 1998). It is likely that GATA-2 and GATA-3 engage in similar cross-regulatory interactions in cells specified for the T cell lineage, although this has yet to be shown directly. These interactions are consistent with an ancestral GATA-1/2/3 type of factor with autoregulatory functions. After gene duplication both the original and the duplicate genes would possess the ability to autoregulate and cross-regulate each other. These modules could then be modified to enhance the use of each factor in different situations, leading to stabilization of one factor or the other. These types of cross-regulatory functions are also observed for the Pax-2/5/8 factors (PFEFFER et al. 1998).

Members of both the GATA-1/2/3 and the GATA-4/5/6 family are present in *Drosophila*, based on the amino acid sequence of the DNA-binding domains (PANCER et al. 1999), indicating a duplication event prior to the divergence of the protostome and deuterostome lineages. Recent characterization of two GATA factors in the sea urchin, one from each subfamily, indicates intriguing expression patterns in these animals (PANCER et al. 1999). SpGATAe, which belongs to the GATA-4/5/6 family, is found in the developing embryo but not in the coelomocytes, which constitute the blood of the sea urchin. The SpGATAc factor, which belongs to the GATA-1/2/3 family, is expressed at restricted times in the embryo, and is the sole GATA factor in coelomocytes. This GATA factor therefore represents a descendant of the ancestral GATA-1/2/3 factor which was recruited into the vertebrate hematopoietic system. The expression pattern of this gene indicates that this recruitment may have preceded the divergence of the vertebrates from the other deuterostomes, if the coelomocyte lineages are derived from a common ancestor of the vertebrate hematopoietic cells in the deuterostome lineage. Much more study of sea urchin coelomocytes will be necessary to determine their relationship to vertebrate blood cells, if any, but this observation certainly warrants further investigation.

6 Differential Expression of Related Genes in Different Contexts: Ets Family Member Divergence Between the Lymphoid Lineages

Many Ets family members are known to be critical for normal T and B cell development and function. Ets family members are present in all metazoans studied thus far, including sponges and cnidarians, and certain subfamilies predate the divergence of the protostomes and deuterostomes (LAUTENBERGER et al. 1992; DEGNAN et al. 1993). The phylogenetic and genomic distribution of the Ets/Erg subfamily members (Fig. 3a) indicates two major duplication and diversification events (MACLEOD et al. 1992). The first, resulting in the Ets-1/Ets-2 ancestor and the Erg/Fli-1 ancestor, preceded the echinoderm/chordate split, as representatives of each are found in the sea urchin (CHEN et al. 1988; QI et al. 1992). The second duplication, of Ets-1 and Ets-2, and Erg and Fli-1, appears to have occurred via a chromosomal translocation event, and preceded the divergence of the amphibian lineage. We have studied the expression of these four closely related Ets family members in B and T cell development in the mouse, and find distinct, cell type-specific patterns of expression (Fig. 3b; M.K.A., E.V.R., unpublished results). The differences we see in lineage-specific expression indicate that these factors have acquired differences in their *cis*-regulatory sequences which are sensitive to the context of these different cell types and strongly suggests that they may play a role in the transcriptional networks associated with lymphoid lineage divergence at the developmental level. Further studies are needed to ascertain whether these factors are expressed in these same patterns in other vertebrates.

PU.1 belongs to a different Ets subfamily with a very divergent DNA binding domain than the Ets/Erg subfamily. The only other known member of the PU.1 subfamily is Spi-B. PU.1 is essential for hematopoietic stem cell survival, B cell development, and myeloid cell differentiation (SCOTT et al. 1994; MCKERCHER et al. 1996), while Spi-B is essential for B cell function (SU et al. 1997). The PU.1 subfamily appears thus far to be restricted to the vertebrates, although the divergent

a

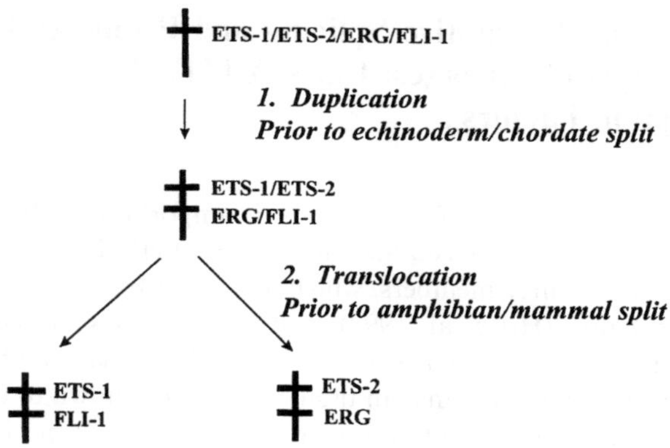

b

	DN-T	DP-T	SP-T	preNK	preB	B	GM
ETS-1	*	*	*	*	*	*	*
FLI-1	*	*	*	*		*	
ETS-2	*	*	*			*	*
ERG	*				*		*

Fig. 3. a Evolutionary history of the Ets/Erg subfamily. Phylogenetic and genomic analyses of the members of the Ets/Erg subfamily indicate that a common precursor to all four mammalian genes existed in a common ancestor of the protostomes and deuterostomes (*black cross*). This locus underwent tandem duplication in an ancestor of the echinoderms and chordates, resulting in the presence of both an Ets-1/Ets-2-like gene and an Erg/Fli-1-like gene in the sea urchin. After the divergence of the echinoderms and the chordates but before the divergence of the amphibians from the vertebrate lineage, a chromosomal translocation event occurred which resulted in the emergence of four genes, Ets-1, Ets-2, Erg, and Fli-1. **b** These genes diverge in their expression patterns in hematopoietic development in the mouse, resulting in different combinations of factors depending on developmental stage and lineage, and indicating alterations in their *cis*-regulatory sequences

nature of its Ets domain may have allowed it to escape methods such as cross-hybridization and degenerate PCR commonly used to isolate other Ets family members in more distantly related species. PU.1 and Spi-B are expressed in overlapping but distinct patterns in hematopoietic cell types, and are completely restricted to the hematopoietic compartment, even during ontogeny (Su et al. 1996). Spi-B can be expressed from either of two promoters (Chen et al. 1998) and in multiple splice isoforms (Ray-Gallet et al. 1996; M.K.A., E.V.R., unpublished results). Only five representatives of this subfamily have been sequenced thus far: human PU.1 (Ray et al. 1990), mouse PU.1 (Klemsz et al. 1990), chicken PU.1 (Kherrouche et al. 1998), human Spi-B (Ray et al. 1992), and mouse Spi-B (Chen et al. 1998). The essential role of PU.1 in mammalian hematopoiesis (Simon 1998), however, suggests that it is widespread throughout the vertebrate lineage. Furthermore, the high level of restriction of these factors to specific hematopoietic lineages will facilitate their use as molecular markers if these expression patterns are conserved throughout the vertebrates.

7 Divergence of Alternative Splicing and Promoter Usage in the Lymphoid Lineages: Class A bHLH Transcription Factors

Another transcription factor family which is important to both T and B cell development is the class A basic helix-loop-helix (bHLH) family. This family in mammals consists of three members, which are called E2A, E2-2, and HEB by the human nomenclature (Hu et al. 1992), and also includes the *Drosophila* daughterless gene product (Cronmiller et al. 1988). The class A bHLH factors are expressed quite widely in mammalian tissues. They are known to operate in muscle and nervous system development in cooperation with class B specific factors, and in early hematopoiesis with the class B bHLH factors Scl-1/Tal-1 and Tal-2. The Id proteins, which contain the HLH but not the DNA binding basic domain, act as repressors by binding and sequestering the class A bHLH proteins (Benezra et al. 1990). In muscle development, the ratio of Id to class A bHLH factor expression is critical to the action of the factor in either activation or repression of target genes. The same mechanism may be operating in T and B cell development (Bain and Murre 1998), although here intraclass A dimers seem to be used instead of class A/class B heterodimers to drive cell type specific gene expression. In lymphocytes we have found that the situation is also complex due to the use of alternative promoters and different splice isoforms for all three factors (Murre et al. 1989; Nielsen et al. 1992; Skerjanc et al. 1996; M.K.A., E.V.R., unpublished observations), making competition possible between different products of the same class A bHLH genes. In any case B cell development is dependent on all three class A bHLH coding genes in a dose-dependent manner (Zhuang et al. 1996). E2A knockout mice also exhibit alterations in T cell development, as do HEB knockout mice (Zhuang et al. 1996; Bain et al. 1997).

An analysis of the domain structures and DNA sequences of E2A, E2-2, and HEB indicate that an E2-2/HEB common precursor diverged from an E2A-like gene, and acquired an alternative promoter before diverging into E2-2 and HEB (Fig. 4). It is unclear when E47 and E12, the two splice isoforms of E2A, emerged, although it appears to be prior to the amphibian/mammal split, since *Xenopus* has homologs of both E12 and E47 (Wulbeck et al. 1994). This unusual situation seems to have resulted from a tandem duplication of the bHLH domain exon of E2A, leading to the alternative use of the E12 exon or the E47 exon in each splice isoform (Murre et al. 1989). We predict that the duplication leading to the emergence of thc three separate class A bHLH genes occurred around the time of the emergence of the vertebrates, although this has yet to be determined. It is known that *Xenopus* has homologs of all three class A bHLH factors, which places their origins prior to the divergence of the amphibians from the other vertebrate lineages (Shain et al. 1997).

We have found that control of the HEB and E2-2 promoters is lineage-specific and differs between B cells, T cells, and NK cells (Fig. 4). The two forms of HEB

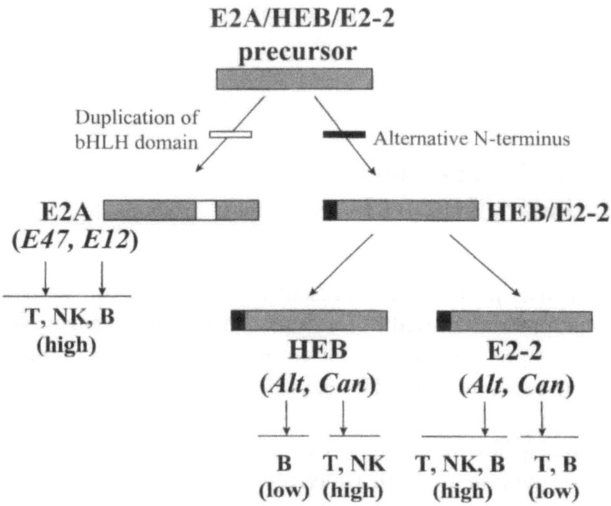

Fig. 4. Evolutionary history of the class A bHLH family. Phylogenetic and genomic analyses of the class A bHLH factors in different metazoan species indicate that a common ancestor to *Drosophila* daughterless and the three mammalian factors E2A, HEB, and E2-2 existed in an ancient metazoan ancestor. Sometime after the protostome/deuterostome divergence, a duplication occurred. One of the resulting factors (E2A) underwent a tandem duplication of the bHLH domain which resulted in the ability to alternatively splice either domain (*E47* or *E12*). The other resulting factor (HEB/E2-2 ancestor) acquired an additional exon (and presumably an additional promoter) at the N-terminal end. Either the alternative (*Alt*) or the canonical (*Can*) 5′ end could presumably be used by alternative splicing. This HEB/E2-2 ancestor duplicated sometime before the divergence of the amphibians from the vertebrate lineage. The expression of the E2A gene and its splice isoforms does not appear to vary significantly between different lymphoid developmental stages and lineages, but the HEB and E2-2 forms are divergent among lymphoid lineages

(as those of E2-2) differ in their 5′ exon as well as their expression patterns. The importance of the correct regulatory control of related factors is illustrated in a study in which the human HEB gene expressed from the mouse E2A promoter is able to rescue the phenotype of the E2A knockout mouse (Zhuang et al. 1998). It seems in this case that either factor is sufficient to do the critical tasks if expressed at the appropriate times and places. This type of redundancy of protein function accompanied by changes in deployment is also seen in the case of the paired, gooseberry, and gooseberry neuro homeobox genes in *Drosophila* (Li and Noll 1994). However, the importance of the domain structures of these factors is highlighted by studies showing that the same bHLH domain can lead to repressive or activating functions in transcription depending on the composition of the 5′ exon in the case of the E2-2 gene (Skerjanc et al. 1996; Liu et al. 1998). The fact that both forms of HEB are expressed concurrently in early T cell precursors (M.K.A., E.V.R., unpublished results) while only one or the other is expressed in mature T or B cells indicates that there may be a functional difference in the different domain structures. Whether the divergence observed during somatic development is indicative of evolutionary heritage or not remains to be seen, but definitely raises questions worth pursuing.

8 Multilineage Priming in Hematopoiesis: A Remnant of Evolutionary History?

In our investigations of the transcription factors present in developmentally early thymocytes which retain developmental potential for several lymphoid lineages, we have found multilineage gene expression of several transcription factor families (M.K.A., E.V.R., unpublished observations). The expression of these genes subsequently diverges as differentiation along various lymphoid lineages progresses. The expression of multilineage genes in uncommitted precursors has also been described in myeloerythroid development. Both transcription factors and lineage-specific effector genes can be expressed in this manner in several hematopoietic lineages (Cross et al. 1994; Voura et al. 1997; Wang et al. 1998) even at the single cell level (Cheng et al. 1996; Hu et al. 1997). It has been suggested that this type of expression may be the result of "priming" of genes of several potential lineage choices, due to a combination of the presence of overlapping transcription factors and open chromatin structure, which is subsequently altered due to the action of both repressive and activating factors in differentiating cells (Cross et al. 1997). This pattern might also suggest that evolutionary duplication of transcription factors and subsequent alterations of the *cis*-regulatory regions of both the original and the duplicate can result in regulatory patterns which are conserved in developmentally early cells but differentially modified in cells committing to one lineage or another. This model would predict the addition of different control modules to pre-existing regulatory instructions in the enhancers of these genes, and it might be possible to test this prediction by analysis of the potentially modular structure of the enhancers of genes which exhibit these patterns. The conservation of combinatorial units of binding sites, regardless of the order of these sites within a module (Magor et al. 1994), might be the key to these systems, and would allow comparison between the enhancer modules of more phylogenetically distant species than if the exact order of sites was needed.

9 Phylogenetic Studies of Lymphoid Lineage Development

One of the problems with drawing evolutionary conclusions from the studies presented here is that they are based primarily on observations in one species, the mouse, in conjunction with what is known about the antigen receptors found in other vertebrates and the transcription factors found in other animals. There could well be a great deal of flexibility in the developmental pathways of vertebrate lymphocytes, analogous to that seen in the mechanisms employed by different vertebrate classes in the diversification of their somatic antigen receptors (LITMAN et al. 1999). In order to test the hypotheses which have been raised here it will be necessary to delve more deeply into the molecular aspects of lymphocyte development in phylogeny than has been heretofore possible. The recent identification of important marker genes for lymphocytes, including Ig, TCRαβ, TCRγδ, RAG-1, TdT, and Ikaros, in divergent vertebrate species (GREENHALGH and STEINER 1995; HANSEN 1997; HANSEN et al. 1997; RAST et al. 1997; TREDE and ZON 1998) is beginning to allow the types of studies necessary to answer these questions. Homologues of these genes are facilitating studies of the sites and pathways of hematopoiesis in teleost fish, as described elsewhere in this volume (HANSEN et al.). Most of these markers are now known in the skate, Raja, and such studies are underway in these cartilaginous fish as well (R. Haire, A. Miracle, and G. Litman, personal communication). Homologues of most of the transcription factors described above should be relatively easy to isolate based on the conservation in the amino acid sequence of the DNA binding domains. It will be very informative to determine whether the transcription factor gene expression patterns shown to be critical in mammalian lymphocyte development are conserved in these divergent vertebrate species. We would expect that if the lineage specificity of these factors is truly a result of evolutionary history, the patterns would indeed be conserved. Recent studies of the SCL and AML-1/CBFα2 factor homologs in zebrafish and *Xenopus* hematopoiesis, respectively, support the validity of this approach (MEAD et al. 1998; TRACEY et al. 1998). One of the most promising regions to look for sites of hematopoiesis and lymphoid development in nonvertebrate chordates might be the gut, since it is known in mammals to be a site of extrathymic T cell development (PAGE et al. 1998) and a secondary or even primary site of lymphoid tissue (gut-associated lymphoid tissue) in most vertebrates (WEILL and REYNAUD 1998). Furthermore, it can easily be identified in animals outside the jawed vertebrates. If the transcription factors known to be instrumental in T or B cell lineage choice are conserved in the cartilaginous fishes, where markers of lymphoid lineages are available, it will probably be most useful to use these genes as markers in adult chordates, where their roles appear to become more specialized than in embryos, where such genes often show broader expression, perhaps indicative of their more ancient roles before the emergence of the lymphocyte developmental program.

10 Conclusions

In summary, it appears in mice that closely related members of transcription factor families diverge in their roles in T and B cell development, and possibly in NK cell development as well. We suggest that these expression patterns reflect an evolutionary divergence in lineage pathways followed by specialization of duplicated factors in each lineage. The proposed genome duplications near the chordate/vertebrate divergence times are correlated very well with the emergence of the lymphocyte morphology and the appearance of the rearranging antigen receptors, Ig and TCR, in phylogeny, and is also correlated with the divergence of several key transcription factor families which play important roles in lymphocyte development. Several different mechanisms appear to have been used in the diversification of the transcription factors, including classic duplication and divergence of amino acid sequence, chromosomal translocation, alternative domain splicing, and alternative promoter usage. In addition, it is apparent from our studies that the regulatory elements controlling these factors have also diverged in a lineage-specific fashion. The evolutionary histories of these factors should be accessible by molecular phylogenetic studies, and in combination with molecular markers of lymphoid development and phenotype should provide insights into the evolution of the transcriptional networks that define the lymphoid lineages themselves.

Acknowledgements. The authors thank the members of the Rothenberg group for their enthusiastic discussion and comments regarding the ideas presented in this review. We also thank Jonathan Rast for his thoughtful critique of these topics. M.K.A. is supported by the Stowers Institute for Medical Research.

References

Agrawal A, Eastman QM, Schatz DG (1998) Transposition mediated by RAG1 and RAG2 and its implications for the evolution of the immune system. Nature 394:744–751

Ardavin C, Wu L, Li CL, Shortman K (1993) Thymic dendritic cells and T cells develop simultaneously in the thymus from a common precursor population. Nature 362:761–763

Bain G, Engel I, Robanus Maandag EC, te Riele HP, Voland JR, Sharp LL, Chun J, Huey B, Pinkel D, Murre C (1997) E2A deficiency leads to abnormalities in alphabeta T-cell development and to rapid development of T-cell lymphomas. Mol Cell Biol 17:4782–4791

Bain G, Murre C (1998) The role of E-proteins in B- and T-lymphocyte development. Semin Immunol 10:143–153

Benezra R, Davis RL, Lockshon D, Turner DL, Weintraub H (1990) The protein Id: a negative regulator of helix-loop-helix DNA binding proteins. Cell 61:49–59

Borrello MA, Phipps RP (1996) The B/macrophage cell: an elusive link between CD5+ B lymphocytes and macrophages [see comments]. Immunol Today 17:471–475

Carlyle JR, Michie AM, Furlonger C, Nakano T, Lenardo MJ, Paige CJ, Zúñiga-Pflücker JC (1997) Identification of a novel developmental stage marking lineage commitment of progenitor thymocytes. J Exp Med 186:173–182

Chen HM, Gonzalez DA, Radomska HS, Voso MT, Sun Z, Zhang P, Zhang DE, Tenen DG (1998) Two promoters direct expression of the murine Spi-B gene, an Ets family transcription factor. Gene 207:209–218

Chen ZQ, Kan NC, Pribyl L, Lautenberger JA, Moudrianakis E, Papas TS (1988) Molecular cloning of the ets proto-oncogene of the sea urchin and analysis of its developmental expression. Dev Biol 125:432–440

Cheng T, Shen H, Giokas D, Gere J, Tenen DG, Scadden DT (1996) Temporal mapping of gene expression levels during the differentiation of individual primary hematopoietic cells. Proc Natl Acad Sci USA 93:13158–13163

Cronmiller C, Schedl P, Cline TW(1988) Molecular characterization of daughterless, a Drosophila sex determination gene with multiple roles in development. Genes Dev 2:1666–1676

Cross MA, Heyworth CM, Dexter TM (1997) How do stem cells decide what to do? Ciba Found Symp 204:3–14

Cross MA, Heyworth CM, Murrell AM, Bockamp EO, Dexter TM, Green AR (1994) Expression of lineage restricted transcription factors precedes lineage specific differentiation in a multipotent haemopoietic progenitor cell line. Oncogene 9:3013–3016

Crozatier M, Valle D, Dubois L, Ibnsouda S, Vincent A (1996) Collier, a novel regulator of Drosophila head development, is expressed in a single mitotic domain. Curr Biol 6:707–718

Czerny T, Bouchard M, Kozmik Z, Busslinger M (1997) The characterization of novel Pax genes of the sea urchin and Drosophila reveal an ancient evolutionary origin of the Pax2/5/8 subfamily. Mech Dev 67:179–192

Dahl E, Koseki H, Balling R (1997) Pax genes and organogenesis. Bioessays 19:755–765

Davidson EH (1994) Stepwise evolution of major functional systems in vertebrates, including the immune system. In: Hoffmann JA, Janeway CA Jr, Natori S (eds) Phylogenetic Perspectives in Immunity: the Insect Host Defense. R.G. Landes Company Austin

Degnan BM, Degnan SM, Naganuma T, Morse DE (1993) The ets multigene family is conserved throughout the Metazoa. Nucleic Acids Res 21:3479–3484

Dubois L, Bally-Cuif L, Crozatier M, Moreau J, Paquereau L, Vincent A (1998) XCoe2, a transcription factor of the Col/Olf-1/EBF family involved in the specification of primary neurons in Xenopus. Curr Biol 8:199–209

Fernandez AS, Pieau C, Reperant J, Boncinelli E, Wassef M (1998) Expression of the Emx-1 and Dlx-1 homeobox genes define three molecularly distinct domains in the telencephalon of mouse, chick, turtle and frog embryos: implications for the evolution of telencephalic subdivisions in amniotes. Development 125:2099–2111

Georgopoulos K, Bigby M, Wang JH, Molnar A, Wu P, Winandy S, Sharpe A (1994) The Ikaros gene is required for the development of all lymphoid lineages. Cell 79:143–156

Greenhalgh P, Steiner LA (1995) Recombination activating gene 1 (Ragl) in zebrafish and shark. Immunogenetics 41:54–55

Hagman J, Belanger C, Travis A, Turck CW, Grosschedl R (1993) Cloning and functional characterization of early B-cell factor, a regulator of lymphocyte-specific gene expression. Genes Dev 7:760–773

Hansen JD (1997) Characterization of rainbow trout terminal deoxynucleotidyl transferase structure and expression. TdT and RAG1 co-expression define the trout primary lymphoid tissues. Immunogenetics 46:367–375

Hansen JD, Strassburger P, Du Pasquier L (1997) Conservation of a master hematopoietic switch gene during vertebrate evolution: isolation and characterization of Ikaros from teleost and amphibian species. Eur J Immunol 27:3049–3058

Holland PW, Garcia-Fernandez J, Williams NA, Sidow A (1994) Gene duplications and the origins of vertebrate development. Dev Suppl 125–133

Hu JS, Olson EN, Kingston RE (1992) HEB, a helix-loop-helix protein related to E2A and ITF2 that can modulate the DNA-binding ability of myogenic regulatory factors. Mol Cell Biol 12:1031–1042

Hu M, Krause D, Greaves M, Sharkis S, Dexter M, Heyworth C, Enver T (1997) Multilineage gene expression precedes commitment in the hemopoietic system. Genes Dev 11:774–785

Kee BL, Murre C (1998) Induction of early B cell factor (EBF) and multiple B lineage genes by the basic helix-loop-helix transcription factor E12. J Exp Med 188:699–713

Kherrouche Z, Beuscart A, Huguet C, Flourens A, Moreau-Gachelin F, Stehelin D, Coll J (1998) Isolation and characterization of a chicken homologue of the Spi-1/PU.1 transcription factor. Oncogene 16:1357–1367

Klemsz MJ, McKercher SR, Celada A, Van Beveren C, Maki RA (1990) The macrophage and B cell-specific transcription factor PU.1 is related to the ets oncogene. Cell 61:113–124

Kondo M, Weissman IL, Akashi K (1997) Identification of clonogenic common lymphoid progenitors in mouse bone marrow. Cell 91:661–672

Lautenberger JA, Burdett LA, Gunnell MA, Qi S, Watson DK, O'Brien SJ, Papas TS (1992) Genomic dispersal of the ets gene family during metazoan evolution. Oncogene 7:1713–1719

Li X, Noll M (1994) Evolution of distinct developmental functions of three Drosophila genes by acquisition of different *cis*-regulatory regions. Nature 367:83–87

Lin H, Grosschedl R (1995) Failure of B-cell differentiation in mice lacking the transcription factor EBF. Nature 376:263–267

Litman GW, Anderson MK, Rast JP (1999) Evolution of Antigen Binding Receptors. Ann Rev Immunol, 17:109–147

Liu Y, Ray SK, Yang XQ, Luntz-Leybman V, Chiu IM (1998) A splice variant of E2–2 basic helix-loop-helix protein represses the brain-specific fibroblast growth factor 1 promoter through the binding to an imperfect E-box. J Biol Chem 273:19269–19276

Macleod K, Leprince D, Stehelin D (1992) The ets gene family. Trends Biochem Sci 17:251–256

Magor BG, Wilson MR, Miller NW, Clem LW, Middleton DL, Warr GW (1994) An Ig heavy chain enhancer of the channel catfish Ictalurus punctatus: evolutionary conservation of function but not structure. J Immunol 153:5556–5563

Martensson A, Martensson IL (1997) Early B cell factor binds to a site critical for lambda5 core enhancer activity. Eur J Immunol 27:315–320

McKercher SR, Torbett BE, Anderson KL, Henkel GW, Vestal DJ, Baribault H, Klemsz M, Feeney AJ, Wu GE, Paige CJ, Maki RA (1996) Targeted disruption of the PU.1 gene results in multiple hematopoietic abnormalities. EMBO J 15:5647–5658

Mead PE, Kelley CM, Hahn PS, Piedad O, Zon LI (1998) SCL specifies hematopoietic mesoderm in Xenopus embryos. Development 125:2611–2620

Minehart PL, Magasanik B (1991) Sequence and expression of GLN3, a positive nitrogen regulatory gene of Saccharomyces cerevisiae encoding a protein with a putative zinc finger DNA-binding domain. Mol Cell Biol 11:6216–6228

Murre C, McCaw PS, Baltimore D (1989) A new DNA binding and dimerization motif in immunoglobulin enhancer binding, daughterless, MyoD, and myc proteins. Cell 56:777–783

Nielsen AL, Pallisgaard N, Pedersen FS, Jorgensen P (1992) Murine helix-loop-helix transcriptional activator proteins binding to the E-box motif of the Akv murine leukemia virus enhancer identified by cDNA cloning. Mol Cell Biol 12:3449–3459

Okuda T, van Deursen J, Hiebert SW, Grosveld G, Downing JR (1996) AML1, the target of multiple chromosomal translocations in human leukemia, is essential for normal fetal liver hematopoiesis. Cell 84:321–330

Page ST, Bogatzki LY, Hamerman JA, Sweenie CH, Hogarth PJ, Malissen M, Perlmutter RM, Pullen AM (1998) Intestinal intraepithelial lymphocytes include precursors committed to the T cell receptor alpha beta lineage. Proc Natl Acad Sci USA 95:9459–9464

Pancer Z, Rast JP, Davidson EH (1999) Origins of Immunity: Transcription Factors and Homologues of Effector Genes of the Vertebrate Immune System Expressed in Sea Urchin Coelomocytes. Immunogenetics 49:773–786

Pebusque MJ, Coulier F, Birnbaum D, Pontarotti P (1998) Ancient large-scale genome duplications: phylogenetic and linkage analyses shed light on chordate genome evolution. Mol Biol Evol 15: 1145–1159

Pfeffer PL, Gerster T, Lun K, Brand M, Busslinger M (1998) Characterization of three novel members of the zebrafish Pax2/5/8 family: dependency of Pax5 and Pax8 expression on the Pax2.1 (noi) function. Development 125:3063–3074

Prasad BC, Ye B, Zackhary R, Schrader K, Seydoux G, Reed RR (1998) unc-3, a gene required for axonal guidance in Caenorhabditis elegans, encodes a member of the O/E family of transcription factors. Development 125:1561–1568

Qi S, Chen ZQ, Papas TS, Lautenberger JA (1992) The sea urchin erg homolog defines a highly conserved erg-specific domain. DNA Seq 3:127–9

Rast JP, Anderson MK, Strong SJ, Luer C, Litman RT, Litman GW (1997) alpha, beta, gamma, and delta T cell antigen receptor genes arose early in vertebrate phylogeny. Immunity 6:1–11

Ray D, Bosselut R, Ghysdael J, Mattei MG, Tavitian A, Moreau-Gachelin F (1992) Characterization of Spi-B, a transcription factor related to the putative oncoprotein Spi-1/PU.1. Mol Cell Biol 12: 4297–4304

Ray D, Culine S, Tavitain A, Moreau-Gachelin F (1990) The human homologue of the putative proto-oncogene Spi-1: characterization and expression in tumors [published erratum appears in Oncogene 1990 Oct;5 (10):1611–2]. Oncogene 5:663–668

Ray-Gallet D, Tavitian A, Moreau-Gachelin F (1996) An alternatively spliced isoform of the Spi-B transcription factor. Biochem Biophys Res Commun 223:257–263

Scott EW, Simon MC, Anastasi J, Singh H (1994) Requirement of transcription factor PU.1 in the development of multiple hematopoietic lineages. Science 265:1573–1577

Seshasayee D, Gaines P, Wojchowski DM (1998) GATA-1 dominantly activates a program of erythroid gene expression in factor-dependent myeloid FDCW2 cells. Mol Cell Biol 18:3278–3288

Shain DH, Neuman T, Zuber MX (1997) Embryonic expression and evolution of duplicated E-protein genes in Xenopus laevis: parallels with ancestral E-protein genes. Genetics 146:345–353

Sigvardsson M, O'Riordan M, Grosschedl R (1997) EBF and E47 collaborate to induce expression of the endogenous immunoglobulin surrogate light chain genes. Immunity 7:25–36

Simon MC (1998) PU.1 and hematopoiesis: lessons learned from gene targeting experiments. Semin Immunol 10:111–8

Skerjanc IS, Truong J, Filion P, McBurney MW (1996) A splice variant of the ITF-2 transcript encodes a transcription factor that inhibits MyoD activity. J Biol Chem 271:3555–3561

Su GH, Chen HM, Muthusamy N, Garrett-Sinha LA, Baunoch D, Tenen DG, Simon MC (1997) Defective B cell receptor-mediated responses in mice lacking the Ets protein, Spi-B. EMBO J 16: 7118–7129

Su GH, Ip HS, Cobb BS, Lu MM, Chen HM, Simon MC (1996) The Ets protein Spi-B is expressed exclusively in B cells and T cells during development. J Exp Med 184:203–214

Takahashi S, Komeno T, Suwabe N, Yoh K, Nakajima O, Nishimura S, Kuroha T, Nagasawa T, Yamamoto M (1998) Role of GATA-1 in proliferation and differentiation of definitive erythroid and megakaryocytic cells in vivo. Blood 92:434–442

Ting CN, Olson MC, Barton KP, Leiden JM (1996) Transcription factor GATA-3 is required for development of the T-cell lineage. Nature 384:474–478

Tracey WD Jr, Pepling ME, Horb ME, Thomsen GH, Gergen JP (1998) A Xenopus homologue of aml-1 reveals unexpected patterning mechanisms leading to the formation of embryonic blood. Development 125:1371–1380

Trede NS, Zon LI (1998) Development of T-cells during fish embryogenesis. Dev Comp Immunol 22: 253–263

Tsai FY, Keller G, Kuo FC, Weiss M, Chen J, Rosenblatt M, Alt FW, Orkin SH (1994) An early haematopoietic defect in mice lacking the transcription factor GATA-2. Nature 371:221–226

Urbanek P, Wang ZQ, Fetka I, Wagner EF, Busslinger M (1994) Complete block of early B cell differentiation and altered patterning of the posterior midbrain in mice lacking Pax5/BSAP. Cell 79: 901–912

Voura EB, Billia F, Iscove NN, Hawley RG (1997) Expression mapping of adhesion receptor genes during differentiation of individual hematopoietic precursors. Exp Hematol 25:1172–1179

Wang H, Diamond RA, Rothenberg EV (1998) Cross-lineage expression of Ig-beta (B29) in thymocytes: positive and negative gene regulation to establish T cell identity. Proc Natl Acad Sci USA 95: 6831–6836

Wang JH, Nichogiannopoulou A, Wu L, Sun L, Sharpe AH, Bigby M, Georgopoulos K (1996) Selective defects in the development of the fetal and adult lymphoid system in mice with an Ikaros null mutation. Immunity 5:537–549

Wang SS, Tsai RYL, Reed RR (1997) The characterization of the Olf-1/EBF-like HLH transcription factor family: implications in olfactory gene regulation and neuronal development. J Neurosci 17:4149–4158

Weill JC, Reynaud CA (1998) Galt versus bone marrow models of B cell ontogeny. Dev Comp Immunol 22:379–385

Weiss MJ, Keller G, Orkin SH (1994) Novel insights into erythroid development revealed through in vitro differentiation of GATA-1 embryonic stem cells. Genes Dev 8:1184–1197

Wu L, Li CL, Shortman K (1996) Thymic dendritic cell precursors: relationship to the T lymphocyte lineage and phenotype of the dendritic cell progeny. J Exp Med 184:903–911

Wulbeck C, Fromental-Ramain C, Campos-Ortega JA (1994) The HLH domain of a zebrafish HE12 homologue can partially substitute for functions of the HLH domain of Drosophila DAUGHTERLESS. Mech Dev 46:73–85

Zhuang Y, Barndt RJ, Pan L, Kelley R, Dai M (1998) Functional replacement of the mouse E2A gene with a human HEB cDNA. Mol Cell Biol 18:3340–3349

Zhuang Y, Cheng P, Weintraub H (1996) B-lymphocyte development is regulated by the combined dosage of three basic helix-loop-helix genes, E2A, E2–2, and HEB. Mol Cell Biol 16:2898–2905

Origin of Receptors

The Phylogenetic Origin of Antigen-Specific Receptors

L. Du Pasquier

Institute for Immunology, Grenzacherstrasse 487, 4005 Basel, Switzerland
e-mail: dupasquier@dial.eunet.ch

1 Introduction

The adaptive immune system of gnathostomes, vertebrates with jaws, is characterized by somatic rearrangement of gene segments to generate genes encoding dimeric antigen-specific receptors (Ig; T-cell receptor, TCR) expressed on lymphocytes (B-cells, T-cells; Tonegawa 1983). Several reviews have recently been published on the evolution of bona fide antigen-specific receptors (Hsu and Steiner 1992; Litman et al. 1999; Marchalonis et al. 1998; Rast and Litman 1998), but their origin remains mysterious. Because of its seemingly abrupt development within the vertebrate lineage (Bernstein et al. 1996; Thompson 1995), this system is an intriguing puzzle for evolutionary theorists. Indeed, while agnathans, vertebrates without jaws, seem to lack most of its specific components, cartilaginous fish seem to have all of them (reviewed in Du Pasquier and Flajnik 1999). Ig and TCR consist of protein domains belonging to the immunoglobulin superfamily (Igsf). Based on length, strand composition, and presence of a V frame, Igsf domains are classified into at least four sets: V (variable), I (intermediary), C2 (constant), and C1 (constant), as shown in Fig. 1 (Harpaz and Chothia 1994; Williams and Barclay 1988). In the simplest cases (Ig light chain, TCRα or γ), somatic rearrangement brings a variable (V) gene segment encoding strands A to F of the V domain in close proximity to the joining segment (J), which encodes the G strand. The number of constant domains varies with the type of receptor, but its structure is always consistent with the presence of the seven strands characteristic of C1. C1 domains have only been reported in gnathostomes, the only vertebrate subphylum in which somatic rearrangement is known to take place. Where do V and C1 domains come from? Is dimerization an ancient feature? What did the ancestral precursor of these antigen-specific receptors look like? The precursor molecule was probably derived from a receptor made of two V1-C1-Tm-Cy polypeptide chains in which the V domain was not generated by somatic rearrangement. V domains generated without rearrangement exist in many metazoa, but are found either alone or in association with C2 domains, shorter than C1 although with 8

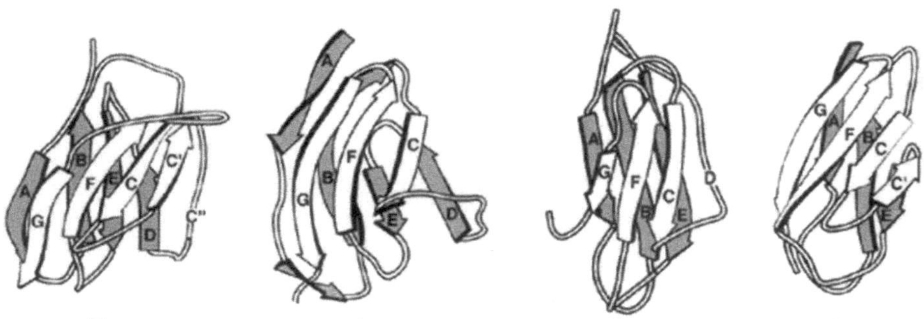

Fig. 1. Imunoglobulin domains. C1, C2, and V (Barclay et al. 1997), I-set (From Harpaz and Chothia 1994)

strands. C2 domains share some of the features of the V domain. Some of these molecules form dimers resembling the TCR, for example, CTX (Chrétien et al. 1996). In gnathostomes, C1 domains are occasionally associated with nonsomatically rearranging V segments that do not serve systematically as lymphocyte specific (Adams et al. 1998), but such molecules have so far not drawn much attention in an evolutionary context. The purpose of this review is to look more closely at all available Igsf members with a V domain not generated by somatic rearrangement and with C1 domain(s) in order to determine whether molecules with the key extracellular structural features of Ig and TCR ancestors have escaped our attention. In attempting to understand the evolutionary paths taken by the various lineages of the Igsf during evolution, I must classify (and occasionally reclassify) such molecules and their genes. The classification of molecules as Igsf members is usually based on amino acid sequences and the strand and loop composition predicted from them (Smith and Xue 1997). Whenever possible I would like to add to these criteria the exon-intron structure and the chromosomal location of these genes.

2 Molecules with V Domains from Invertebrates

2.1 Porifers

Ig superfamily domains are found in a receptor tyrosine kinase of the most ancient metazoa, the porifers *Geodia* (Pancer et al. 1998; Schacke et al. 1994). The domain length and its amino acid composition led to the classification of its domains as C2 domains. Indeed they resemble the C2 domains of many of the *Caenorhabditis elegans*, zebrafish, and *Xenopus* neural cell adhesion molecule in length and in homology (both up to 51% similarity). However, the homology of the most membrane proximal domain with the mammalian immunoglobulin V domain is 34%, rising to 51% if one includes positive replacements. All but one of the key amino acids for the V frame are present (G in the A′, B strand region is missing at the expected position, but there is one in the preceding position). The domain is short in the A′, B, and C strands area, otherwise the homology with a V domain is remarkable (Blumbach et al. 1999; Fig. 2: homology with V_λ and a bovine immunoglobulin VH). Chou-Fassman or Garnier-Robson strand predictions confirm the V nature of these molecules. If they are not real V, they certainly belong to the I set (Harpaz and Chothia 1994). This suggests that Igsf domains with V features were among the most ancestral molecules of this class.

More recently a second Ig domain has been described, preceding that above (Blumbach et al. 1999). The molecule would then be reminiscent of a mouse tyrosine kinase which also possesses two Ig domains and three fibronectin domains (tk19–2). Neither of these mammalian I-set genes contains an intron (Mark et al. 1994). Interestingly, some signal regulatory proteins (SIRPs, see below) have been

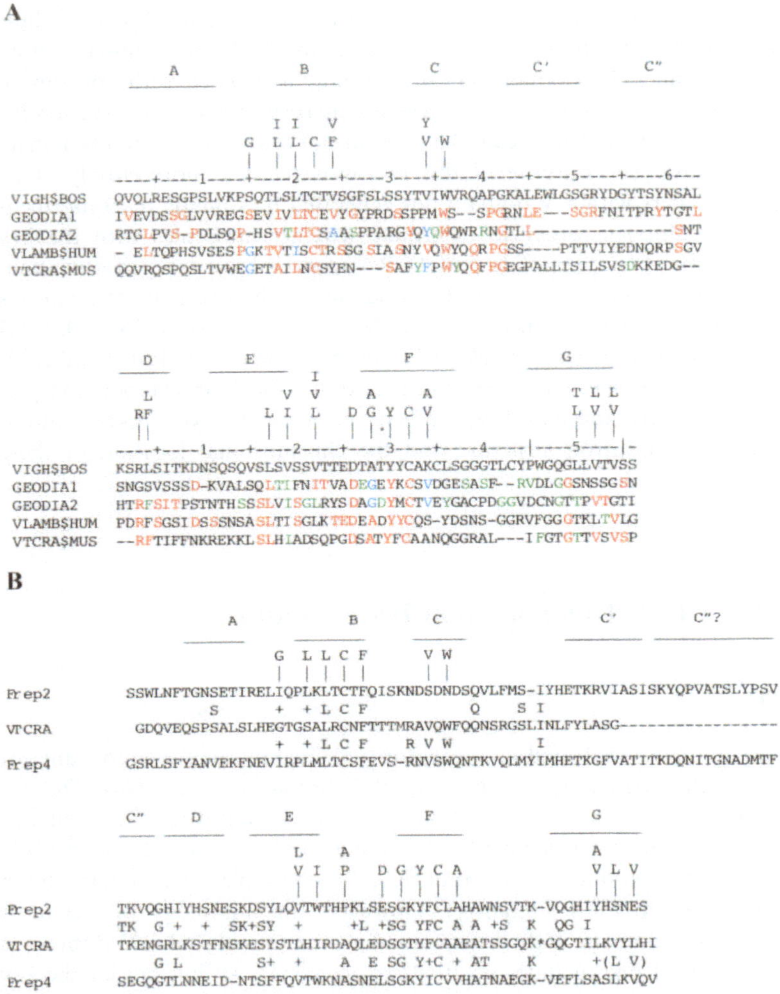

Fig. 2. A V or I-set domains of the porifer *Geodia* have been aligned with mammalian Ig heavy-chain Ig light-chain and murine TCRα sequences. The vertical bars above the alignments indicate, the V frame residues according to Harpaz and Chothia (1994) with the corresponding amino acid at this position. In order not to charge the figure only the appropriate selection among the V-frame residues at each position are indicated. For instance, in strand B the last amino acid can be any of the neutral or hydrophobic amino acids, here an F. *Red*, Residues shared with VIGH$BOS; *green*, residues shared with human V λ; *cyan*, acceptable V-frame residues when not already outlined; *VIGH$BOS*, bovine Ig heavy-chain variable region (AF015498); *GEODIA1*, first Ig domain of *Geodia* tyrosine kinase (X98340); *GEODIA2*, second Ig domain of *Geodia* tyrosine kinase (X98340); *VLAMB$HUM*, variable region of human Ig light-chain λ (X98219); *VTCRA$mus*, variable region of mouse TCRα (Y11102). **B** Comparison of the mollusc *Biomphalaria* fibrinogen-related proteins (FREP) V domains with TCRα from mouse. Same symbols in a for the V frame residues. +, A conservative replacement; *letters*, identities; *asterisk*, junctional diversity has been removed; *brackets*, residues that match the V frame. The strand attribution has been obtained with the lasergene program. Accession numbers; Frep2, U82479; Frep4, U82478; *VTCRA*, Variable region of mouse TCRα, X02968

shown to inhibit signaling through receptor tyrosine kinases and are expressed on myeloid cells and neuronal cells (ADAMS et al. 1998). One wonders whether similar interactions could take place in sponge, and whether the reported polymorphism of the receptor tyrosine kinase of sponges is in fact recognized. Elucidation of the ligand of the sponge molecule would be very interesting.

2.2 Cnidarians

Receptor tyrosine kinase KTK90 of the diploblastic *Hydra viridis* also contains five external Igsf domains some of which with V frame residues (accession no. AAB03389) but their length seems to be more compatible with I-set characteristics than with true V domains.

2.3 Nematodes (Triploblastic Acoelomates)

Many Ig domains from *C. elegans* have been found in intracellular molecules such as twitchin (a molecule binding myosin), titin and hemicentin (VOGEL and HEDGECOCK 1998, AF 074901) a molecule required for hemidesmosome-mediated cell adhesion, and many sequences of this type have been analyzed previously (HARPAZ and CHOTHIA 1994). Moreover, many undefined but homologous sequences are present in cosmid clones (e.g., celf12f3 595) and should be analyzed further.

So far, no obvious homology with V domain using TCR Vδ, Vα, or Ig V regions as "probes" in computer Blast searches has been detected. All the molecules detected are too short to correspond to a real V architecture even though several show the typical V-frame and therefore fit the I-set definition. No involvement of any of these molecules in the immune responses of the worm has been reported. The hemicentin has been mentioned because a gene with some significant homology to it has been found in the class III region of the mouse MHC (ROWEN 1998; accession AF109905). Short stretches of this sequence match segments from the constant region of CTX, a MHC-linked molecule in *Xenopus*, but this does not go beyond the usual Igsf features. It would be interesting to see if hemicentin is present on the MHC class III synteny group found in *C. elegans* (see Kasahara, this volume).

2.4 Molluscs (Triploblastic Coelomates)

Fibrinogen-related proteins (FREPs) from the snail *Biomphalaria* represent a diversified family of genes coding for soluble molecules with a distal domain homolog to TCR V region and apparently involved in the defense against parasites (ADEMA et al. 1997). The homologies are strongest in the B, E, F, and G strands, with substantial divergence in A, C, D, E strands. Yet the key residues in those strands are conserved (Fig. 2) and the Garnier-Robson prediction of β-strand gives a typical V domain distribution of the A, A′, B, C, C′, C″, D, E, F, and G strands with cysteines in the B and F strands, as expected in a canonical V domain.

As in *Geodia*, the glycine residue between strands A and B is missing. The best vertebrate homology is obtained with TCR, and the alignment is presented in Fig. 2. Southern blot data indicate that several genes are present in the germ line (Adema et al. 1997).

In molluscs another Ig superfamily member, molluscan defence molecule (MDM), with five Igsf C2 domains has been discovered in *Lymnaea* and seems to be down-regulated during parasitic infections (Hoek et al. 1996). MDM is similar to adhesion molecules with a strong homology to NCAM and neuroglian genes from various species of invertebrates. Although lacking a proper V domain, one should consider this molecule because of its possible involvement in immune responses (Hoek et al. 1996). It also has some homology with an arthropod molecule involved in defense, hemolin (see below).

The most distal external domain of MDM show moderate homology with the nurse shark Igsf receptor V domain (Greenberg et al. 1995) mainly in the D, E, and F strands [54% positive residue matches with one new antigen receptor (NAR) sequence U18715]. The four other domains are closer to NCAM domains and more resemble C2. Some V frame residues are absent from all domains (e.g., G in strand A).

2.5 Arthropods (Triplobastic Coelomates)

Many insect molecules with Ig domains have been identified, most without V domains but with C2 domains, such as the adhesion molecule titin. In principle, we do not want to expand on these. Among them peroxidasin is conspicuous (Nelson et al. 1994) because of the strong homology of its four Ig domains to the constant region of the *Xenopus* CTX a thymocyte receptor (Chrétien et al. 1998).

Other Igsf insect molecules may be involved in immunity, such as hemolin (Sun et al. 1990), mentioned above. It is found in silkworms (Lindstrom-Dinnetz et al. 1995; Shin et al. 1998). Overall, as well as domain by domain, it shows a strong homology to neuroglian of *Drosophila* (Grenningloh et al. 1990; Hortsch and Goodman 1991). It is made up of four Ig domains, so far considered as C2, but its third domain has been analyzed by Harpaz and Chotia (1994) and found to belong to the I-set because of its V frame.

Among arthropods, two molecules have easily recognizable distal V domains (Du Pasquier 1989; Seeger et al. 1988), lachesin and amalgam, both discovered in insects, and are built according to the scheme V-C2-C2. Amalgam is encoded by a single exon whereas lachesin, studied in *Drosophila* and *Schistocerca*, is similar to *Drosophila* amalgam (V-C2-C2; Karlstrom et al. 1993). It functions in neurite outgrowth and other cell surface recognition events. In addition, as in Porifers, Cnidarian and *Drosophila* receptor tyrosine kinase has Igsf external domains including one V-like domain (Pulido et al. 1992) with homology to the distal domain of hemolin as well as to the V domains of mouse butyrophilin, poliovirus receptor and *Xenopus* CTX (see below).

In all, there is evidence for V-like structures of the I-set type and also for real V domain genes without introns in several phyla of invertebrates. However, among the I-set members the degree of homology to V can vary. Some molecules are closer to the canonical V domains of vertebrates than others. The *Geodia* molecules are a striking example of this and confirm that the V frame is a very ancient feature in metazoa. However, a V-C1 architecture reminiscent of a true TCR or Ig has not been encountered in any of the invertebrate phyla. Perhaps vertebrates themselves will be found to contain in their genome, genes or gene fragments that have at least retained the architecture of early lymphocyte receptors even though they have evolved in other directions.

3 Vertebrate Molecules with V Domains Not Generated by Somatic Rearrangement

3.1 V Without C1 Domains

The molecules described in invertebrates with nonrearranging V domains have counterparts in vertebrates. The closest homologs have V domains associated with a C2 or an I-set domain 1 (Fig. 3). Many different V lineages have developed among these and are characterized by their exon-intron organization and strand properties. In some cases the exon-intron organization and the splicing laws will provide useful tags. (References for this section are too numerous to be listed in the present context. They can be found in Barclay et al. 1997.)

3.1.1 V Without Intrachain Disulfide Bridges

A good example is the CD2 family in which the V domain is encoded by a single exon whenever described. Related members include CD48, CD58, 2B4, Ly9, and CD150. These V domains are always associated with I-set or C2 domains, never with C1. The carcinoembryonic antigen V domain also belongs to this type (CD66 a to e) as well as the second V domain of CD4.

3.1.2 V With a Noncanonical Intrachain Disulfide Bridge

CD33 is the prototype of this category with the second cysteine involved in the disulfide bridge at position 84. This position is in fact similar to that of the second cysteine in the first V domain of CD4. The family contains several genes all present together with CD33 (Peiper et al. 1988) on human chromosome 19q13: CD22 and the myelin-associated glycoprotein (Barton et al. 1987).

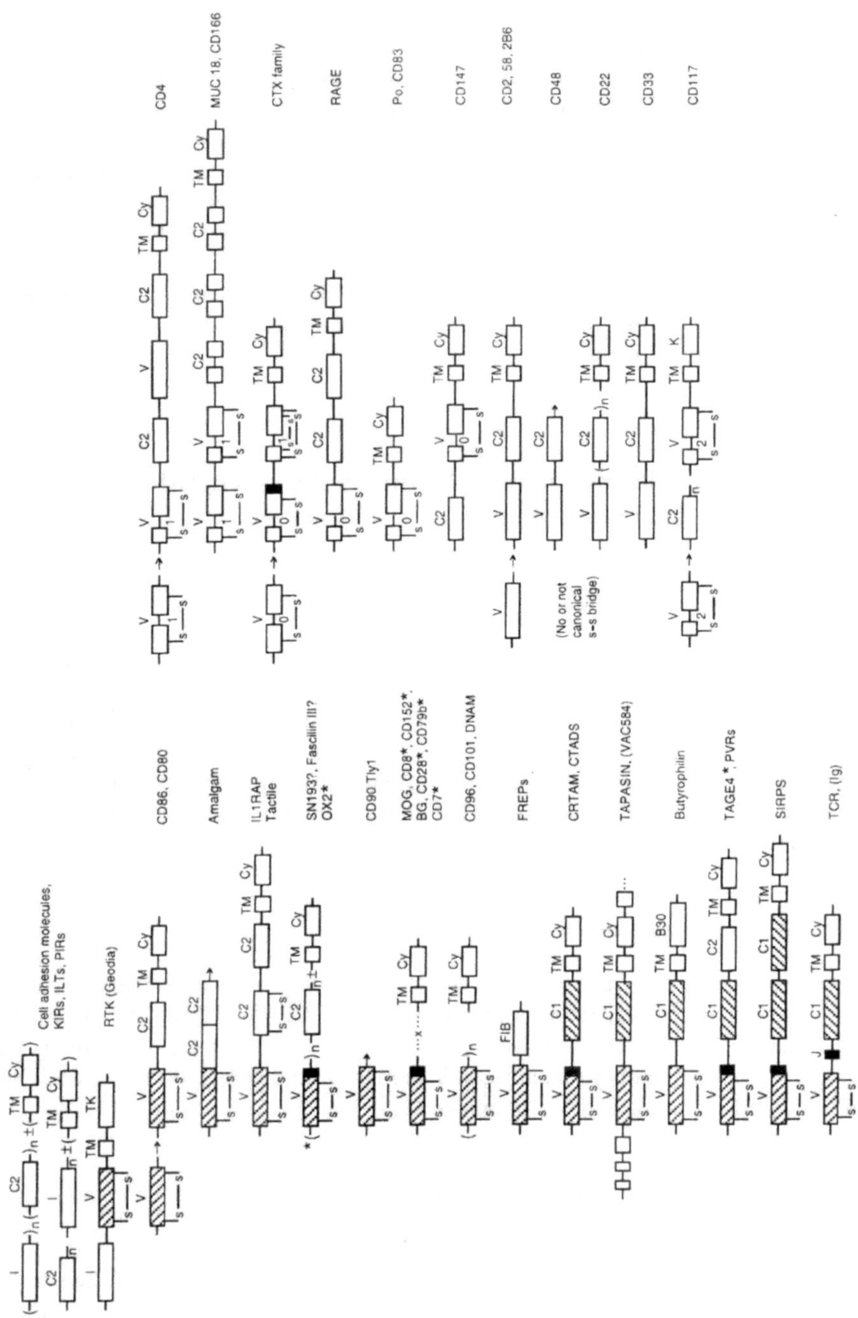
Cell adhesion molecules, KIRs, ILTs, PIRs
RTK (Geodia)
CD86, CD80
Amalgam
IL1RAP
Tactile
SN193?, Fasciclin III?
OX2*
CD90 Thy1
MOG, CD8*, CD152*, BG, CD28*, CD79b*
CD7*
CD96, CD101, DNAM
FREPs
CRTAM, CTADS
TAPASIN, (VAC584)
Butyrophilin
TAGE4 *, PVRs
SIRPS
TCR, (Ig)
CD4
MUC 18, CD166
CTX family
RAGE
Po, CD83
CD147
CD2, 58, 2B6
CD48
CD22
CD33
CD117
(No or not canonical s-s bridge)

3.1.3 V Encoded by Two Exons with a Type 1 Splicing Law

V domains of this category can be encountered alone (CD79b) or in association with other V and C2 domains (as in the case of the first V domain of CD4 (VC2VC2) and the V domains of Muc18 (VVC2C2C2). CD166 has the same architecture as Muc 18, but its genomic organization is unknown.

3.1.4 V Encoded by Two Exons with a Type 2 Splicing Law

This is a case of CD117 or C-kit in which the V domain is close to the membrane and not distal. The molecule is expressed by mast cells, acute myeloid leukemic cells, and some bone marrow cells and is another receptor tyrosine kinase (Yarden et al. 1987).

3.1.5 V Domains Encoded by Two Exons with a Type 0 Splicing Law

This is found in the CTX family members from amphibians to mammals, where the V domain is associated with a special C2 domain with an additional disulfide bridge (Chrétien et al. 1998). CTX defines a subset of genes and molecules the function of which is largely unknown but apparently related to cell proliferation (Robert et al. 1997a,b). In the immune system CTX members have been found in frogs and birds on cortical thymocytes (Chrétien et al. 1996; Katevuo et al. 1998), but so far, although many mammalian homologs have been discovered, none shows a similar distribution (Chrétien et al. 1998). The fact that the molecule bears a resemblance to TCR (Chrétien et al. 1996) but is not generated via somatic rearrangement places it at the crossroad between adhesion molecules and antigen specific receptors. Several mammalian homologs are expressed in the genital and gastrointestinal tract but not on lymphocytes. I describe here a new human member reconstituted from an high throughput genomic sequences (HTG) of chromosome X and called provisionally for this reason CTXhumx (see sequence of the V domain in Fig. 11a). This gene seems to be the closest relative to the cortical thymocyte markers CTX and chicken thymocyte antigen (ChT1). The level of duplication of this gene in

Fig. 3. Molecules with V domains. The various architectures of the Igsf members and genes with V domains not generated by somatic rearrangement compared to TCR. The molecules are classified in function of the intron-exon organization or the composition, i.e., their association with I-set C2 or C1 domains. Only examples are given. The V domain sequences of these molecules can be found again in the alignment of Fig. 11a. *Asterisk*, molecules with a G strand resembling a J segment with a diglycine bulge. *Black*, J segment-like sequences. In the CTX family only some members have a diglycine. The splicing law within V domain is indicated by a 0, 1, or 2 between the schematized exons. *KIR*, Killer inhibitory receptors; *RTK*, receptor tyrosine kinase; *Il1RAP*, Il1 receptor accessory molecule; *FREP*, fibrinogen-related protein; *Tage4*, tumor associated glycoprotein E4; *SIRP*, signal regulatory protein (Kharitonenkov et al. 1997); *CTX*, cortical thymocyte marker of *Xenopus* RAGE receptor for advanced glycosilation end products; *MUC18*, a marker of tumor progression in human melanoma; *DNAM*, DNAX accessory molecule-1 expressed on peripheral lymphocytes (Shibuya et al. 1996); *ILT*, Ig-like transcripts (Cella et al. 1997); *PVR*, poliovirus receptor; *TM*, transmembrane; *Cy*, cytoplasmisc domain; *MOG*, myelin/oligodendrocyte glycoprotein (Linington and Lassmann 1987); *PIR*, paired Ig-like receptors (Kubagawa et al. 1997)

Fig. 4. Constant domains. **A** CRTAM constant region domain compared to C1 containing molecules Tage4 has one C1 (C2 Tage) and one C2 (C3 Tage) domain. Constant domains have been aligned against a C1 domain of the bovine MHC that gave the best blast homology with the SIRP and Tage4 sequences. By length and residue composition the SIRPs molecules appear clearly related to C1 domains. In Tage4 the third domain lacks amino acids in the D strand region being close to a C2 domain whereas the second molecule resembles a C1. **B** Comparison of tapasin constant domain (CTAP6) with the gene discovered on chromosome 12 (CTAP12). Above each sequence or group of sequences the residue most often associated with the C1 and the C2 type have been represented (Barclay et al. 1997). Tapasin is clearly on the C1 side. *Red*, C1 domain characteristic residues; *green*, C2 domain characteristic residues; *black*, Ig domain residues. *BovMHC*, Bovine MHC class I haplotype AW10 (M69206); *SIRP C1*, *C2*, first and second constant domains of SIRP (Y10376); C_3 Tage, C_2 Tage, first and second constant domains of murine Tage4 (U 35836) mouse Tage4; *Ctap*, human tapasin constant Igsf domain (Y13582). **C** C1 domain found on chromosome 11 (AC406) compared to fish MHC Ig domain

mammals and the loose linkage of many of them from *Xenopus* to mammals with the MHC or MHC paralogous regions suggests the conservation of an ancient association and therefore a long-lasting relationship with immunologically relevant genetic regions (see below).

The family contains lymphocyte receptors (CTX, ChT1) and also many molecules expressed on tissues of endodermic origin such as A33 and others of widespread tissue distribution such as the coxsackie virus receptors (CAR). CTX, ChT1, and the coxsackie virus receptors contain ITAM (tyrosin based activation VxYxxV/l motifs that were not initially recognized). If splicing law type 0 between the two half V domain exons can be considered a tag, other members join the family: the myelin protein P0 (Lemke and Axel 1985), the antigen presenting cell molecule CD83, and basigin (CD147) in which the V domain as in C-kit is close to the membrane and not distal. The comparison of linkage groups to which these single

Table 1. Chromosomal locations of Igsf members

	1q21–25	6p21–23	9q33-34	11p13	11q22-24 11q23-24	12p12–13	19q13	19q13	X
V type o splic.	Po[a] A33?	CD83 RAGE			EVA CTH V1235a + b		CD147		CTX humX q21-2
V no intron		Butyro. tapasin MOG	Tapasin-like (V2320)		CRTAM PVR-like	Tapasin-like (AC5840)		PVRs MAG V1335	HS159 q14-?
C_1	CD1 MR1	Tapasin class I class II Butyro.	?	AC406	PVR-like CRTAM	Tapasin-like (AC5840)		FcRec neonate PVRs, (TAGE 4 homo?)	
Complement		C4	C5			α_2-Macro	C3		
MHC I, III + paralogues	+	+	+			+	+	+?	
Other genes				RAGs	$CD_3\gamma, \delta, \varepsilon$	CD4 LAG-3		KIRs	

When not specifically mentioned, sequence and accession numbers of the members can be found in Fig. 4. The positions have been obtained from the Gene Map 1999. Sequences of V and c1 domains of tapasinlike (AC5840) VHS159, V1335 which correspond to sequence identified from HTGs in this paper can be found in Fig. 11. Another MHC class I-like molecule, the zinc α_2-glycoprotein has a bona fide C1 domain. Its gene maps to chromosome 7q22.1 in human (see Kasahara, this volume), but the linkage group does not as yet contain other immunologically relevant genes.

[a] Other genes related to P0 have been identified from HTGs (see Du Pasquier et al. 1999).

domain molecules belong is a further argument to unify the family since many members belong to the MHC and paralogous regions (Table 1). In phylogenetic analysis the V of P0, RAGE (VLASSARA et al. 1987), basigin (MIYAUCHI et al. 1990), and CD83 do indeed cluster with the CTX family members.

CTX bears some interesting similarity to TCR because of its apparent dimeric state and the presence of a diglycine bulge in the G strand that makes it very reminiscent of a J segment (CHOTHIA et al. 1985; DU PASQUIER and CHRÉTIEN 1996). Given the difference in the gene organization of TCR and CTX, this situation could be the result of convergence. The family also contains monomeric molecules.

3.1.6 Canonical V Domains Encoded by Genes Without Intron

In principle this feature should be shared by the ancestor of TCR or Ig V domain genes. In the real TCR ancestor a single fragmentation led to separation of the gene segment coding for the J part from the rest of the V gene.

3.1.6.1 V Alone or Associated with a Non-Igsf Domain

Several molecules of this type are listed in Fig. 3. Many contain a non-Igsf segment. They can be integral membrane proteins or GPI anchored. Some form dimers (CD152, CD8, CD28) a feature often associated with the presence of a diglycine bulge in the G strand, as noted above.

This category contains genes that have a surprisingly high homology with primitive vertebrate T-cell receptors. For instance, Dora (BATES et al. 1998) expressed on dendritic cells and other cells of the myeloid lineage is 42% in human, similar to the shark TCRδ-like sequence (RAST et al. 1997) and also shows a diglycine bulge.

3.1.6.2 V-C2 Associations

In teleost fish the SN193 type molecule (fugu; RAST et al. 1995) is described in detail by YODER and LITMAN (this volume). The V domain of this molecule is clearly of the immune receptor type. In phylogenetic analysis it always clusters with Ig and TCR V sequences.

In human molecules such as CD80 and CD86 (both VC2) their genes are located on the same chromosome 3q13–23.

Tactile, a novel human T-cell activation antigen (CD96; WANG et al. 1992) shows significant similarity to *Drosophila* amalgam, the melanoma antigen MUC-18, a V-V-C2-C2-C2 molecule (see previous section), members of the carcinoembryonic antigen family (CD166), the poliovirus receptor, and NCAM. The protein has three Ig domains defined as three V domains, a long serine/threonine/proline-rich region typical of an extensively O-glycosylated domain, a transmembrane domain, and a 45-residue cytoplasmic domain. Tactile may be involved in adhesive interactions of activated T and natural killer cells during the late phase of the immune response. The distal V domain of human tactile preferentially matches the distal V domain of *Drosophila* amalgam and of another homolog, the poliovirus receptor (V-C2-C2). It also matches the myelin protein P0, another V domain of

A

```
                      A                 B                          C              C'
                 ___________       _________                  ____________    ___________
                          |   |  |  |  |                          |  |            |
                 ----+----1----+----2----+----3----+----4----+----5----+----6----+----
TAPASIN$CH      VALAVLTHTPTLRARVGSPIHLHCAFA-------------APPSS---FV--LEW-RHQNRGAGRVLLAYD
TAPASIN$6       'V'T''''''AP'V'L'QDAL'DLS''YMPPTSEAASSL''G'PP--'G--'''-'R'HL'K'HL'''AT
TAGE4           GEI''QVL-SNSTGFL'GSTV'''SL'--------------SKDNV--TITQ'T'-MKRDPDGSHPSV'VF
TAPASIN$12      'EFQ'M'Q'QS'SFLL''SAS'D'G'S------------M''GLDL---IS-V'''-L'HK'R'QLVYSWT
VPVR            TGDV'VQAPTQVPGFL'DSVT'P'YLQ--------------V'NMEVTH'SQ'T'A''GES'SMAVFHQT

                     C''            D           E               F                G
                 __________     ________    ___________     _________      _____________
                                   ||              | |   |    |  |  |  |  |       |  |  |
                 ----+----1----+----2----+----3----+----4----+----5----+----6----+----
TAPASIN$CH      SSTARAPRAHPGAELLLGTRDGDG-----VTAVTLRLARPSPGDEGTYICSVFLPH-GHTQTVLQLHVF
TAPASIN$6       PGLNGQMP'AQEGAVAFAAW'D'EPWGPWTGNG'FW'P'VQ'FQ'''LATIH''Y-LQG'VT'E'A'Y
TAGE4           HPKKGPSISD'ERVKF'VAKVYED-----LRNAS'AISNLRVE'''I'E'QIATFPTDSKSANVW'K''
TAPASIN$12      AGQGQ'V'KGAT----'EPAQLGM-----ARDAS'T'PGLTIQ''''''QITTSL-YRA'QII''NIQ
VPVR            QGP-SYSESKRLEFVA--ARL'AE-----LRNAS''MFGLRVE'''N'T'LFVTFPQ'SRSVDIW'R'L
```

B

```
             ----+----1----+----2----+----3----+----4----+----5----+----6----+----7----+--
TAPASIN$6   -VVLTVLTHT-PAPRVRLGQDALLDLSFAYMP-----PTSEAASSLAPGPPPFGLEWRRQHLGKGH--------LLL
TAPASIN$9   VSPLGPPSPVSPPPPANAEPSPERRRPQSVVGTRDPGPGSEMRTSWAQGPPPWPLAWQGLGGAKGARLSLRGADLLL
            ...|.......|.|...|......................|.||...|.|.||||..|.|......||.........|||

             ----+----1----+----2----+----3----+----4----+----5----+----6----+----7----+---
TAPASIN$6   AATPGLNG----QMPAAQEGAVAFAAWDDDEPWGPWTGNG---TFWLPRVQPFQEGTYLATIHLPYLQGQVTLELAVY
TAPASIN$9   AAGPALGRLEGRHPGAENGGTVCFAALQP---WAPWPGAGWHARLWLGWVQPLEQGTVVLGQLIPTDV--VTLV----
            ||.|.|..____...|...|.|.|||...___||.|.|.|___..||..|||...||.......|...--|||.
```

Fig. 5. A Alignment of tapasin V domains sequences from chicken and human with related sequences from genes identified on chromosome 9 and 12 in human. *Dots*, identity with chicken tapasin V domain; *red*, residues common to human tapasin; *green*, residues in common to Tage4. Tapasin 12, as with chicken tapasin but unlike the tapasin of chromosome 6 has canonical cysteins. *Bars* V frame residues (see Fig. 2 for more explanations). **B** Chromosome 9 V-like sequence has been separated from A to minimize the number of gaps in a multialignment display. As in the tapasin of chromosome 6, this sequence lacks the canonical cysteins. *Tapasin$ch*, Chicken tapasin (AJ005071). *Tapasin$6*, Human MHC linked tapasin (chromosome 6; 13582); *Tapasin$12*, human tapasin homologue reconstructed from HTG AC005840 and belonging to chromosome 12p13 region; *VPVR*, variable domain of the human poliovirus receptor (P15151); *Tapasin$9*, reconstructed V domain of a human tapasin homologue from HTG AC002320 and belonging to chromosome 9q34

mammals with the CTX type of gene organization. The two other Ig domains are more difficult to classify. The second one bears no significant homology to any specific Ig domain. The third domain best matches the C2 domain of Muc 18 and RAGE and the membrane proximal domain of the mouse tumor-associated glycoprotein 4 (Tage4) or poliovirus receptors, all members of a newly defined V-C1-C2 type (Fig. 4). DNAX accessory molecule-1 (DNAM-1), a human T-cell protein, is also made of two external Igsf domains (Shibuya et al. 1996). The two domains seem to belong to the V family (as was proposed for tactile). The first one is closer to Tage4 and poliovirus receptor (PVR) molecules while the second is closer to a TCR or butyrophilin V region or to I-set molecules.

3.2 Molecules with Igsf C1 Domains

None of the previous molecules seems to be directly related to what could have been a complete ancestor of Ig and TCR because of major differences in gene organization and lack of C1 domains.

In fact, in addition to the above combinations found across practically all phyla, gnathostomes show smaller numbers of molecules with V and C1 domains. This suggests a lower level of duplication, compatible with the hypothesis that these genes appeared more recently in the history of vertebrates. They are not necessarily lymphocyte specific. They may not even be cell surface molecules but have strong homologies with Ig TCR and MHC domains for the C1 domains, and also the V domains of chondrichthyan species (Figs. 4–7). Ancient duplicates of such genes could have generated a TCR-like receptor since their basic structure is already V-C1 and that some of them can form dimers. In several of the genes encoding these molecules it looks as if a gene segment consisting of a core unit of one V-C1 polypeptide has been duplicated and shuffled into various combinations with other gene segments.

3.2.1 Molecules with a VC1 Associated with Another Set of Gene Segments

3.2.1.1 VC1C2: Poliovirus Receptors and Tage4

These molecules were selected by the BLAST (Tblastn) searches when looking for homologs of invertebrate molecules such as lachesin and amalgam. These were previously classified as composed of one V and two C2 domains, but I have found the homology of the so-called second C2 domains with MHC and Ig heavy-chain constant regions (Fig. 4) to be sufficiently striking to question this early classification and to suggest in fact a VC1C2 structure. Systematically the first C domain gave better homology with MHC Ig domains and the second one better homologies with the C2 domain of adhesion molecules.

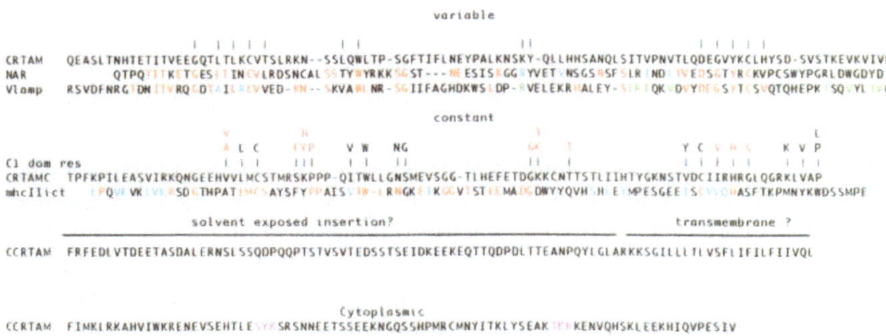

Fig. 6. Comparison of CRTAM to the shark NAR sequence and to the limbic-associated membrane protein lamp (*Vlamp*). Note the long DE-rich region before the transmembrane region. *Red*, residues common to CRTAM; *green*, residues common to NAR and lamp; *magenta*, phosphorylarion sites; *Bars*, V-frame residues. See Fig. 5 for more explanations

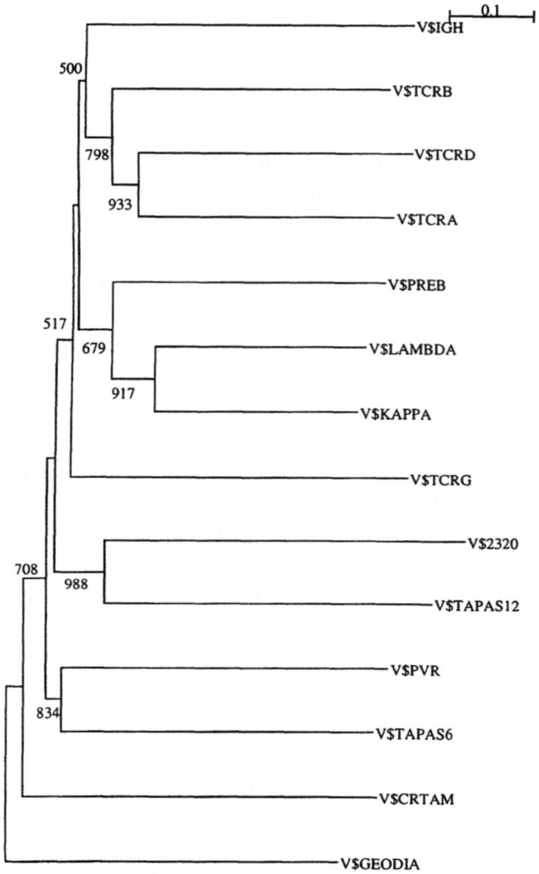

Fig. 7. V domain dendrogram showing the probable paralogous nature of the tapasin-related V sequences. The phylogenetic tree was generated by the neighbor-joining method with the Clustalx program. Boostrap values are given when they are above 500/1000. This tree is not a complete phylogenetic analysis. It is meant to be a useful way of grouping the sequences and to complement the data obtained from gene structure and chromosomal locations. Note the early separation of all V lineages from CRTAM and the grouping of all the V region genes that undergo somatic *rearrangement*. Accession numbers are given in Fig. 4. *V$Ighhhum*, variable region of human heavy chain; *V$TCR* αβγδ, variable region of human TCR. *V$KAPPA*, *V$LAMBDA*, variable regions of human light-chain; *V$preB*, V domain of the preB-cell receptor; *V2320*, human chromosome 9 variable domain with homology to tapasin (see corresponding section in the text); *V$Tapas12*, V domain of a second tapasinlike gene from human chromosome 12 (accession AC005840); *V$CRTAM*, V domain of CRTAM, novel immunoglobulin supergene family member expressed by class I MHC restricted T cells; *V$PVR*, V domain of the poliovirus receptor; *V$TAPAS6*, V domain of the human tapasin situated in the MHC on chromosome 6; *V$Geodia*, sponge tyrosine kinase V domain. Bootstrap values around and above 500/1000 are indicated

Tage4, a molecule identified in rodents, is at the crossroad of many invertebrate and vertebrate molecules. Tage4 is a recently discovered member of the Igsf expressed in rat colon and mammary carcinoma (Chadeneau et al. 1996).

In Blast searches, the Tage4 V domain generates more significant homologies than any other member with a similar structure. Poliovirus receptors and herpes virus receptors are closely related to Tage4. CTX, EVA (epithelial vascular antigen, a molecule with a V domain and splicing law type 0 on the same linkage group as one of the human homologs of CTX family members), V of RAGE and V of Vcam also show good homologies. Tage4 first domain is an unambiguous V domain similar to P0 and Ig λ V region ($P < 0.001$). Its second domain is closer to C1 by Blast search and by Garnier-Robson prediction (Fig. 4). It has better homologs among class II, Ig domains of *Xenopus* and zebrafish than among C2 domains. It would be interesting to know the organization of this gene since many of the related molecules have a different exon-intron organization.

The third domain, another assumed C2 domain, is indeed closer by structure to C2 for instance those of *C. elegans* hemicentin precursor; ($p = 0.009$).

3.2.1.2 X-V-C1: Tapasin

Tapasin, a molecule involved in the MHC class I assembly (Ortmann et al. 1997; Sadasivan et al. 1996) is encoded by a MHC-linked gene in human (Herberg et al. 1998) and chicken (S. Milne, J. Kaufman, S. Beck, locus entry: GenBank CAA18963). The second exon after the leader encodes a structure that has no strong homology with any known gene family (scheme in Fig. 3). The fourth exon encodes an Igsf V domain that was not recognized as such until recently because in mammals it does not contain the famous disulfide bridge tag. However, it is contained in chicken, which helped to identify the gene family (Frangoulis et al. 1999). This V domain is encoded by a gene segment without intron. The fifth exon is a recognized Igsf C1 domain gene (Frangoulis et al. 1999) and is followed by transmembrane and cytoplamic exons. It is also homologous to the C1 domain of Tage4/PVR group.

In addition, I have found two new genes or pseudogenes related to tapasin in the human genome. One (AC005840) with a V (including a disulfide bridge as in chicken; see Fig. 11A) and a recognizable C1 domain (Fig. 4–5) associated with the V is located on chromosome 12 in the newly described MHC paralogous region close to the complement homolog α_2-macroglobulin (Kasahara 1999). Both V and C domains remind one more of the chicken tapasin than the human tapasin present on chromosome 6.

The other one (AC002320) (Fig. 5B) is on chromosome 9q34 in the vicinity of a MHC class III paralogous region which includes C5, another complement component. I was able to reconstruct this V tapasin-like segment from two "exons" or gene fragments separated from each other by 90kb and in the opposite orientation but on the same HTG. For this reason it looks like a pseudo gene unless mistakes are still in the HTGs. In summary, Tapasin gene seems to host a gene architecture very much like the one anticipated for an ancestor of the T-cell receptor.

3.2.1.3 VC1 B30-2

A VC1 core can be associated after its transmembrane region to a domain quite different from that found in Ig TCR and MHC. This is a case of the MHC-linked butyrophilins forming a family of at least four VC1 molecules (Henry et al. 1998) where the cytoplasmic domain is a B30-2, a characteristic sequence of about 170 amino acids apparently able to bind xanthine oxidase (Ishii et al. 1995).

3.2.1.4 VC1C1 and SIRPs

This category contains intracellular SIRPs, also known as the P84 neural adhesion molecule (Kharitonenkov et al. 1997), with homologs such as BIT and SHPS-1 (Yamao et al. 1997). These molecules act as signal inhibitor regulatory molecules via immunoreceptor tyrosine-based inhibitory motifs present in their cytoplasmic tail.

The homology of SIRP C domain with the C1 domains of MHC or Ig heavy chain is strong and unequivocal. This was recognized by Adams et al. (1998). The first C1 domain is most homologous to the immunoglobulin heavy-chain C region; the second more resembles MHC class I. In other words, they are both C1 but differ substantially from each other, and the SIRP gene could be the result of an exon shuffling rather than of a tandem duplication. The V domain is close to that of the shark TCR HFU09531. These molecules have many features of Ig or TCR, including the diglycine bulge in the G strand. In human SIRP is encoded on chromosome 20. Similar architectures can be encountered among shark Ig isotypes (Greenberg et al. 1996).

3.2.2 Molecules with a VC1 Architecture Similar to that of TCR

CRTAM, a class I restricted T-cell associated molecule (Kennedy 1998; accession AF001622), is made of a V-C1 core followed by an intermediary sequence of about 75 amino acids (D-, E-rich) with no obvious homology to any domain. This hydrophilic segment is reminiscent one of the solvent-exposed region of the TCRβ found in mammalian TCRβ chains but not in cold-blooded vertebrates (Bentley et al. 1995; Chrétien et al. 1997). The molecule is an integral membrane protein with a transmembrane and a cytoplasmic region. Structurally it resembles the best relative of an Ig or TCR ancestor except for its long cytoplasmic tail. The CRTAM V domain ends by a G strand which is not unlike the J segment of *Xenopus* σ light-chain isotype, where the diglycine bulge is replaced by a di-serine. The CRTAM V domain is also a homolog to a nervous system limbic associated membrane protein of the VC2C2 type (Lamp) and also to the primitive receptor V region of the shark NAR (32% identities and 60% positive replacements). It seems to have retained several features of a primordial nonrearranging receptor. Its cytoplasmic tail contains Camp PKC and CK2 phosphorylation sites. In humans it is encoded by a gene in the TCR CD3γ δ ε linkage group. This molecule and its homologs are represented in Fig. 4 and 6. The C domain of CRTAM and CTADS (its chicken homolog described by Ruble and Foster 1997; accession AF035677) have best homologs among MHC class II Ig domains or constant domains of Ig L chains and clearly belongs to the C1 set.

3.2.3 C1 Alone

The only known C1 domain existing as a single domain is the β_2-microglobulin, the gene of which is on chromosome 15 q21–q22.2 (Manolov et al. 1979). Using the tapasin C1 sequences as a "probe" in further Blast searches, I have identified one interesting sequence from a human HTG corresponding to no product and no expressed sequence tag so far in the databases. This is a C1 domain, the closest relative of which is a fish MHC sequence from *Ictalurus* (C$AC 406 in Fig. 11B). It is located next to RAG1 and 2. This sequence has the characteristics of a "fossil" fragment and is clearly more related to bird and cold-blooded vertebrate MHC class I and II C1 sequences than to anything else. However, on chromosome 11 it does not belong to the CRTAM linkage group but represents a potentially interesting association between a receptorlike molecule and the rearranging machinery. The fact that no expressed sequence tag corresponds to this sequence reinforces the idea that this gene or gene fragment could be a left over with an essentially phylogenetic value.

4 Evolutionary Scenarios

4.1 Relationship to Dimerization

The families of molecules with the various architectures defined above may have evolved in different directions depending on their exon shuffling and splicing law possibilities.

Genes with a V-C1 core, especially the ones with a G strand that resemble a joining segment, are structurally more related to the genes of an ancestor of Ig or TCR at introduction of the somatic rearrangement. It is unknown whether this introduction, accompanied by the split in the V domain gene between the ABCC′C″DEF and the G=J strand had already occurred in a gene encoding a dimeric lymphocyte receptor. It is a likely possibility given the multiple examples of dimers among Igsf members including those with more primitive features such as CTX with its C2 domains. However, dimerization does not seem to be a prerequisite since molecules with primitive characteristics, such as the NAR of the shark, can bind in the absence of light chain (Roux et al. 1998). Dimerization is encountered in TCR and Ig, which suggests that all molecules with such an assembly form combining sites with the loop equivalent to CDR1, 2, and 3. For example, the dimeric CD28 and CD152 interact with monomeric CD80 and CD86b via a highly conserved motif in the CDR3-like loop (Peach et al. 1994). However, other types of interaction are possible (Bork et al. 1994), and the functional selective pressures exerted on molecules that appear similar may in fact be very different.

4.2 Linkage Groups

After selecting putative candidates a few DNA sequences remain as possible relatives of the primordial receptor genes. These can be roughly mapped in hu-

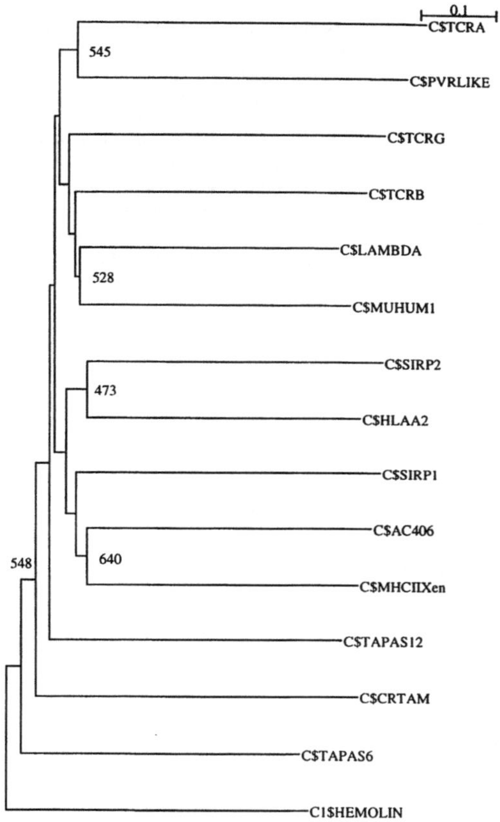

Fig. 8. Dendrogram of the constant region of the C1 type. *C$TCRA* αβγδ Constant domains of TCR; *C$PVRlike*, C1 domain of the PRV-like gene of the chromosome 11q linkage group. Note the clustering of MHC class I, II and SIRPs sequences and their separation from the C1 domains of Ig and TCR and from tapasins and CRTAM. *C$LAMBDA*, Human light-chain C1 domain; *C$ MUHUM1*, first C1 domain of the human IgM gene; *C$SIRP1, 2*, first and second domains of the signal regulatory protein; *C$HLAA2*, human MHC class I Ig domain; *C$AC406*, human chromosome 11 Ig domain; *MHCIIF8Ig*, MHC class II Ig domain; *C$TAPAS12*, C1 domain of a Tapasin related gene on human chromosome 12; *C$CRTAM*, C1 domain of CRTAM (see legend to Fig. 8). *C$TAPAS6*, C1 domain of human tapasin; *C1HEMOLIN*, first domain of silkworm hemolin. Bootstrap values around and above 500/1000 are indicated

mans, and most of them reside in regions related to the adaptive immune system (except SIRP on chromosome 16). This is probably too much to be only a coincidence; however, it is difficult to decide which came first and to build a scenario (Table 1):

- CRTAM-CD3 linkage. This gene is the most similar to a putative prerearrangement receptor and is flanked by genes for the CD3 complex γ, δ, ε also in the vicinity of a member of the CTX family.

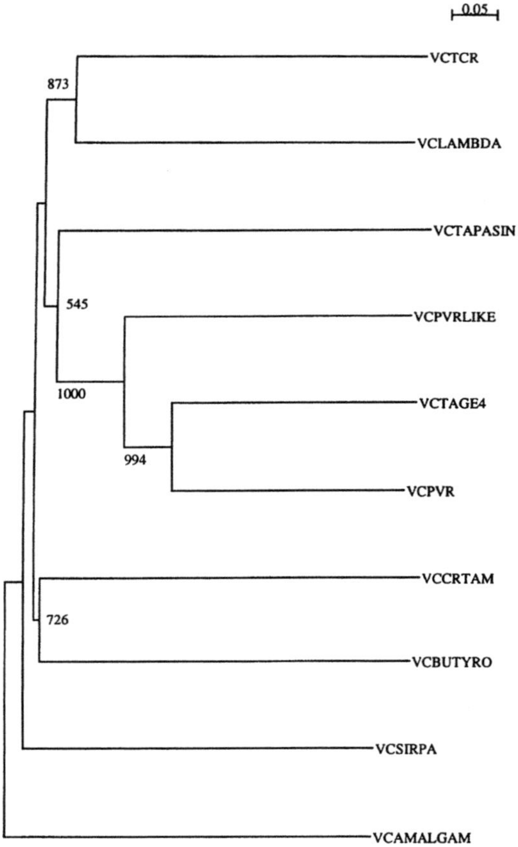

Fig. 9. Dendrogram of the VC1 regions taken together VC stands for V plus C1 domain. *VC Butyro*, VC1 domains of human butyrophilin; *VC Tage4*, V and C1 domain of mouse Tage4; *VCAmalgam*, V and first C domain of amalgam from *Drosophila*. Note the early separation of the CRTAM/butyrophilin family from the tapasin and Ig/TCE. Other abbreviations as in Figs. 7, 8. Bootstrap values around and above 500/1000 are indicated

- Ca406. A C1-RAG linkage type, this gene fragment is not expressed and is most similar to the cold-blooded vertebrate MHC II Ig domain while Ig heavy chain is close to RAG1 and 2.
- Tapasin or tapasinlike genes–MHC paralog linkage. With their VC1 domain these genes form a family with paralogs on chromosome 6p21, 9q34, 12p12–13,

Fig. 10. Origin and evolution of Igsf members with V and C1 domains. A hypothesis taking into account the genetic linkages and the homologies presented in this chapter. *1*, The sequence of the genes segment VHS159 is close to CTX family sequence even though it is not coded for bysplit exons. It could resemble a gene ancestor common to several types of V. *2*, A first duplication creates two types of V that remain linked. *3*, C1 is created by a duplication that leads to C2 and C1 and the complex is now duplicated. *4*, The lineage leading to CRTAM, Ig, and TCR is born (*right*), and the introduction of RAGs determines the lineage of the somatic rearranging of receptors. Probably around the same time the lineage leading to tapasin is linked to class III complement components. Tapasin, or a related ancestor, donates C1 to peptide binding domains α1 and α2 and this generates class I (for MHC paralogies see Kasahara, this volume). These linkages are preserved in paralogous groups. *5*, The origin of SIRPS and PVR is not clear. The similarity of PVR with tapasin and of SIRPs C1 domain with MHC Ig domains makes possible a link with the left part of the scheme

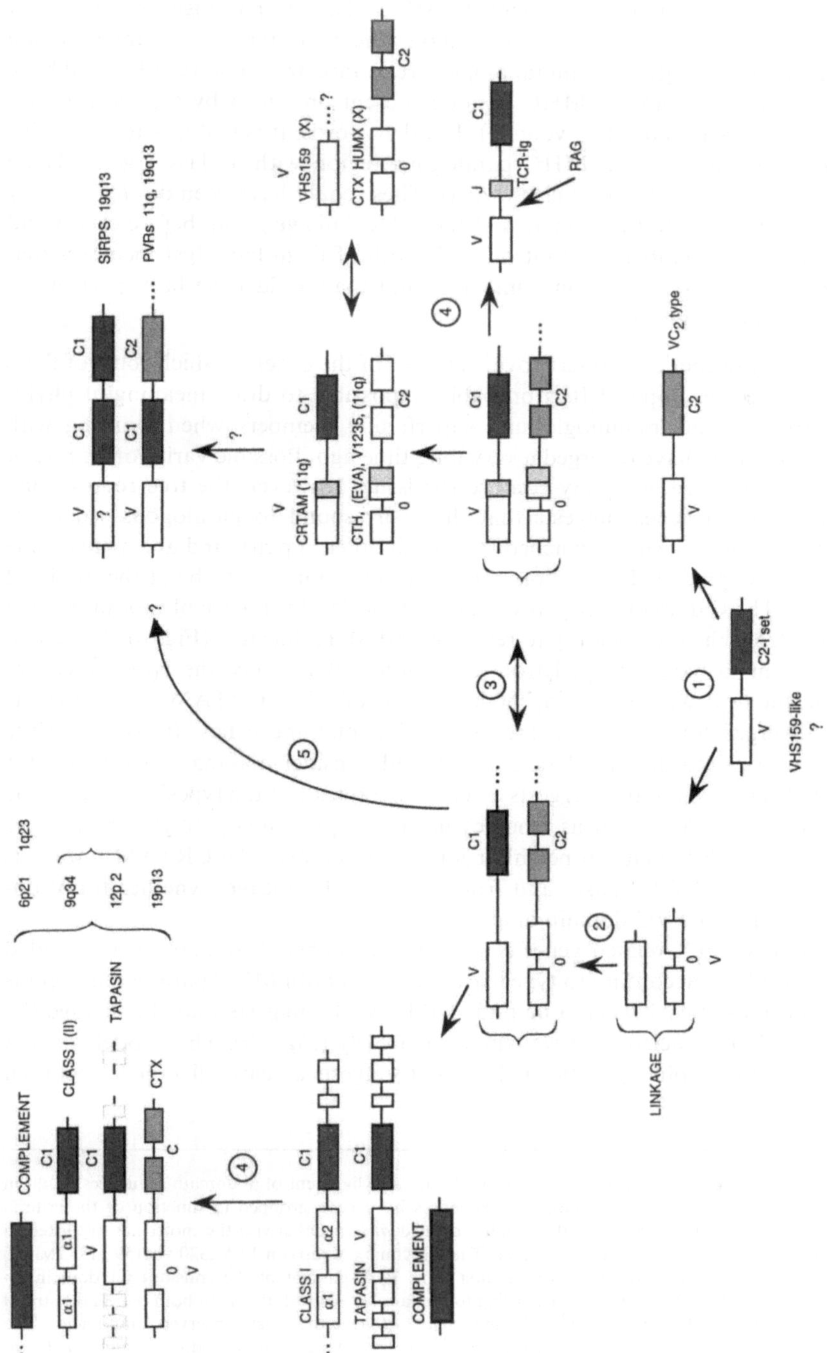
COMPLEMENT
CLASS I (II)
TAPASIN
CTX
6p21
9q34
1q23
12p 2
19p13
CLASS I
LINKAGE
SIRPS 19q13
PVRs 11q, 19q13
CRTAM (11q)
CTH, (EVA), V1235, (11q)
VHS159 (X)
CTX HUMX (X)
TCR-Ig
RAG
VC2 type
C2-I set
VHS159-like
?

and 19q13. All regions correspond to MHC class III paralogs except that of chromosome 19q13. However, the region where the tapasin distant homolog (PVR) and many genes of immunological relevance are found (9q13) could have been generated from the MHC paralog present on 19p13 by a pericentric inversion (see Kasahara, this volume). In other words, it is still possible that this region corresponds to a MHC paralogous region with a class I gene. There remain two regions without class I or II. These could have been deleted, or they could correspond to the remnants of the MHC linkage group before class I and class II were assembled and introduced. Both of them have Igsf members that seem to have diverged a long time ago, and that could have been part of the original linkage group.

Does phylogenetic analysis provide an idea of the order in which some of these different molecules appear? It is probably impossible to draw meaningful phylogenetic trees of the immunoglobulin superfamily members when working with members that may have diverged a very long time ago. Possible variation in rate of evolution, saturation, etc. may confuse the issue. However, the tree representing tapasin V related genes suggests that they correspond to paralogous sequences (Fig. 7). The various genes branched from each other in pairs, and at a similar time in evolution as predicted if a whole genome duplication was to be at the origin of the family. The C domain tree gives a similar time for divergence of tapasin-related sequences from the ones leading to receptor and MHC lineages (Fig. 8). These two lineages seem to have diverged from each other later, hence the hypothesis presented in the next section. Phylogenetic analysis of C1 of CRTAM argues that its lineage diverged long ago from the other C1 containing genes. In assuming the creation of a primordial V-C1 unit one should consider making a tree with VC1 segments (Fig. 9). This also suggests an early divergence of the types found with the C domain alone. A consensus emerges in all the preliminary phylogenetic trees made simply with the clustal neighbor joining programs. The CRTAM lineage is distinct from the PVR/tapasin and from the TCR Ig lineages whether for V domain, C domain, or VC domain analysis.

Curiously, CTX related genes characterized by their V domain gene encoded by 2 exons spliced according to type 0 are present in both MHC paralogous regions and in chromosome 11 where one finds CRTAM. It suggests that the linkage (V-C1)-(CTX-like) is ancient and was duplicated early (Fig. 10). This model is again consistent with the phylogenetic analysis that suggests an early divergence between

---

Fig. 11 can be found at the end of the book as fold-out. **A** Alignment of V domain sequences found in invertebrates and vertebrates molecules. The sequences have been grouped in function of the criteria described in Fig. 6 and according to the families defined *at left*. Otherwise the molecules are listed in alphabetical order with their accession number. The CTXhumx V tapasin 12 V2320 vhs159 v1235 v1335 are reconstructed and presented here for the first time. **B** Alignment of the constant C1 domain sequences. Sequences have been grouped according to the families defined *at left*. In both panels the strand composition is drawn *above the sequence*. *Various shades of gray*, the most conserved amino acids. The color code for the amino acids for which there is a consensus which is indicated *at the bottom*. *Right*, Accession numbers

the groups formed by CRTAM as opposed to the one formed by tapasin MHC, IgF, and TCR. In this group the earliest separation is that from MHC tapasin from Ig, TCR, and SIRPS. The model is consistent with the finding that CTX family members do not disappear easily in evolution and hence remain on ancient linkage groups and can be found in large numbers. In the polyploid amphibian *Xenopus ruwenzoriensis*, CTX is the only gene so far of which the 12 copies expected from the dodecaploid nature of the species have in fact been found (Du Pasquier et al. 1999). The finding of CTX members on chromosome 21 and X will require further analysis to determine whether other Igsf members are linked to these genes.

The present review does not consider the natural killer receptor family. Encoded by genes present on human chromosomes 19q13 (Table 1; Barclay et al. 1997), this family of molecules could also be considered as an ancient lineage. They have some relationships with the emergence of antigen specific receptors because they already form a coevolving unit with the MHC, and because they are Igsf members. Yet these molecules do not have well characterized V nor C1 domains compared to those discussed here, and therefore these do not seem to be on a direct line to the somatically rearranging receptor genes with the V-C1 building block.

4.3 Have Igsf Members Preceded Class I and II in the MHC?

If one accepts the idea that the adaptive immune system developed in a darwinian context, where variation precedes selection, the first important thing that shaped the system was the introduction of variation via somatic rearrangement in a V-C1 gene segment. It is not impossible from the above that this gene belonged to the primitive MHC linkage group, still devoid of class I and II.

This created a situation in which selection was to be developed to sort the various products now clonally expressed on lymphocytes. This could have been followed by the recruitment of the MHC peptide binding region gene in the immune system from another gene via a scenario similar to the one imagined by Flajnik et al. (1991). The C1 domain could have been borrowed from a tapasin-like gene. We now know that HSP70 may no longer be the closest relative to the peptide binding domain of the MHC molecule (Flajnik et al. 1991). On the other hand, one should perhaps pay attention to the second exon of tapasin the one preceding the V domain. Its relationship to other protein family members is mysterious, and some think it has some features of an Igsf member (Ortmann et al. 1997). The presence of all these elements in a single location might have been an advantage either for the purpose of expression or because of linkage disequilibrium. In this model, RAG should have invaded the vertebrate genome at the time of the second set of duplication leading on the one side to the antigen receptor and MHC, on the other side to tapasin (Fig. 10). The phylogenetic analysis of the C1 domain (Fig. 9) suggests an ancient separation of the tapasin-MHC lineage from the PVR-CRTAM and Ig-TCR lineages. The primitive lineage now represented by CRTAM (VC1 with a non rearranging V gene) seems to have diverged long before the tapasin group and perhaps in species that have not undergone whole genome duplications. It is therefore puzzling that no C1 domain sequence has been reported in any invertebrate or agnathan. At present we must therefore assume that searches have been unsuccessful, or that the transition C2 to C1 or I-set to C1 occurred in vertebrates. Given the distances separating the hagfish from the lampreys, it would be worthwhile investigating more thoroughly the latter for the presence of C1 domains.

5 Conclusion

This survey reminds us that V domains are very ancient and exist in the most primitive Metazoa. V domains diversified in many different directions, often materialized by different exon-intron organizations. C1 domains are probably more abundant in mammals than has been previously thought, although they remain so far unique to gnathostome vertebrates. A model (Fig. 10) is presented accounting for many of the linkages and homologies described. This suggests that altogether there were several VC1 gene segments in primitive vertebrates serving as possible "targets" when the rearranging machinery was introduced. One was affected, leaving the others to evolve in different directions with their primordial architecture still visible in their modern descendants.

Acknowledgements. We thank C.M. Steinberg and M.F. Flajnik for critical reading of the manuscript. I thank Lucy Trippmacher and Catherine Forbes for help in the preparation of the manuscript and the Figures. The Basel Institute for Immunology was founded and is supported by F. Hoffmann–La Roche Ltd., Basel, Switzerland.

References

Adams S, van der Laan LJ, Vernon-Wilson E, Renardel de Lavalette C, Dopp EA, Dijkstra CD, Simmons DL, van den Berg TK (1998) Signal-regulatory protein is selectively expressed by myeloid and neuronal cells. J Immunol 161:1853–1859

Adema CM, Hertel LA, Miller RD, Loker ES: 1997 A family of fibrinogen-related proteins that precipitates parasite-derived molecules is produced by an invertebrate after infection. Proc Natl Acad Sci USA 94:8691–8696

Barclay AN, Brown MH, Law SKA, McKnight AJ, Tomlinson MG, van der Merwe PA (1997) The leucocyte antigen facts book. Academic Press, London, San Diego

Barton DE, Arquint M, Roder J, Dunn R, Francke U (1987) The myelin-associated glycoprotein gene: mapping to human chromosome 19 and mouse chromosome 7 and expression in quivering mice. Genomics 1:107–112

Bates EE, Dieu MC, Ravel O, Zurawski SM, Patel S, Bridon JM, Ait-Yahia S, Vega FJr, Banchereau J, Lebecque S (1998) CD40L activation of dendritic cells down-regulates DORA, a novel member of the immunoglobulin superfamily. Mol Immunol 35:513–524

Bentley GA, Boulot G, Karjalainen K, Mariuzza RA (1995) Crystal structure of the β chain of a T cell antigen receptor. Science 267:1984–1987

Bernstein RM, Schulter SF, Shen S, Marchalonis JJ (1996) A new high molecular weight immunoglobulin class from the carcharhine shark: Implications for the properties of the primordial immunoglobulin. Proc Natl Acad Sci USA 93:3289–3293
Blumbach B, Diehl-Seifert B, Seack J, Steffen R, Müller IM, Müller EG (1999) Cloning and expression of novel molecules belonging to the immunoglobulin superfamily from the marine sponge Geodia cydonium: putative cell recognition molecules. Immunogenetics: 49:751–763
Bork P, Holm L, Sander C (1994) The immunoglobulin fold. Structural classification, sequence patterns and common core. J Mol Biol 242:309–320
Cella M, Dohring C, Samaridis J, Dessing M, Brockhaus M, Lanzavecchia A, Colonna M (1997) A novel inhibitory receptor (ILT3) expressed on monocytes, macrophages, and dendritic cells involved in antigen processing. J Exp Med 185:1743–1751
Chadeneau C, LeCabellec M, LeMoullac B, Meflah K, Denis MG (1996) Over-expression of a novel member of the immunoglobulin superfamily in Min mouse intestinal adenomas. Int J Cancer 68: 817–821
Chothia C, Novotny J, Bruccoleri R, Karplus M (1985) Domain association in immunoglobulin molecules. The packing of variable domains. J Mol Biol 186:651–663
Chrétien I, Courtet M, Marcuz A, Katuevo K, Vainio O, Du Pasquier L (1998) CTX, a *Xenopus* thymocyte receptor, defines a molecular family conserved throughout vertebrates. Eur J Immunol 28: in press
Chrétien I, Marcuz A, Fellah J, Charlemagne J, Du Pasquier L (1997) The T cell receptor β genes of *Xenopus*. Eur J Immunol 27:763–761
Chrétien I, Robert J, Marcuz A, Garcia-Sanz JA, Courtet M, Du Pasquier L (1996) CTX, a novel molecule specifically expressed on the surface of cortical thymocytes in *Xenopus* Eur J Immunol 26:780–791
Du Pasquier L (1989) Evolution of the immune system. In: WE Paul (ed) Fundamental Immunology, pp 139–165, Raven Press Ltd., New York
Du Pasquier L, Chrétien I (1996) CTX, a new lymphocyte receptor in *Xenopus* and the early evolution of Ig domains. Res Immunol 147:218–226
Du Pasquier L, Courtet M, Chrétien I (1999) Duplication and MHC linkage of the CTX family of genes in *Xenopus* and in mammals. Eur J Immunol 29:1729–1739
Du Pasquier L and Flajnik MF (1999) Origin and evalution of the vertebrate immune system. In: WE Paul (ed) Fundamental Immunology pp 605–650 Lippincolt-Raven, Philadelphia
Flajnik MF, Canel C, Kramer J, Kasahara M (1991) Which came first, MHC class I or class II? Immunogenetics 33:295–300
Frangoulis B, Park I, Guillemot F, Severac V, Auffray C, Zoorob R (1999) Identification of the tapasin gene in the chicken major histocompatibility complex [In Process Citation]. Immunogenetics 49: 328–337
Greenberg AS, Avila D, Hughes M, Hughes A, McKinney EC, Flajnik MF (1995) A new antigen receptor gene family that undergoes rearrangement and extensive somatic diversification in sharks. Nature 374:168–173
Greenberg AS, Hughes Al, Avila D, McKinney CE, Flajinik MF (1996) A novel "chimeric" antibody class in cartilaginous fish: IgM may not be the primordial immunoglobulin. Eur J Immunol 26: 1123–1129
Grenningloh G, Bieber AJ, Rehm EJ, Snow PM, Traquina ZR, Hortsch M, Patel NH, Goodman CS (1990) Molecular genetics of neuronal recognition in Drosophila: evolution and function of immunoglobulin superfamily cell adhesion molecules. Cold Spring Harb Symp Quant Biol 55:327–340
Harpaz Y, Chothia C (1994) Many of the immunoglobulin superfamily domains in cell adhesion molecules and surface receptors belong to a new structural set which is close to that containing variable domains. J Mol Biol 238:528–539
Henry J, Mather IH, McDermott MF, Pontarotti P (1998) B30.2-like domain proteins: update and new insights into a rapidly expanding family of proteins [In Process Citation]. Mol Biol Evol 15:1696–1705
Herberg JA, Sgouros J, Jones T, Copeman J, Humphray SJ, Sheer D, Cresswell P, Beck S, Trowsdale J (1998) Genomic analysis of the Tapasin gene, located close to the TAP loci in the MHC. Eur J Immunol 28:459–467
Hoek RM, Smit AB, Frings H, M VJ, de Jong-Brink M, Geraerts WPM (1996) A new Ig-superfamily member, molluscan defence molecule (MDM) from *Lymnaea stagnalis*, is down-regulated during parasitosis. Eur J Immunol 26:939–944
Hortsch M, Goodman CS (1991) Cell and substrate adhesion molecules in Drosophila. Ann Rev Cell Biol 7:505–557

Hsu E, Steiner LA (1992) Primary structure of immunoglobulin through evolution. Curr Op Struct Biol 2:422–431

Ishii T, Aoki N, Noda A, Adachi T, Nakamura R, Matsuda T (1995) Carboxy-terminal cytoplasmic domain of mouse butyrophilin specifically associates with a 150-kDa protein of mammary epithelial cells and milk fat globule membrane. Biochim Biophys Acta 1245:285–292

Karlstrom RO, Wilder LP, Bastiani MJ (1993) Lachesin: an immunoglobulin superfamily protein whose expression correlates with neurogenesis in grasshopper embryos. Development 118:509–522

Kasahara M (1999) The chromosomal duplication model of the major histocompatibility complex. Immunol Rev 167: in press

Katevuo K, Boyd R, Bean A, Dunon D, Imhof B, Göbel TWF, Vainio O (1999) ChT1, a new IgSF member inhibits thymocyte differentiation at or before $CD3^{-}$ $CD4^{+}CD8^{+}$ stage: J Immunal. 162:5685–5694

Kharitonenkov A, Chen Z, Sures I, Wang H, Schilling J, Ullrich A (1997) a family of proteins that inhibit signalling through tyrosine kinase receptors. Nature 386:181–186

Kubagawa H, Burrows PD, Cooper MD (1997) A novel pair of immunoglobulin-like receptors expressed by B cells and myeloid cells. Proc Natl Acad Sci USA 94:5261–5266

Lemke Gand Axel, R (1985) Isolation and sequence of a cDNA encoding the major structural protein of peripheral myelin. Cell 40:501–508

Lindstrom-Dinnetz I, Sun SC, Faye I (1995) Structure and expression of Hemolin, an insect member of the immunoglobulin gene superfamily. Eur J Biochem 230:920–925

Linington C, Lassmann H (1987) Antibody responses in chronic relapsing experimental allergic encephalomyelitis: correlation of serum demyelinating activity with antibody titre to the myelin/oligodendrocyte glycoprotein (MOG). J Neuroimmunol 17:61–69

Litman GW, Anderson MK, Rast JP (1999) Evolution of antigen binding receptors. Annu Rev Immunol 17:109–147

Manolov G, Manolova Y, Kieler J (1979) Cytogenetic investigation of assignment of locus for beta-2-microglobulin in K562 leukemia and Namalwa and Daudi Burkitt lymphoma cells. Cytogenet Cell Genet 25:182

Marchalonis JJ, Schluter SF, Bernstein RM, Shen S, Edmundson AB (1998) Phylogenetic emergence and molecular evolution of the immunoglobulin family. Adv Immunol 70:417–506

Mark MR, Scadden DT, Wang Z, Gu Q, Goddard A, Godowski PJ (1994) rse, a novel receptor-type tyrosine kinase with homology to Axl/Ufo, is expressed at high levels in the brain. J Biol Chem 269:10720–10728

Miyauchi T, Kanekura T, Yamaoka A, Ozawa M, Miyazawa S, Muramatsu T (1990) Basigin, a new, broadly distributed member of the immunoglobulin superfamily, has strong homology with both the immunoglobulin V domain and the β-chain of major histocompatibility complex class II antigen. J Biochem Tokyo 107:316–323

Nelson RE, Fessler LI, Takagi Y, Blumberg B, Keene DR, Olson PF, Parker CG, Fessler JH (1994) Peroxidasin: a novel enzyme-matrix protein of Drosophila development. EMBO J 13:3438–3447

Ortmann B, Copeman J, Lehner PJ, Sadasivan B, Herberg JA, Grandea AG, Riddell SR, Tampe R, Spies T, Trowsdale J, Cresswell P (1997) A critical role for tapasin in the assembly and function of multimeric MHC class I-TAP complexes. Science 277:1306–1309

Pancer Z, Skorokhod A, Blumbach B, Müller WE (1998) Multiple Ig-like featuring genes divergent within and among individuals of the marine sponge Geodia cydonium. Gene 207:227–233

Peach RJ, Bajorath J, Brady W, Leytze G, Greene J, Naemura J, Linsley PS (1994) Complementarity determining region 1 (CDR1)- and CDR3-analogous regions in CTLA-4 and CD28 determine the binding to B7–1. J Exp Med 180:2049–2058

Peiper SC, Ashmun RA, Look AT (1988) Molecular cloning, expression, and chromosomal localization of a human gene encoding the CD33 myeloid differentiation antigen. Blood 72:314–321

Pulido D, Campuzano S, Koda T, Modolell J, Barbacid M (1992) D*trk*, a *Drosophila* gene related to the *trk* family of neurotrophin receptors, encodes a novel class of neural cell adhesion molecule. EMBO J 11:391–404

Rast JP, Anderson MK, Strong SJ, Luer C, Litman RT, Litman GW (1997) α, β, γ, and δ T cell antigen receptor genes arose early in vertebrate phylogeny. Immunity 6:1–11

Rast JP, Haire RN, Litman RT, Pross S, Litman GW (1995) Identification and characterisation of T-cell antigen receptor-related genes in phylogenetically diverse vertebrate species. Immunogenetics 42: 204–212

Rast JP, Litman GW (1998) Towards understanding the evolutionary origins and early diversification of rearranging antigen receptors. Immunol Rev 166:79–86

Robert J, Brown DM, Du Pasquier L, Cohen N (1997a) Antibody cross-linking of the thymocyte-specific cell surface molecule CTX causes abnormal mitosis and multinucleation of tumor cells. Exp Cell Res 235:227–237

Robert J, Chrétien I, Guiet C, Du Pasquier L (1997b) Cross-linking CTX, a novel thymocyte-specific molecule, inhibits the growth of lymphoid tumor cells in *Xenopus*. Mol Immunol 34:133–143

Roux KH, Greenberg AS, Greene L, Strelets L, Avila D, McKinney EC, Flajnik MF (1998) Structural analysis of the nurse shark (new) antigen receptor (NAR): molecular convergence of NAR and unusual mammalian immunoglobulins. Proc Natl Acad Sci USA 95:11804–11809

Sadasivan B, Lehner PJ, Ortmann B, Spies T, Cresswell P (1996) Roles for calreticulin and a novel glycoprotein, tapasin, in the interaction of MHC class I molecules with TAP. Immunity 5:103–114

Schacke H, Rinkevich B, Gamulin V, Muller IM, Muller WE (1994) Immunoglobulin-like domain is present in the extracellular part of the receptor tyrosine kinase from the marine sponge *Geodia cydonium*. J Mol Recognit 7:273–276

Seeger MA, Haffley L, Kaufman TC (1988) Characterization of amalgam: a member of the immunoglobulin superfamily from Drosophila. Cell 55:589–600

Shibuya A, Campbell D, Hannum C, Yssel H, Franz-Bacon K, McClanahan T, Kitamura T, Nicholl J, Sutherland GR, Lanier LL, Phillips JH (1996) DNAM-1, a novel adhesion molecule involved in the cytolytic function of T lymphocytes. Immunity 4:573–581

Shin SW, Park SS, Park DS, Kim MG, Kim SC, Brey PT, Park HY (1998) Isolation and characterization of immune-related genes from the fall webworm, Hyphantria cunea, using PCR-based differential display and subtractive cloning. Insect Biochem Mol Biol 28:827–837

Smith DK, Xue H (1997) Sequence profiles of immunoglobulin and immunoglobulin-like domains. J Mol Biol 274:530–545

Sun SC, Lindstrom I, Boman HG, Faye I, Schmidt O (1990) Hemolin: an insect-immune protein belonging to the immunoglobulin superfamily. Science 250:1729–1732

Thompson CB (1995) New insights into V(D)J recombination and its role in the evolution of the immune system. Immunity 3:531–539

Tonegawa S (1983) Somatic generation of antibody diversity. Nature 302:575–581

Vlassara H, Valinsky J, Brownlee M, Cerami C, Nishimoto S, Cerami A (1987) Advanced glycosylation endproducts on erythrocyte cell surface induce receptor-mediated phagocytosis by macrophages. A model for turnover of aging cells. J Exp Med 166:539–549

Wang PL, O'Farrell S, Clayberger C, Krensky AM (1992) Identification and molecular cloning of tactile. A novel human T cell activation antigen that is a member of the Ig gene superfamily. J Immunol 148:2600–2608

Williams AF, Barclay AN (1988) The immunoglobulin superfamily domains for cell surface recognition. Ann Rev Immunol 6:381–405

Yamao T, Matozaki T, Amano K, Matsuda Y, Takahashi N, Ochi F, Fujioka Y, Kasuga M (1997) Mouse and human SHPS-1: molecular cloning of cDNAs and chromosomal localization of genes. Biochem Biophys Res Commun 231:61–67

Yarden Y, Kuang WJ, Yang-Feng T, Coussens L, Munemitsu S, Dull TJ, Chen E, Schlessinger J, Francke U, Ullrich A (1987) Human proto-oncogene c-kit: a new cell surface receptor tyrosine kinase for an unidentified ligand. EMBO J 6:3341–3351

Evolution of Receptors

Immunoglobulin Isotypes: Structure, Function, and Genetics

E. Bengtén[1], M. Wilson[1], N. Miller[1], L.W. Clem[1], L. Pilström[2], and G.W. Warr[3]

1 Introduction

Immunoglobulin (Ig) classes (in mammals, IgM, IgA, IgD, IgG, IgE) are defined by the isotypes of heavy (H) chains (μ, α, δ, γ, and ε). Each isotype is in turn distinguished by unique structures in its constant region domains. These different structures confer distinctive functions on the Ig classes. When two or more Ig classes are very similar, as occurs with the four different types of IgG found in man

[1] Department of Microbiology, University of Mississippi Medical Center, Jackson, MS 39216-4505, USA
[2] Division of Microbiology, Department of Medical Biochemistry and Microbiology, BMC, Uppsala University, 75123 Uppsala, Sweden
[3] Department of Biochemistry and Molecular Biology, Medical University of South Carolina, Charleston, SC 29425-211, USA

and mouse, they are usually termed subclasses. Each isotype is encoded by a distinct gene and multiple heavy chain isoforms can be produced by alternative pathways of RNA processing, such as the secreted (sIg) and membrane (mIg) forms of all H chains, or the full-length and truncated H chain isoforms of certain avian antibodies. Allelic variation in the constant (C) regions gives rise to allotypes. The different types of light (L) chains (in mammals, κ and λ) are also typically referred to as isotypes. This system of classification of Igs was developed from studies of man and his immunological understudy, the mouse, and has proven useful not only in these two species, but also in other mammalian species. Although the classification of mammalian Ig classes and isotypes is quite clear, the situation with Igs from nonmammalian vertebrates is not. For example, is the shark molecule referred to as IgM really IgM? Should we call the predominant low molecular weight Ig in chickens IgG or IgY? This chapter discusses the ways in which these and similar questions have been approached.

Most workers interested in the phylogeny of immunity prefer systems of classification and nomenclature that reflect their best judgement concerning evolutionary descent and relationships. In the case of the nonmammalian vertebrate Ig genes one is faced with the task of distinguishing the evolutionary lineages of individual H and L chain genes by studying extant phylogenetically diverse species. Although all vertebrate Ig H chain genes (and all Ig L chain genes) are clearly related, this task is difficult because multiple gene duplications and losses have undoubtedly occurred during vertebrate evolution. This means that inferences concerning the orthologous or paralogous nature of Ig genes in widely distant species can be difficult to make. However, many features of Ig genes and their encoded proteins can potentially offer insight into their evolutionary relationships at either the functional analogy or structural homology levels. Some of the commonly used features are listed below.

Protein
- Primary amino acid sequence
- Conservation of critical residues
- Amino acid composition
- Presence/absence of hinge
- Degree of polymerzation
- Presence in serum and/or secretions

Gene
- Nucleotide sequences
- Exon Structure
  - Exon number
  - Presence of hinge exons
  - TM exon number
- Organization
  - Multicluster or translocon
  - Order of IgH C region genes
  - Transcriptional orientation

- Expression
  - Site of expression
  - Alternative RNA processing pathways
  - Class switch recombination
  - V(D)J recombination

2 Immunoglobulin Features

2.1 Analyses of Primary Sequences

Without doubt, primary amino acid and nucleotide sequences are the molecular properties most frequently used for phylogenetic analyses. There are several good reasons for this, the most obvious being that such sequences represent data at the ultimate fundamental level. In addition, sequence data are amenable to quantitative comparisons and to phylogenetic tree constructions that depict their relationships. The relationship that such trees are most often used to investigate is relatedness by evolutionary descent. Given the unequivocal nature of sequence data, one might think that deduction of evolutionary relationships should be quite easy. However, in fact, this area is extremely (some might say excessively) complicated, and "molecular systematics" has become a discipline of its own. Problems arise at many stages in the process of building phylogenetic trees from Ig sequence data. First, before tree construction begins, the sequences under consideration must be aligned. Unfortunately, Ig molecules have probably undergone high rates of sequence changes during evolution (McLaughlin and Dayhoff 1972; Sitnikova and Nei 1998). These changes include not only amino acid substitutions but sequence additions and/or deletions as well. These additions and deletions can involve entire domains or even multiple domains. Tree-building programs are sensitive to the data received with the nature of the alignment influencing the nature of the deduced tree. One alignment problem includes whether or not to assign "weights" to amino acid exchanges. Most programs can be instructed to give higher weights to functionally significant or rare amino acids and the resulting alignments are then typically anchored by residues such as Cys or Trp. Another alignment problem involves the significance of the gaps that must usually be introduced to optimize alignments. Questions to be considered include the magnitude of the penalty for each missing residue, and whether the penalty should become greater as the introduced gap lengthens. Second, once alignments have been generated, there are numerous programs for constructing trees, each using very different methods and assumptions. Distance (Thompson et al. 1994), parsimony-based (Swofford 1993), and neighbor-joining (Saitou and Nei 1987) methods are among the most frequently encountered. Within each tree-building program there are also usually multiple parameters that can be set, each one of which can influence the topology of the resulting trees. For example, one obvious question is whether a gap in an alignment should be treated as missing data or as a new character, i.e., a 5th base or a 21st amino acid. Optimistically the above brief discussion illustrates why molecular systematics has evolved into its own separate discipline. It can be argued that when tree-building analyses are conducted with Igs, it would be instructive to compare the results produced by different methods, in particular to ascertain the effects of changing the full range of parameters in the aligning and tree-building programs. However, none of the published analyses of Ig evolution have exhaustively compared the various molecular systematic approaches that can be applied to a given data set; nearly every research group has its own set of preferred analytical methods. In fact, sequence-based analysis of evolutionary relationships is very complicated (and/or equivocal), particularly when the sequences under comparison have relatively low similarities (approx. 25% or less). Not only are exact relationships hard to discern at such levels (KLEIN 1998), but many tree-building programs suffer from the problem of long branch attraction (discussed in LAKE and MOORE 1998) which can, at its worst, generate artifactual relationships between distant sequences. Thus a good case can be made for looking to other aspects of Igs (at both the protein and gene levels) for evolutionarily relevant information. Characteristics such as locus organization and exon structure (in contrast to sequence alignments) are not usually susceptible to quantification or computer analyses, and their values are therefore open to debate. Nevertheless, one can argue that such qualitative information is a valuable addition when analyzing primary sequence data. Consequently our own approach to interpretation of the history of the Igs (see below) has been to include all available data, including trees, where appropriate or useful, in our analysis.

2.2 Other Properties of the Protein

While primary amino acid sequences are most often used for tree-building analyses, they contain other information that can be phylogenetically informative. For example, the Cys residues that form the intradomain disulfide bonds in Igs are not the only highly conserved residues; there are also other invariant residues which are often diagnostic for a particular domain (KABAT et al. 1991). Other sequence-related information comes from the transmembrane segment of the receptor forms of Igs. For example, a short cytoplasmic tail with a typical C-terminal sequence of Lys-Val-Lys is characteristic of μ and δ chains. In contrast the cytoplasmic tails of the other Ig membrane forms have sequences that diverge greatly, especially in length (MAGOR K et al. 1994; MUßMANN et al. 1996a). Another example of protein comparisons, of historic interest, is the use of global amino acid composition as a measure of phylogenetic relatedness, although it is somewhat more prone to inaccurate measurement and lacking information on the order of the residues. This approach (MARCHALONIS and WELTMAN 1971) was of use before the era of molecular biology which has subsequently rendered it obsolete.

Ig domain structure and organization can also provide useful information. While individual H chains always have a single V domain, the number of C domains they possess is variable ranging from as few as two to as many as eight (CHENG et al. 1982; GREENBERG et al. 1996; BERNSTEIN et al. 1996; WILSON et al. 1997). Thus, while the number of C region domains of an Ig might be a characteristic for drawing conclusions concerning relatedness, taken in isolation it could easily be misleading.

Ig H chains have a certain degree of flexibility which aids in multivalent binding of antigen(s). Mammalian IgG, IgA, and IgD contain a distinct region, between the first and second C region domains, that permits a high degree of segmental flexibility. This region is referred to as the hinge; it consists of a short sequence containing a preponderance of distinctive amino acids (e.g., Cys, Pro and Gly) and is thought to adopt an extended conformation (BURTON 1987; SCHUMAKER et al. 1991). Mammalian IgG and IgA each have H chains with only three C domains. In the case of IgG, the domain that most likely would originally have been Cγ2 (in a four C domain precurser molecule) is believed to have collapsed and become the hinge (YAMAWAKI-KATAOKA et al. 1981). In the case of IgA, the original Cα2 domain has also apparently been lost and the hinge is thought to represent an N-terminal extension of the present Cα2 domain (i.e., the domain that would have originally been Cα3; TUCKER et al. 1981a). The different genetic features underlying distinct hinges in IgA and IgG are discussed below. Other Ig protein properties, such as degree of polymerization and physiological role, are less useful in determining evolutionary relatedness. For example, most Igs are divalent molecules with a H_2L_2 structure. Some Igs such as IgM and IgA can exist as polymers, of structure $(H_2L_2)_n$, where n varies from species to species, or even within a species (for examples drawn from fish IgM, see WARR 1995). Because the degree of polymerization of an Ig can be influenced by many factors, such as the presence or absence of J chain or mutation of individual amino acids, it should not be considered a major property on which to base evolutionary conclusions. The antibodies, as soluble defense molecules, exert their effector functions both internally (e.g., in blood plasma and tissue fluids) and externally (in body secretions). The classes of mammalian Ig show considerable divergence of function in this respect. For example, IgA is considered to be the secretory Ig found in mucus and other body secretions, whereas all other classes of Ig remain confined mostly to the internal body fluids. However, using the sites in which Igs are found as clues to their evolutionary relationships is problematic. Whereas the secretory Igs of birds and mammals appear to be homologous IgA molecules, the Igs involved in secretory immunity in teleosts and amphibians are clearly not IgA (as discussed below).

2.3 Genetic Properties

In addition to nucleotide sequence analyses (primarily used for phylogenetic tree-building) other informative data can be obtained from cDNA and genomic sequences. In many cases the most powerful analysis can be made from comparisons of cDNA and genomic sequences, where intron/exon boundaries and patterns of alternative pre-mRNA processing can be clearly defined.

The manner by which specific genes are structured, and whether overall locus organization is conserved from species to species are also of considerable interest. There is much evidence for similar (or even identical) patterns of gene colinearity between species. One rationale for the *Fugu* genome project was the assumption that this species possesses a genome equivalent in organization to that of more highly evolved vertebrates but with very much shorter introns (BRENNER et al. 1993; APARICIO and BRENNER 1997; ELGAR et al. 1997). However, from studies of other genes, as well as Igs, it is now obvious that the genomes of various vertebrates show, despite many similarities, some differences in their genomic organizations. These differences may be of functional significance in their own right, but in addition they may provide insight into how the genome evolved. At a higher level of resolution than the whole genome, it seems reasonable to use the organization of genes, or even the arrangement of exons within genes, as indicators of phylogenetic relatedness.

Within the IgH locus, dramatic differences in organization are found throughout the vertebrates (Fig. 1). Perhaps the greatest difference lies between the elasmobranchs and all other vertebrates. The typical vertebrate IgH locus conforms to what has been termed the translocon structure, in which there is one large locus containing all VH genes in the 5′ region, followed by D, J, and then C region genes at the 3′ end. In contrast, the IgH loci of elasmobranchs exhibit a multicluster arrangement, in which units of (V-D-D-J-C) are repeated hundreds of times in the

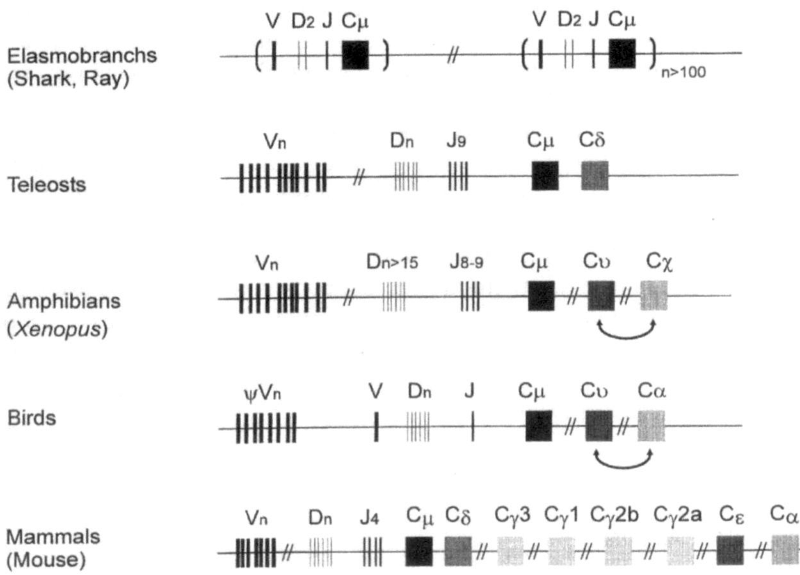

Fig. 1. IgH chain gene organization in the different classes of vertebrates. (Modified from DU PASQUIER and FLAJNIK 1999) Arrows indicate that the relative gene position in the germline is unknown

genome (KOKUBU et al. 1988; SHAMBLOTT et al. 1989; reviewed in LITMAN et al. 1999). There also appear to be increases in the number and complexity of C_H-region genes during vertebrate evolution, even when the number of functional VH and JH genes has shrunk to as few as one of each, as in chickens (REYNAUD et al. 1989). The order of genes encoding the various isotypes in the IgH locus may be equally informative. C_H region genes occupy characteristic positions in the locus: the μ gene is always observed to be the 5′-most C_H gene and the δ gene, when present, is found immediately 3′ of μ. Likewise, the α gene is always found as the most 3′ C_H gene in the mammalian IgH locus (HONJO and MATSUDA 1995); it is not yet known whether the avian α gene occupies a similar position.

Additionally, the transcriptional orientation of Ig gene segments can vary. Although all the genes within an IgH locus are typically in the same transcriptional orientation, it has recently been observed that in one avian species (the duck) two of the C_H region genes (υ and α) are in opposite transcriptional orientations (MAGOR et al. 1999). In addition to having significance for aspects of function such as class-switch recombination, this organization may reflect some reorganizations of the IgH locus specific to the avian lineage.

When the C_H region genes from different vertebrates are examined at the exon level much variation is found. Both the number of exons in a gene and the structure of the exons themselves, can change markedly from species to species. For example, as mentioned above, mammalian α and γ chains contain three C_H region domains, with the presumptive fourth domain being replaced by a hinge (Fig. 2). However, as discussed above, the origins of the two hinges appear very different, as revealed by studies of the manner in which the hinge is encoded (TUCKER et al. 1981a). Other informative exons in the Ig genes are those that encode the transmembrane (TM) and cytoplasmic (CYT) regions of the membrane-bound (receptor) forms. All of these segments, except those of Cα, are encoded by two exons. The membrane-spanning hydrophobic TM1 segment is approximately 26 residues in length and is relatively conserved among all H chains. For example, it contains functionally important residues that make up the conserved antigen receptor transmembrane (CART) motif. This motif is found in all known antigen receptor transmembrane domains and these amino acids have been shown, at least in mammals, to be important for the assembly and signaling of the lymphocyte antigen receptor (CAMPBELL et al. 1994). TM2 segments which encode CYT regions are more variable in sequence and size, ranging from as few as three to as many as 27 residues. The value of these exons in defining Ig evolutionary relationships has been noted by several groups (MAGOR K et al. 1994; WARR et al. 1995; MUßMANN et al. 1996a; MARCHALONIS et al. 1998). For example, duck υ TM exons have characteristic features of γ and ε genes; the TM1 and TM2 exons of υ, γ, and ε are similar in both length and coding sequence, with the hydrophobic region of TM1 being the most conserved. In fact the TM exons of duck υ are very similar in sequence to the γ TM exons (89% overall amino acid identity with human γ1; MAGOR K et al. 1994). While the TM1 and TM2 segments of *Xenopus* υ contain conserved residues found in avian υ and mammalian γ and ε, overall they, as with the TM1 and TM2 segments of *Xenopus* mIgX, are most similar to the TM segments of *Xenopus* mIgM (MUßMANN et al. 1996a).

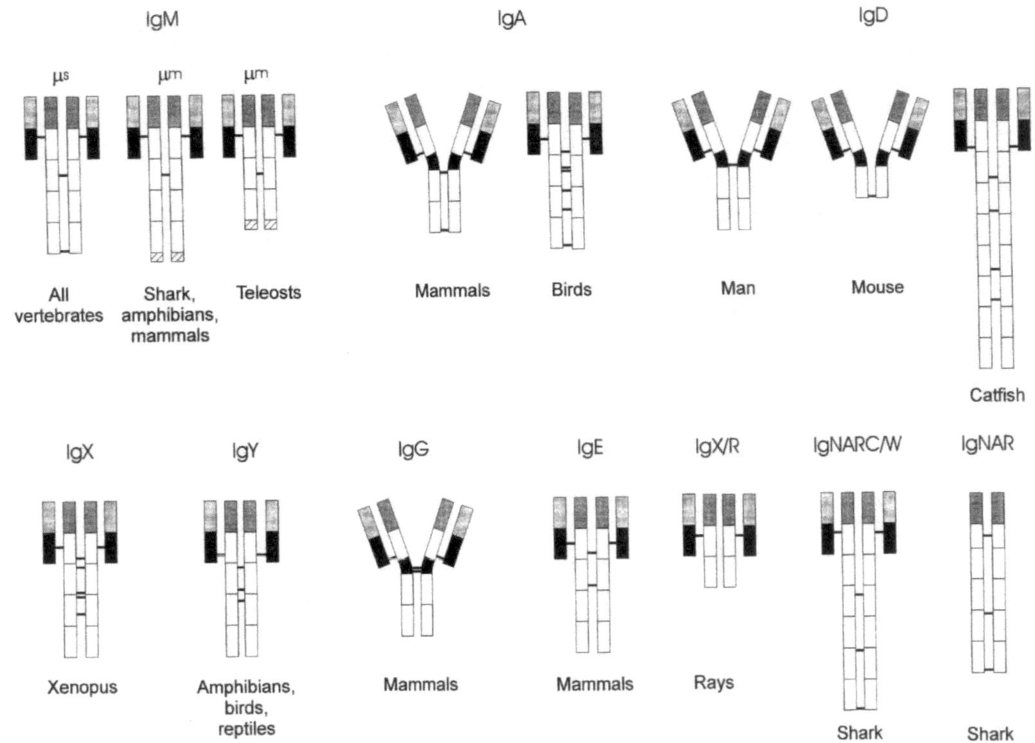

Fig. 2. Schematic showing the various vertebrate Ig classes (for references, see text). *Gray boxes*, V regions; *angles*, hinges; *hatched boxes*, TM regions of µm; *bold lines*, predicted interchain disulfide bonds; Igs are shown in monomeric (H_2L_2) structure only

Gene expression is another useful criterion for phylogenetic consideration. The expression of all Ig loci is dependent on both the initial productive V(D)J rearrangement and the correct processing to mRNA of a complex primary transcript. It is not unusual for an IgH transcript to encode up to four different H chains. For example, following initial productive V(D)J rearrangement in an IgH locus, the primary transcript can be processed by alternative pathways to generate mRNAs encoding the secreted and membrane forms of μ and δ (PETERSON 1994). Expression by alternative RNA processing rather than by class-switch recombination in the IgH locus was one of the features that prompted the designation of a novel C_H region gene of the channel catfish as a homolog of δ (discussed below; WILSON et al. 1997).

In the following sections the phylogenetic distribution of the vertebrate Ig isotypes is examined. Various aspects of the structure of the Igs as proteins as well as aspects of their gene structure and expression (see above) are considered in reaching conclusions about their evolutionary relationships. While it is possible that a satisfactory classification of the Igs of nonmammalian vertebrates will never be achieved, an effort to define their presumed evolutionary relationships as the basis for a more rational nomenclature has been made in this review.

3 Heavy-Chain Isotypes

3.1 IgM

IgM serves as the predominant B cell antigen receptor, and during B cell development μ H chains are the first isotype expressed. IgM is also the first antibody to appear in the serum during a primary immune response and in neonatal development. It is presently considered to be the only class of Ig universally found in all vertebrates that possess immunoglobulins (DU PASQUIER and FLAJNIK 1999). While IgM is not the major class of serum Ig in birds and mammals, an isotype considered to be homologous to IgM is the most prevalent serum Ig in elasmobranchs and bony fish. These "lower vertebrate" molecules can clearly be reasonably classified as IgM based upon H (μ) chain properties: (a) they all exhibit a five-domain structure (one V and four C), (b) show sequence similarities, (c) have membrane forms that are similar, and (d) are encoded by the most 5′ H chain gene (except in elasmobranchs). As a result some investigators suggested that IgM may have been the primordial Ig. In all vertebrates, serum IgM molecules are usually found as polymers, frequently as pentamers $(\mu_2L_2)_5$, although tetramers $(\mu_2L_2)_4$ characterize the teleosts (Fig. 3). However, caution must be used in terms of drawing evolutionary conclusions from such gross architectural observations. Specifically monomers (μ_2L_2) of IgM are frequently found in the elasmobranchs; in fact in several species of sharks monomeric IgM is the predominant serum Ig (CLEM et al. 1967; CLEM and SMALL 1967), albeit in certain species of rays the

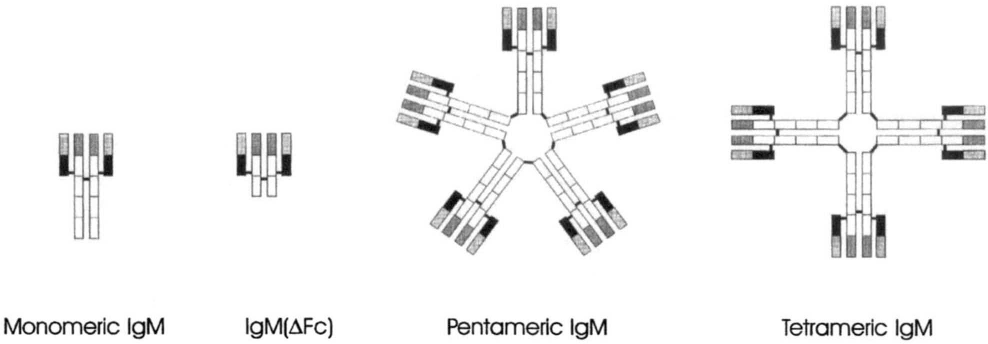

Fig. 3. Schematic illustrating a variety of different IgM structures. Positions and number of interchain disulfide bonds are variable between different species. (Modified from WARR 1995)

pentameric form predominates (KOBAYASHI et al. 1984). Assigning evolutionary significance to such observations must be tempered by the findings of monomeric IgM in normal human cord blood (PERCHALSKI et al. 1968) and in serum from patients suffering from lupus (HARISDANGKUL et al. 1975; McDOUGAL et al. 1975). The occasional finding of what appears to be monomeric or $(Fab)_2$-like IgM forms in certain teleosts (CLEM 1971) is another example of potential, albeit not understood, significance. Although the derivations of these various IgM forms are not clear, the limited available data indicate that they are likely directly related via presecretory events, i.e., the shark extracellular pentamer does not appear to dissociate to the monomer nor does the monomer associate to the pentameric form (SMALL et al. 1970).

All vertebrate IgM molecules are heavily glycosylated and in nonmammalian species where only genetic sequence information is available, N-linked glycosylation sites are readily identified. The carbohydrate contents vary from approx. 10% in mammals (reviewed in FRAZER and CAPRA 1999) to as high as 16% in teleosts (WILSON and WARR 1992). The μ chains associate with each other and with L chains via disulfide bonds. As in all Ig isotypes, the amino acids responsible for the Ig fold, such as the Cys residues that form the intrachain disulfide bonds and the invariant (required) Trp residues, are the most highly conserved. However, the overall degree of similarity between the inferred amino acid sequences of IgM molecules from different vertebrates is low. In general the regions between the first Cys to the first Trp of each domain are the most conserved. Across species, Cμ4 domains are the most highly conserved (approx. 37% at the amino acid level) followed by Cμ1 domains (HSU 1994; MARCHALONIS et al. 1998). Although IgM molecules do not contain a distinct hinge region as defined by sequence (BURTON 1987), they are flexible, a situation probably involving residues at the Cμ1/Cμ2 and Cμ2/Cμ3 boundaries. Certainly IgM molecules from the complete spectrum of vertebrates, including sharks and bony fish, exhibit proteolytic susceptibility reminiscent of the hinge region in other Ig classes (KLAPPER et al. 1971; VAN GINKEL et al. 1991).

Low amounts of IgM are also found in secretions in mammals where they seemingly participate in mucosal immunity. However, in mammals IgA is the predominant secretory antibody, although IgM can perform the secretory functions in IgA deficiencies (BRANDTZAEG et al. 1968). IgM molecules (like IgA) can be transported across mucosal surfaces in association with J chain and bound to the secretory component in epithelial cells (KERR 1990). It has been proposed that the role of IgM in secretory immunity developed early in evolution (HSU 1994). This notion is based upon the observation that IgM antibodies are found in the mucus of normal and immune fish skin (reviewed WILSON and WARR 1992; ROMBOUT and JOOSTEN 1998) and in *Xenopus* bile (MUßMANN et al. 1996b). To date, no additional isotypes other than μ and δ (WILSON et al. 1997) have been identified in bony fish. Thus the finding of non-serum-derived IgM in fish secretions (LOBB 1987) argues strongly for the existence of an IgM-based secretory immune system in fish. However, in *Xenopus*, where there are clearly three antigenically distinct H chain isotypes, IgX (which is most homologous to *Xenopus* IgM) is now considered to be a functional analog of IgA (MUßMANN et al. 1996b). IgX is found in the bile and large numbers of IgX-positive or IgM-positive cells are found lining the intestinal gut epithelium. Thus it appears that in different vertebrates the secretory immune system can involve different Ig isotypes, a situation that seemingly provides little if any information of obvious evolutionary consequence.

As mentioned above, the μ gene is the most 5′ C_H region in the IgH locus, and it is comprised of four C_H (Cμ1 to Cμ4) and two TM exons. The two forms of μ H chain, secreted (μs) and membrane (μm), are produced by alternative pre-mRNA processing pathways of a single transcript. The Cμ4 domains (μs) of serum IgM molecules possess hydrophilic C-termini while the membrane forms (μm) of μ possess typical hydrophobic C-terminal regions that can associate with the B cell membrane. In mammals the μm C-terminus allows the μ chain to insert into the B cell plasma membrane and interact with Ig-accessory molecules, CD79a and b, involved in signal transduction following antigen binding (PLEIMAN et al. 1994; BUHL and CAMBIER 1997).

The pathways of RNA processing that generate the μs and μm forms have, perhaps surprisingly, seemingly undergone distinctive changes during vertebrate evolution (reviewed in ROSS et al. 1998). The generation of μm in mammals, birds, amphibians, and elasmobranchs involves splicing the TM1 exon into a cryptic donor splice site (U/C/GG↓GUA̧AA) within the Cμ4 exon (PETERSON and PERRY 1989). In contrast, the channel catfish produces the membrane form of the μ chain by splicing the TM sequences directly to the donor splice site at the 3′ boundary of the Cμ3 exon since a cryptic donor splice site within Cμ4 is lacking (see Fig. 2; WILSON et al. 1990). This finding was subsequently confirmed in other teleost species (BENGTÉN et al. 1991; HORDVIK et al. 1992; LEE et al. 1993). Thus production of teleost μs and μm involves competition for the Cμ3 donor splice site and should involve regulatory mechanisms different from those in other vertebrates (ROSS et al. 1998). This different processing pathway for μm in bony fish has an interesting phylogenetic history. Holostean fish (such as gar and bowfin), which are considered to have been the evolutionary precursors of the teleosts (WILSON et al. 1995a,b), utilize, remarkably, both the "mammalian" and the atypical teleostean pathways for the production of μm mRNA. Additionally, the bowfin produces a third μm message by splicing TM1 to a cryptic splice site within its Cμ3 exon (WILSON et al. 1995b). Hence it seems that, as the teleosts emerged, a single pathway for the production of μm message was selected (ROSS et al. 1998). A cryptic Cμ4 splice site is also found in the sturgeon, a modern descendant of the chondrostean precursors to the holosteans. However, at this time, the μm pre-mRNA splicing pattern of sturgeon has not been determined (LUNDQVIST et al. 1998). There appears to be no functional significance to the loss of the Cμ4 domain in the teleost μm chain since the membrane IgM of teleost B cells appears to be effective as an antigen receptor (MILLER et al. 1985; VAN GINKEL et al. 1994; RYCYZYN et al. 1996). Since teleosts are deficient in the ability to process many pre-mRNAs (ROSS et al. 1998; BENTANCOURT et al. 1993) this may have been a factor influencing the evolution of genes other than the Igs.

3.2 IgA

IgA is the predominant Ig isotype participating in mucosal immunity in mammals and as such can be considered an important component of the first line of defense. IgA exists in two forms, secretory and serum. The secretory form is typically a dimeric molecule which includes J chain and secretory component (SC) whereas serum IgA is predominantly monomeric (Fig. 4). In mammals the majority of the IgA synthesized is of the secretory form.

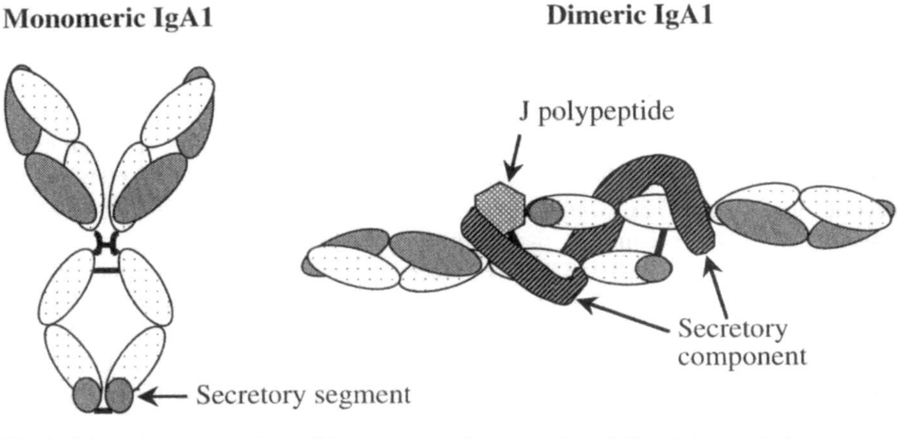

Fig. 4. Schematic representations of the structures of monomeric and dimeric human IgA1. In monomeric IgA1 L chain domains and secretory segments of the H chains are *gray*. The structures of IgA depicted here are based on those proposed by KERR (1990)

To date the only unequivocal homologs of mammalian IgA are found among the avian species. In the early 1970s a third class of Ig (in addition to IgM and IgY) was identified in domestic chickens. This antibody was proposed to be IgA since it was (a) found in both serum and secretions, (b) antigenically distinct from chicken IgM and IgY, (c) expressed in epithelial associated lymphoid tissues, and (d) capable of binding purified human secretory component (reviewed in MANSIKKA 1992). A full-length chicken Ig α cDNA sequence (MANSIKKA 1992) indicated that, unlike mammalian α, which contains three constant region domains, chicken α contains four C_H-domains and no distinctive hinge region. When analyzed via domain by domain amino acid comparisons, chicken α4 aligned best with human α3 (41%), chicken α3 with human α2 and chicken α1 with human α1. These results tend to confirm the speculation of TUCKER et al. (1981b) on the origins of the hinge in mammalian IgA, as does the fact that the chicken α2 showed only low relatedness to any of the three human α domains (<29%). Overall the deduced amino acid sequence of the chicken α H chain shows 36% and 35% identity to human IgA1 and IgA2, respectively. Identities to human μ (29%), δ (24%) or γ2 (33%) are lower. Recently α chain cDNA from the white Peking duck has been sequenced (MAGOR et al. 1998). Duck α, as expected, is comprised of four domains and comparisons of inferred amino acid sequences of individual domains reveal duck α most closely resembles chicken α (50%). The duck α cDNA was shown to encode the Ig previously identified as "IgX" isolated from duck bile (NG and HIGGINS 1986). This molecule, assuming that J chain and SC are present, would likely form a tetramer, $(\alpha_2L_2)_4$, akin to that observed for IgA in the bile and secretions of chickens (WATANABE and KOBAYASHI 1974).

While α homologs have not been demonstrated in ectothermic vertebrates, it is clear, as discussed previously, that these animals possess functional analogs of IgA. IgM molecules in teleosts and *Xenopus*, IgX in *Xenopus*, and IgY in axolotls (see below) all are found in secretions and hence are likely involved in mucosal immunity.

3.3 IgD

IgD is a class of immunoglobulin previously believed to be recently evolved since it had been found only in primates and rodents. In man, δ chains are composed of three C domain encoding exons, while in mouse there are only two δ C_H exons. Delta chains of both species have distinct hinges encoded by one or two exons (in mouse and human, respectively), located between δ1 and δ2 and characterized by highly charged (acidic and basic) amino acids and the presence of carbohydrate. The hinges of mouse and human δ are, however, different in length; the human δ hinge is almost twice the length of the mouse δ hinge with all the charged residues being at the carboxy-terminus instead of distributed equally throughout (TUCKER 1981b). Since the N-terminal regions of the two hinges show homology, it was suggested that the human δ hinge may have resulted from a duplication of the mouse δ hinge and then subsequently mutated (PUTNAM et al. 1981). Although the biochemistry and genetics of IgD have been definitively described in the mammals, the functional significance of the molecule is poorly understood (THORBECKE and LESLIE 1982; NITSCHKE et al. 1993; ROES and RAJEWSKY 1993; LUTZ et al. 1998). Recently a complex homolog of the δ chain has been identified at the cDNA level in the catfish (Fig. 2; WILSON et al. 1997). The catfish δ is unique in that it is expressed as a chimeric molecule containing a rearranged VDJ, the Cμ1 exon, and seven novel IgH chain exons, some of which show sequence homology to mammalian δ by parsimony-based (PAUP) analyses. It seems reasonable to suggest that the presence of Cμ1 is required for catfish IgD to covalently bind L chains since the δ1 exon does not contain an appropriate Cys for L chain binding. Genetic mapping and sequence analyses of the catfish δ gene show the δ1 exon to be only 1580bp 3′ of the TM2 exon of μ (WILSON et al. 1997; MILLER et al. 1998). There are no class-switch sequences in this short intron and transfection studies have shown this region to contain the catfish Ig H chain enhancer (MAGOR B et al. 1994). This finding suggests that catfish δ message, as with that of mammalian δ, is produced by alternative mRNA splicing, rather than class switching. Further evidence for alternative mRNA splicing is provided by a catfish clonal B cell line, 3B11 which expresses both μ and δ messages each of which utilizes an identical VDJ rearrangement. Other catfish clonal B-cell lines (MILLER et al. 1994) express only μ (WILSON et al. 1997). Recently, by genomic mapping and sequencing, similar δ-like genes have been identified in the lungfish (OTA et al. 1997), pufferfish (T. Ota, personal communication), and Atlantic cod (T. Jørgenson, personal communication). While no cDNA information on this gene in the pufferfish is yet available, it is interesting to note that the lungfish Cμ1 exon is not included in δ transcripts (T. Ota, personal communication).

We propose that the sequence relatedness, location immediately downstream of the μ locus, and coexpression (by alternative mRNA processing pathways) with μ in some B cells is consistent with catfish δ being a homolog of mammalian δ. However, catfish δ differs from mammalian δ in several ways. It contains 7 δ exons, not including the secreted and TM regions, and the inclusion of Cμ1 makes it an isotypically chimeric molecule. In addition, its predicted amino acid sequence suggests a molecule with reduced flexibility due to the lack of a hinge region. The function of putative IgD in the catfish is still unknown. Western blot analyses using monoclonal (mAbs) antibodies specific for the C-terminal sequence of the catfish putative δ secreted peptide and Cμ1 have revealed that δ is found in normal catfish serum (MILLER et al. 1998 and unpublished observations). It will be interesting to determine whether δ is also present on the surface of catfish lymphocytes or in the serum of the other fish species in which the gene has been described.

3.4 IgY

IgY is a low molecular weight, monomeric antibody of wide distribution. It is found in amphibians, reptiles, and birds (reviewed in WARR et al. 1995). Avian IgY antibody was originally considered to be an IgG (and is still referred to as such by some authors), but in 1969 LESLIE and CLEM presented physiochemical evidence indicating that it is likely a separate isotype, a situation subsequently confirmed by molecular biological approaches. In contrast to IgG, the υ chain of IgY has four C_H region domains and lacks a distinct hinge (as is typical of all nonmammalian antibodies). However, it does possess regions at the boundaries between the first and second and the second and third domains that could confer flexibility (WARR et al. 1995). While it is clear that IgY serves as a functional analog of IgG in many species (HIGGINS and WARR 1993), it also appears to serve other functions as well. For example, IgY serves as a secretory antibody in the axolotl (FELLAH et al. 1992), and it is also transported effectively into the yolk-sac of oviparous animals. IgY also exists in some species of turtle and in anseriform birds (ducks, geese, and their relatives), in a truncated form which is designated IgY(ΔFc) to denote the absence of the two C-terminal υ domains (MAGOR et al. 1992). IgY(ΔFc) production can occur concurrently with the full-length IgY; it results from the presence of a small terminal exon lying between the second and third υC exons, although the evolutionary origins and significance of IgY(ΔFc) are not clear (WARR et al. 1995; PARHAM 1995). However, it seems that IgY, IgG, and IgE are related in that IgY appears to have been the Ig that was ancestral to both mammalian IgG and IgE. Chicken IgY can mediate hypersensitivity reactions (FAITH and CLEM 1973): it is interesting to speculate that the development of IgG and IgE from IgY may have involved a separation of physiological functions previously combined in IgY (see Fig. 5 and below).

3.5 IgG and IgE

IgG and IgE are isotypes found thus far exclusively in the mammals. Because of their clinical importance they have been subjected to intensive studies and hence many excellent reviews describing their structure, function and genetics are available (e.g., BURTON 1987; BRÜGGEMANN 1987; HONJO and MATSUDA 1995). Thus only a few points relevant to their presumed evolution are discussed in this brief review. Recently cDNAs encoding IgG, IgA, and IgE have been identified from the nonplacental mammal, the short-tailed opossum (AVESKOGH and HELLMAN 1998). These genes conform to the general H chain gene structures of IgG, IgA, and IgE in the placental mammals, and the authors suggest that the major changes in Ig structure occurred before the separation of mammals from the early reptile lineages. Numerous investigators have suggested the likely derivation of both IgG and IgE from an IgY-like ancestor (PAVARI et al. 1988; MAGOR K et al. 1994; WARR et al. 1995; MUßMANN et al. 1996a). This notion is based on several lines of evidence, including similarities in sequence (especially in the TM region), domain structure, patterns of RNA splicing to encode the membrane forms and function. In the case of function IgY is the predominant serum antibody in most amphibians, reptiles, and birds, whereas this role falls to IgG in the mammals. IgE is the predominant antibody that mediates immediate hypersensitivity reactions in mammals: from an evolutionary perspective there would appear to be an advantage in restricting immediate hypersensitivity reactions to an Ig class present in very low concentration (e.g., IgE) as opposed to those present at relatively high levels in the blood, such as IgG or IgY. However, this logic may be circumvented by the findings

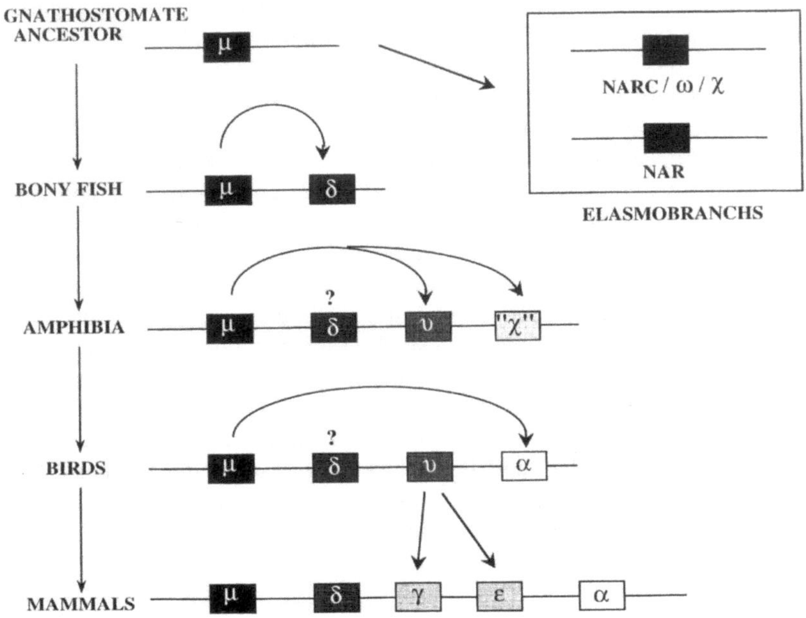

Fig. 5. Scheme for the evolutionary development of Ig C region genes by a series of gene duplications. A duplication of μ, forming δ, occurred early in vertebrate evolution, before divergence of the tetrapod and bony fish lineages. In the amphibian lineage υ and then later X were produced by duplications from μ, with υ becoming the progenitor for both γ an ε as the mammals emerged. α was formed by another duplication of μ that preceded the divergence between the avian and mammalian lineages. *Question marks*, no evidence of δ has yet been found in the amphibians and birds. The NARC/W/NAR H chain isotypes are assumed to have derived from duplications of μ

that IgG and IgY can mediate such reactions in mice (HAZENBOS et al. 1998) and chickens (FAITH and CLEM 1973), respectively.

In the case of structure, IgG achieves segmental flexibility through possession of a hinge, whereas flexibility in IgE is associated with less-specialized regions at the Cε1/Cε2 boundary and, perhaps more importantly, at the Cε2/Cε3 boundary (SLATTERY et al. 1985; PUMPHREY 1986). These regions are called switch regions (HELM et al. 1991). Similar switch regions of the molecule appear to be associated with segmental flexibility in IgM (BURTON 1987) and probably IgY (WARR et al. 1995).

3.6 IgNARC, IgW, IgX(R), and IgNAR

The second class of Ig discovered in cartilagenous fish was first called IgR, now termed IgX (but not related to *Xenopus* or duck IgX). It was isolated as a distinct (non-μ) low molecular weight serum Ig in longnose skate (KOBAYASHI et al. 1984, 1988). More recently IgX/IgR encoding H chain genes were isolated at both the cDNA and genomic levels from the clearnose skate (HARDING et al. 1990a,b). Skate IgX genes, as with other elasmobranch antigen receptor chain genes, are arranged in clusters; *in situ* hybridization studies have shown that IgM and IgX clusters are found unlinked on multiple chromosomes (ANDERSON et al. 1994). The skate IgX H chain cDNA is composed of one V and two C_H region domains; the second C_H-encoding exon ending with a Cys-rich terminal segment.

Several large multidomain Ig H chain cDNAs have also been isolated in sharks. These molecules (IgNAR, and IgNARC, from the nurse shark and IgW from the sandbar shark) appear to be related by sequence analysis to the IgX of rajiformes (GREENBERG et al. 1995, 1996; BERNSTEIN et al. 1996; LITMAN et al. 1999). They have been proposed as candidates for the primordial antibody. IgNARC and IgW H chains are composed of one rearranged V and six C_H region domains. Domain by domain comparisons of the six H chain C region exons of NARC and ω show them to be equivalent (SCHLUTER et al. 1997; MARCHALONIS et al. 1998). The NARC/ω H exons exhibit typical Ig H chain structures, i.e., residues that form the Ig fold and available free Cys residues for forming the disulfide bonds in the H chain dimer and between the H and the L chains. Phylogenetic analyses using Clustal W show that the first two domains of NARC H chain (as well as ω H chain) are homologous to IgX, while the four C-terminal domains are homologous to NAR (described below). cDNAs for both secreted and membrane-bound forms of NARC have been isolated. Southern blots of shark genomic DNA indicate the presence of only a few IgNARC genes arranged in a cluster organization. Immunoprecipitation experiments using L chain specific mAbs with shark splenocyte lysates suggest that only one of the shark L chain isotypes associates with IgNARC (GREENBERG et al. 1996). Interestingly, studies by ANDERSON et al. (1994) have shown that long IgX transcripts approximating the size of IgNARC/W are produced in the skate. Recently, these authors used differential screening of cDNA libraries to isolate the long form of the skate IgX gene and they propose that the two different IgX forms are produced by alternative splicing (ANDERSON et al. 1999).

The additional novel Ig isotype identified in the nurse shark has been termed IgNAR. Sequencing analyses show IgNAR H chain cDNAs to be composed of a V domain produced by rearrangement and five C_H domains. Both secreted and transmembrane forms have been isolated at the cDNA level. To date, IgNAR protein has only been found in shark serum (demonstrated by immunoprecipitation experiments using an mAb that reacts with a stretch of five amino acids in the fifth NAR C domain). IgNAR is unusual in that (a) its V regions appear more TCR-like than Ig-like, and (b) its CH1 exon is short with no available Cys residues that would allow for L chain binding (GREENBERG et al. 1995). This lack of L chain binding has been confirmed by electron microscopic studies (ROUX et al. 1998). NAR V regions undergo extensive somatic hypermutation (DU PASQUIER et al. 1998; DIAZ et al. 1998), in contrast to what is observed with the V regions of shark IgM genes, where the mutation rate is low (HINDS-FREY et al. 1993). It is not yet established whether the newly described Igs of the elasmobranchs (IgNAR/NARC/W/X) are IgM-derived or represent more ancient (primordial?) classes of Ig (KLEIN 1998).

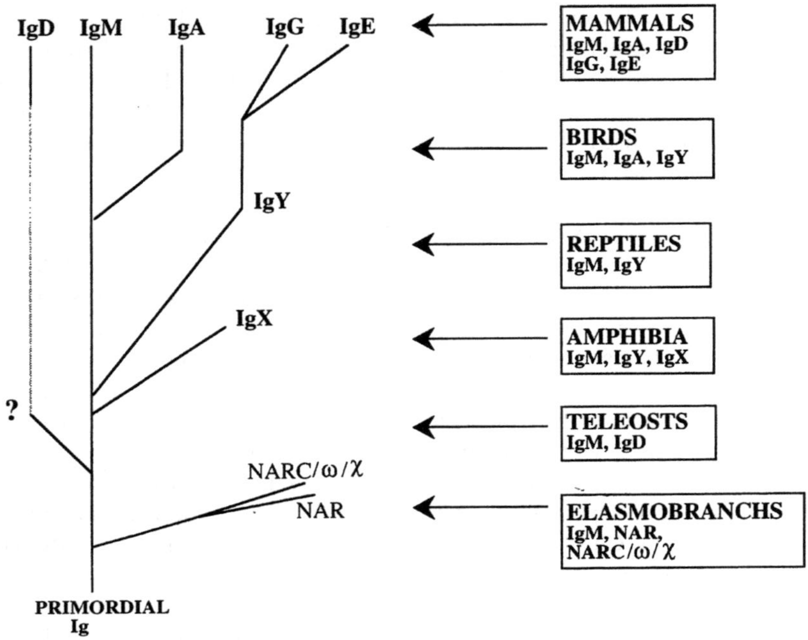

Fig. 6. Distribution and evolutionary relationships of Ig classes in the vertebrates. *Faint line* (*with question mark*), the hypothetical relationship between the IgD of teleosts and mammals

However, it is clear (especially after reviewing the previous sentence) that the nomenclature of these particular elasmobranch Igs requires attention.

3.7 Summary

Conclusions about the evolutionary relationships of vertebrate Ig H chains are likely to remain controversial. However, the available data presented and discussed in this review are arguably compatible with the scheme of Ig evolution illustrated in Fig. 6. The most radical aspect of this scheme is the early emergence of IgD (by the time of the actinopterygian fish). Further investigations in species representing a variety of vertebrate classes will test this suggestion.

4 Immunoglobulin Light Chains

The L chains of Ig molecules are single polypeptides of 25–30kDa, consisting of 210–220 amino acids folded into separate V and C domains, each containing the classical Ig fold. In most cases L chains associate with H chains via a disulfide bridge between their C domain and the C1 domain of the H chains; an exception is some allotypes of human IgA2, where the H and L chains are held together noncovalently (CHINTALACHARUVU and MORRISON 1996). The antigen binding region of the Ig molecule is stabilized by hydrophobic interactions between the VL and VH domains. Insights into the evolution of the various vertebrate L chains have benefited from studies of L chain genes in a wide variety of extant species. These studies suggest that most vertebrates (above the phylogenetic level of the cyclostomes) with the exception of the birds express two or more L chain isotypes. In teleosts and elasmobranchs the L genes are clustered and each cluster can be viewed as a separate locus resulting in at least 50 L chain loci in teleosts and 100 L chain loci in elasmobranchs. An issue that arises involves defining the benefits of maintaining several L chain loci. One answer to this question might involve the notion that since VL genes normally do not rearrange until after a productive H chain rearrangement has occurred, several L chain loci simply may enhance the probability for a productive rearrangement in each B cell. However, this is not a compelling answer in that other mechanisms, such as secondary gene rearrangements to rescue abortive recombinations, are present (FEDDERSON and VAN NESS 1985). Another possible answer is that several L chain loci each expressing different V domains may increase the overall antigen combining site repertoire for the animal.

The L chains in humans and mice, κ and λ, were initially classified by serology (EDELMAN and GALLY 1962; WU and KABAT 1970). While this isotypic classification is suitable in other mammals, it is not particularly applicable to nonmammalian vertebrates. What criteria can be applied to nonmammalian L chains to help distinguish likely from less likely evolutionary relationships? In contrast to H chain C regions, L chain C regions do not possess special effector functions that make them distinctive. The function of L chains is to various degrees to contribute to the antigen combining site, even at the phylogenetic level of the elasmobranchs (SHANKEY and CLEM 1980a,b). However, in some cases, as in certain camelid IgGs, the L chains are not needed for antigen binding (HAMERS-CASTERMAN et al. 1993). There is also insufficient information available to assign the different L chain isotypes distinct syntenic locations in the genome that could facilitate their classification. Consequently as considered below there are three criteria for considerations of L chain phylogeny/evolution, i.e., gene organization, differences in heptamer-nonamer signals, and homology at nucleotide and amino acid levels.

4.1 Mammalian Light Chains

The mammalian κ L chain locus has a translocon type organization, whereas the λ locus in all mammalian species studied to date has a hybrid "cluster/translocon" organization, i.e., multiple V genes followed (3′) by a variable number of duplicated J-C genes. Recently λ chains have been cloned and characterized from a marsupial, the short-tailed opossum (LUCERO et al. 1998). This locus encompasses at least three diverse Vλ families, each of which seems to be related to distinct Vλ families present in placental mammals. Phylogenetic analyses comparing the opossum Cλ with mammalian Cλs (LUCERO et al. 1998) and other studies using mammalian Cλ regions (HAYZER 1990) suggest that duplications of J-Cλ have occurred independently in each species. Hence J-C duplication appears to be a mammalian λ L chain feature that may or may not be characteristic of other vertebrates' λ-like L chain genes.

It is of importance to note that the recombination signal sequences (RSS) are of different orientation in mammalian κ and λ loci. The V segments in the κ locus are followed by heptamer/nonamer motifs separated by a 12-bp nucleotide spacer with J segments being preceded by heptamer/nonamer motifs separated by a 23-bp nucleotide spacer. The converse occurs in the λ locus, i.e., the heptamer/nonamer motifs of the V genes are separated by 23 nucleotides and the J heptamer/monomer motifs are separated by 12 nucleotides (TONEGAWA 1983).

Mammalian κ and λ L chains are expressed at different levels in various species and the mechanism(s) involved in this regulation is poorly understood. In mouse serum the κ/λ chain molar ratio is 95/5, in man 60/40, and in pig approximately 50/50. In contrast the λ chain dominates in cats, dogs, horses, and cattle with more than 90% of the plasma cells expressing λ chains (ARUN et al. 1996). It has been assumed that the level of L chain expression in an animal directly reflects the number of functional VL elements associated with a specific locus (reviewed in MAX 1999). However, since the number of κ and λ V elements in horses are similar, it has been suggested that the temporal sequence in which the horse L chains rearrange may be important (HOME et al. 1992; FORD et al. 1994).

4.2 Avian Light Chains

In contrast to other vertebrates, birds express only one L chain isotype (KUBO et al. 1970). The L chain locus of the domestic chicken contains only one functional V gene that rearranges to a single functional J segment. This rearrangement occurs in B cell progenitors during embryogenesis (between days 10 and 15) prior to migration to the bursa. Imprecise joining, somatic point mutation and gene conversion from a pool of at least 25 pseudo-VL genes are used to achieve diversity (REYNAUD et al. 1987). Recently chicken-derived mAbs have been developed and have demonstrated that antigen specific antibodies produced by Ig gene conversion display both high affinity and specificity; the H and L chains each contribute to the binding site (MICHAEL et al. 1998). Based on the orientation of RSS and sequence homology, the chicken L chain gene is clearly related to mammalian Cλ genes (REYNAUD et al. 1983, 1987). For example, the chicken Cλ chain exhibits 59% identity at the amino acid level with mouse Cλ1, compared to 41% identity with mouse Cκ. While the chicken has been the most extensively studied avian species, it has been shown that a variety of other birds (quail, pigeon, turkey, cormorant, and hawk) also utilize a single VL for gene rearrangements (MCCORMACK et al. 1989). A modification of this pattern is observed in the Muscovy duck, where two functional VL genes are found in the germline (MCCORMACK et al. 1989). It has been suggested that a second avian L chain isotype is unnecessary in chickens, since nonproductive rearrangements are rarely observed due to selective amplification of cells containing successful L chain rearrangements during development of the bursa (McCormack and Thompson 1990).

4.3 Amphibian Light Chains

Three L chain isotypes have been identified at the genetic level in *Xenopus*, i.e., ρ, σ, and type III (Schwager et al. 1991; Haire et al. 1996). The designations ρ and σ were originally chosen to distinguish them from κ and λ. However, the *Xenopus* ρ locus is now considered κ-like, because it is organized in a translocon fashion with multiple Vs and Js associated with typical κ-type RSS and one C region gene. (Zezza et al. 1992; Stewart et al. 1993). Phylogenetic tree analyses show that *Xenopus* Vρ sequences always group with mammalian Vκ chain sequences (Zezza et al. 1992; Sitnikova and Nei 1998). Likewise, the Cρ sequence groups with (or close to) mammalian Cκ sequences (Haire et al. 1996). *Xenopus* σ L chains are encoded at a separate locus, containing multiple V segments of two distinct families, two J segments and two C region genes (Schwager et al. 1991). Sigma V and C sequences do not exhibit strong relatedness to either κ or λ (V and C) sequences and σ J segments are unusual in that two Ser (or Arg) residues replace the diglycine bulge which is conserved in all IgH, IgL, and T cell receptor J segments. The third *Xenopus* L chain, type III, is considered λ-like based on phylogenetic analyses. In contrast to ρ and σ, which both express limited VL heterogeneity, the type III locus contains multiple V segments comprised of at least six families, two J segments and two distinct C region genes (Haire et al. 1996). The genomic organization and the architecture of the RSS in type III have not been reported. Studies have shown that rearrangements of the ρ locus begin by day 8 after fertilization, and rearrangement of the σ locus begins by day 13 (Mußmann et al. 1998). However, it is not yet known when (or if) type III is expressed in tadpoles. Immunoprecipitation experiments using mAbs to distinct L chain forms (25, 27, 29kDa) have demonstrated that each of the three *Xenopus* H chain isotypes, μ, υ, and χ, are associated with the 25 and 29kDa L chains. In contrast the 27kDa L chain form is rarely associated, if at all, with υ chains (Hsu et al. 1991). The L chain isotype specificities of the mAbs have not been reported. In the bullfrog the C region of the major L chain expressed in the serum has been sequenced at the protein level. The amino acid identities are 40%, 28%, and 29% to *Xenopus* ρ, σ, and type III, respectively. These L chains are not covalently associated with H chains due to the replacement of a Pro residue with a Cys residue at amino acid position 119. This Cys allows for the formation of an unusual intrachain disulfide bond to Cys 214, in lieu of the usual interchain disulfide bridge between the H and L chains (Mikoryak and Steiner 1988).

4.4 Teleost and Chrondrostean Light Chains

At least two different immunoglobulin L chain forms have been detected by antigenic analyses of several species of ray-finned fish; for example, channel catfish (Lobb et al. 1984), rainbow trout (Sanchez et al. 1995), and white sturgeon (Adkison et al. 1996). In channel catfish mAbs specific to L chains designated as F and G have demonstrated the F:G protein ratio in serum to be 60:40 (Lobb et al. 1984). At the genomic level, channel catfish F and G, as well as the isotypes designated as L1 from rainbow trout and Atlantic cod have been characterized (Ghaffari and Lobb 1993, 1997; Daggfeldt et al. 1993). In contrast to the H chains, teleost L chains have a clustered gene organization with multiple repeats of $(V\text{-}J\text{-}C)_n$. One interesting feature of the channel catfish G locus is that each cluster contains two V and two J segments. The most 5′ J segment may or may not be functional, because the heptamer of its RSS differs from the consensus sequence by one nucleotide and the spacer between the heptamer and nonamer consists of 21 nucleotides instead of the usual 23 (Ghaffari and Lobb 1997). In the catfish germline the components of L chain G clusters are closely linked and the V regions are in the opposite transcriptional orientation, suggesting that rearrangement of this locus occurs by inversion rather than deletion (Ghaffari and Lobb 1993). Quite similar observations apply to catfish F and cod L1 clusters (Ghaffari and Lobb 1997; Bengtén et al. 1997). Each of these L chain isotypes have κ-type RSS and their V sequences group with *Xenopus* Vρ sequences (Sitnikova and Nei 1998). Therefore they may be considered κ-like. In contrast, a second rainbow trout isotype, L2, has C regions that share 29% amino acid identity with trout L1C regions (Partula et al. 1996). The L2 Vs and Cs group with both *Xenopus* Vσ and Cσ regions in phylogenetic trees. These genes have a clustered organization with κ-type RSS; however, unlike other teleost L chains, their V regions appear to be in the same transcriptional orientation as the J and C genes (S. Timmusk, personal communication). A second L chain isotype, L2, has been identified in Atlantic cod. This gene has λ-type RSS, and its C region sequences are clearly different from those L chain isotypes described in other species (N. Wermestam, personal communication). Hence it appears that in teleosts there are several L chain isotypes that differ in genomic organization and RSS orientation.

Interestingly, chondrosteans, believed to represent an evolutionary branch between elasmobranchs and teleosts, do not have L chain gene clusters. The L chain loci of the Siberian sturgeon and the sterlet contain multiple V genes upstream of at least seven J segments (in the sturgeon) and at least two C region genes. In the sturgeon the RSS are of κ-type and the V region sequences group with V regions from *Xenopus* ρ and teleost κ-like V region sequences. However, the C regions are most similar to the L chain C regions from cartilaginous fish. The presence of a κ-like organization in the L chain locus of the chondrosteans suggests that the clustered organization of Ig L chain loci in bony fish and sharks (see below) arose from two distinct events (Lundqvist et al. 1996).

Recently the activities of L chain enhancers in the Atlantic cod L1 locus have been studied in channel catfish B cell lines. As in mouse (Hagman et al. 1990) and human λ genes (Blomberg et al. 1991) the Atlantic cod L1 enhancer is located downstream of the C region and the J-C intron does not show enhancer activity. Interestingly, λ enhancers from mouse and man, in contrast to κ enhancers, showed strong enhancer activity in the channel catfish cell lines. This suggests that the factors involved in λ transcription are likely present in catfish B cells. Taken together these findings suggest that although the Atlantic cod L1 genes are κ-like, the location and perhaps the function of the enhancers are more λ-like (BENGTÉN et al. 1997; Bengtén, unpublished observations).

4.5 Chondrichthean Light Chains

Three L chain isotypes (types I, II, and III) have been identified in chondrichthyes, and all have cluster organizations. Unlike the situation with teleost VL genes, chondrichthean VL genes do not appear to have reversed transcriptional orientation. Type I L chains from the horned shark and little skate show strong sequence relationships to each other, although their genomic organizations of these genes are different. In horned sharks each of the type I L chain clusters contain separate V, J, and C region genes $(V\text{-}J\text{-}C)_n$ (SHAMBLOTT and LITMAN 1989), while in little skates the V and J are joined in the germline $(VJ\text{-}C)_n$ (ANDERSON et al. 1995). In addition, whereas the orientations of the RSS in the horned shark type I loci are of κ-type, their V region sequences group with rainbow trout L2 and *Xenopus* σ VL region sequences (PILSTRÖM et al. 1998). Hence they cannot definitively be classified as either κ or λ.

Type II L chains from the sandbar shark (HOHMAN et al. 1993), horned shark, little skate, and ratfish (RAST et al. 1994) are all germline joined (VJ-C). Thus RSS orientation cannot be used as a criterion for classification with type II L chains. Several investigators consider the chondrichthean type II sequences to be more similar to avian and mammalian λ sequences than to κ sequences (HOHMAN et al. 1992; DU PASQUIER and FLAJNIK 1999). Type III L chains from the nurse and horned sharks are considered more κ-like (RAST et al. 1994). For example, the type III V genes in the nurse shark show up to 60% amino acid identity with mammalian Vκs (GREENBERG et al. 1993). Type III L chain clusters contain separated V, J, and C region genes (V-J-C) with the RSS orientation being of the κ-type. Furthermore a small repertoire of main chain conformations (canonical structures identified in κ hypervariable regions) responsible for forming the antigen binding site in man and mouse are also present in the nurse and horned shark type III V regions (BARRE et al. 1994). This clearly suggests that κ-like features were present during the divergence of the sharks and that they have been subsequently preserved during evolution.

Although the chondrichthean VL genes exhibit κ- and λ-like features, the L chain C regions in cartilaginous fish always group together in phylogenetic analyses. One interpretation of this finding is that the different L chain loci of cartilaginous fish developed before the separation of κ and λ (i.e., at a time when κ- and λ-like features were already present), such that some of the "dual" features might have been preserved in the VL regions, possibly by antigenic selection pressure.

4.6 Conclusion

Based on the above considerations vertebrate L chains can be arbitrarily grouped into three different lineages. The first is based upon a putative relationship to mammalian λ chains, the second to mammalian κ chains, and the third shows no close relationship to either κ or λ chains (Fig. 7). These relationships are based on amino acid sequence similarities of the V and C regions and the RSS orientation. Cluster versus translocon gene organization is not an informative characteristic in the L chain gene classification scheme. It cannot be determined unequivocally when during evolution the three groups diverged; the different L chains probably have continued to evolve significantly since separation of the different lineages. Similarly the internal relationships between the three L chain groups is not clear. It is important to note that the three groups do not constitute three isotypes, rather three lineages from which different isotypes probably have originated. This raises a question about a common or unifying nomenclature for nonmammalian L chains. When an orthologous relationship can be readily inferred, for example with chicken λ, isotype names are easy to assign. However, problems using the above criteria that generated the proposed pattern occur (Fig. 7). For example, if the proposed

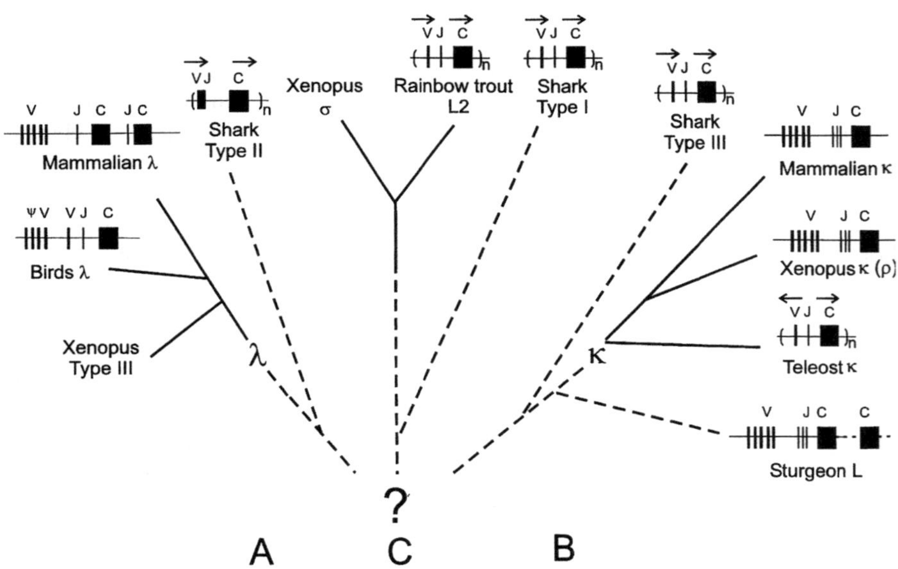

Fig. 7. Representation of the relationships of three vertebrate L chain lineages. *Lineage A*, the vertebrate light chains exhibiting relationships to mammalian λ chains; *lineage B*, those that exhibit relationships to mammalian κ chains; *lineage C*, those that do not exhibit clear relationships to either mammalian λ or κ chains. *Solid line* (*within a lineage*), the relationship shown is based on V and C sequence comparisons and the RSS orientation (when known). *Dashed line*, the relationship shown is based on V sequence comparisons only and the RSS orientation (when known); *arrows*, transcriptional orientation. Distances between branch points are arbitrary and do not represent evolutionary distances. Schematics of the germline gene organization are shown when known

pattern holds true, it can be rationalized that the channel catfish has developed two isotypes, F and G, from the same κ-like lineage. However, sequence analyses show that F and G have approximately equal identity to each other and to κ and λ. Hence, whether sequence comparisons along with RSS orientation are sufficient to be used to designate nonmammalian L chain isotypes as κ or λ is debatable.

Acknowledgements. Research in our laboratories has been supported by grants from NIH (AI-19530), NSF (MCB9807531), USDA (NRICGP9602942), Swedish Council for Forestry and Agricultural Research (grant no. 40.0484/96), EU (AIR2-CT94-1334 and FAIR-CT95-0666).

References

Adkison MA, Basurco B, Hedrick RP (1996) Humoral immunoglobulins of the white sturgeon, *Acipenser transmontanus*: partial characterization of and recognition with monoclonal antibodies. Dev Comp Immunol 20:285–298

Anderson M, Amemiya C, Luer C, Litman R, Rast J, Niimura Y, Litman G (1994) Complete genomic sequence and patterns of transcription of a member of an unusual family of closely related, chromosomally dispersed Ig gene clusters in *Raja*. Int Immunol 6:1661–1670

Anderson MK, Shamblott MJ, Litman RT, Litman GW (1995) Generation of immunoglobulin light chain gene diversity in *Raja erinacea* is not associated with somatic rearrangement, an exception to a central paradigm of B cell immunity. J Exp Med 182:109–119

Anderson MK, Strong SJ, Litman RT, Luer CA, Amemiya CT, Rast JP, Litman (1999) A long form of the skate IgX gene exhibits a striking resemblance to the new shark IgW and Ig NARC genes. Immunogenetics 49:56–67

Aparicio S, Brenner S (1997) How good a model is the *Fugu* genome? Nature 387:140

Arun SS, Breuer W, Hermanns W (1996) Immunohistochemical examination of light chain expression (lambda/kappa ratio) in canine, feline, equine, bovine and porcine plasma cells. Zentralbl Veterinarmed A 43:573–576

Aveskogh M, Hellman L (1998) Evidence for an early appearance of modern post-switch isotypes in mammalian evolution; cloning of IgE, IgG and IgA from the marsupial *Monodelphis domestica*. Eur J Immunol 9:2738–2750

Barre S, Greenberg AS, Flajnik MF, Chothia C (1994) Structural conservation of hypervariable regions in immunoglobulins in evolution. Natl Struct Biol 1:915–920

Bengtén E, Leanderson T, Pilström L (1991) Immunoglobulin heavy chain cDNA from the teleost Atlantic cod (*Gadus morhua* L.): nucleotide sequences of secretory and membrane form show an unusual splicing pattern. Eur J Immunol 21:3027–3033

Bengtén E, Daggfeldt A, Magor BG, Warr GW, Pilström L (1997) Organization and transcription enhancers of Ig L chain genes in Atlantic cod. Dev Comp Immunol 21:162

Bernstein RM, Schluter SF, Shen S, Marchalonis JJ (1996) A new high molecular weight immunoglobulin class from the carcharhine shark: implications for the properties of the primordial immunoglobulin. Proc Natl Acad Sci USA 93:3289–3293

Bentancourt OH, Attal J, Theron MC, Puissant C, Houdebine LM (1993) Efficiency of introns from various origins in fish cells. Mol Mar Biol Biotech 2:181–188

Blomberg BB, Rudin CM, Storb U (1991) Identification and localization of an enhancer for the human lambda L chain Ig gene complex. J Immunol 147:2354–2358

Brandtzaeg P, Fjellanger I, Gjeruldsen ST (1968) Immunoglobulin M: local synthesis and selective secretion in patients with immunoglobulin A deficiency. Science 160:789–791

Brenner S, Elgar G, Sandford R, Macrae A, Venkatesh B, Aparicio S (1993) Characterization of the pufferfish (*Fugu*) genome as a compact model vertebrate genome. Nature 366:265–268

Brüggemann M (1987) Genes encoding the immunoglobulin constant regions. In: Calabi F, Neuberger MS (eds) Molecular genetics of immunoglobulin. Elsevier Science, Amsterdam, pp 51–80

Buhl AM, Cambier JC (1997) Co-receptor and accessory regulation of B-cell antigen receptor signal transduction. Immunol Rev 160:127–138

Burton DR (1987) Structure and function of antibodies. In: Calabi F, Neuberger MS (eds) Molecular genetics of immunoglobulin. Elsevier Science, Amsterdam, pp 1–50
Campbell KS, Backstrom BT, Tiefenthaler G, Palmer E (1994) CART: a conserved antigen receptor transmembrane motif. Sem Immunol 6:393–410
Cheng H-L, Blattner FR, Fitzmaurice L, Mushinski JF, Tucker PW (1982) Structure of genes for membrane and secreted murine IgD heavy chains. Nature 296:410–415
Chintalacharuvu KR, Morrison SL (1996) Residues critical for H-L disulfide bond formation in human IgA1 and IgA2. J Immunol 157:3443–3449
Clem LW (1971) Phylogeny of immunoglobulin structure and function. IV. Immunoglobulins of the giant grouper, *Epinephalus itaira*. J Biol Chem 246:9–15
Clem LW, DeBoutaud F, Sigel MM (1967) Phylogeny of immunoglobulin structure and function. Immunoglobulins of the nurse shark. J Immunol 99:1226–1235
Clem LW, Small PA (1967) Phylogeny of immunoglobulin structure and function. I. Immunoglobulins of the lemon shark. J Exp Med 125:893–920
Daggfeldt A, Bengtén E, Pilström L (1993) A cluster type organization of the loci of the immunoglobulin light chain in Atlantic cod (*Gadus morhua* L.) and rainbow trout (*Oncorhynchus mykiss* Walbaum) indicated by nucleotide sequences of cDNAs and hybridization analysis. Immunogenetics 38: 199–209
Diaz M, Greenberg AS, Flajnik MF (1998) Somatic hypermutation of the new antigen receptor gene (NAR) in the nurse shark does not generate the repertoire: possible role in antigen-driven reactions in the absence of germinal centers. Proc Natl Acad Sci USA 95:14343–14348
Du Pasquier L, Wilson M, Greenberg AS, Flajnik MF (1998) Somatic mutation in ectothermic vertebrates: musings on selection and origins. In: Kelsoe G, Flajnik M (eds) Somatic diversification of immune responses. Springer, Berlin Heidelberg New York, pp 199–216
Du Pasquier L, Flajnik MF (1999) Origin and evolution of the vertebrate immune system. In: Paul WE (ed) Fundamental immunology, 4th edn. Lippincott-Raven, Philadelphia, pp 605–650
Edelman GM, Gally JA (1962) The nature of Bence-Jones proteins. Chemical similarities to polypeptide chains of myeloma globulins and normal γ-globulins. J Exp Med 116:207–227
Elgar G, Clark M, Green A, Sanford R (1997) How good a model is the Fugu genome? Nature 387:140
Faith RE, Clem LW (1973) Passive cutaneus anaphylaxis in the chicken. Biological fractionation of the mediating antibody population. Immunol 25:151–164
Feddersen RM, Van Ness BG (1985) Double recombination of a single immunoglobulin kappa-chain allele: implications for the mechanism of rearrangement. Proc Natl Acad Sci USA 82:4793–4797
Fellah JS, Iscaki S, Vaerman JP, Charlemagne J (1992) Transient developmental expression of IgY and secretory component like protein in the gut of the axolotl (*Ambystoma mexicanum*). Dev Immunol 2:181–190
Ford JE, Home WA, Gibson DM (1994) Light chain isotype regulation in the horse. Characterization of Ig kappa genes. J Immunol 153:1099–1111
Frazer JK, Capra JD (1999) Immunoglobulins: structure and function. In: Paul WE (ed) Fundamental immunology, 4th edn. Lippincott-Raven, Philadelphia, pp 37–74
Ghaffari SH, Lobb CJ (1993) Structure and genomic organization of immunoglobulin light chain in the channel catfish. An unusual genomic organizational pattern of segmental genes. J Immunol 151: 6900–6912
Ghaffari SH, Lobb CJ (1997) Structure and genomic organization of a second class of immunoglobulin light chain genes in the channel catfish. J Immunol 159:250–258
Greenberg AS, Steiner L, Kasahara M, Flajnik MF (1993) Isolation of a shark immunoglobulin light chain cDNA clone encoding a protein resembling mammalian kappa light chains: implications for the evolution of light chains. Proc Natl Acad Sci USA 90:10603–10607
Greenberg AS, Avila D, Hughes M, Hughes A, McKinney EC, Flajnik MF (1995) A new antigen receptor gene family that undergoes rearrangement and extensive somatic diversification in sharks. Nature 374:168–173
Greenberg AS, Hughes AL, Guo J, Avila D, McKinney EC, Flajnik MF (1996) A novel "chimeric" antibody class in cartilaginous fish: IgM may not be the primordial immunoglobulin. Eur J Immunol 26:1123–1129
Hagman J, Rudin CM, Haasch D, Chaplin D, Storb U (1990) A novel enhancer in the immunoglobulin lambda locus is duplicated and functionally independent of NF kappa B. Genes Dev 4:978–992
Haire RN, Ota T, Rast JP, Litman RT, Chan FY, Zon LI, Litman GW (1996) A third light chain gene isotype in *Xenopus laevis* consists of six distinct VL families and is related to mammalian λ genes. J Immunol 157:1544–1550

Hamers-Casterman C, Atarhouch T, Muyldermans S, Robinson G, Hamers C, Songa EB, Bendahman N, Hamers R (1993) Naturally occurring antibodies devoid of light chains. Nature 363:446–448

Harding FA, Cohen N, Litman GW (1990a) Immunoglobulin heavy chain gene organization and complexity in the skate, *Raji ernacea*. Nucl Acids Res 18:1015–1020

Harding FA, Amemiya CT Litman RT Cohen N, Litman GW (1990b) Two distinct immunoglobulin heavy chain isotypes in a primitive, cartilaginous fish, *Raja erinacea*. Nuc Acids Res 18:6369–6376

Harisdangkul V, McDougal JS, Knapp M, Christian CL (1975) Naturally occurring low molecular weight IgM in patients with rheumatoid arthritis, systemic lupus erythematosus and macroglobulinemia. I. Purification and immunologic studies. J Immunol 115:216–222

Hayzer DJ (1990) Immunoglobulin lambda L chain evolution: Igλ and Igλ like sequences form three major groups. Immunogenetics 32:157–174

Hazenbos WLW, Heijnen IAFM, Meyer D, Hofhuis FMA, de Lavalette CR, Schmidt RE, Capel PJA, van de Winkel JGJ, Gessner JE, van den Berg TK, Verbeek JS (1998) Murine IgG1 complexes trigger immune effector functions predominantly via FcγRIII (CD16). J Immunol 161:3026–3032

Helm BA, Ling Y, Teale C, Padlan EA, Brüggemann M (1991) The nature and importance of the inter-ε chain disulfide bonds in human IgE. Eur J Immunol 21:1543–1548

Higgins DA, Warr GW (1993) Duck immunoglobulins: structure, functions and molecular genetics. Avian Pathol 22:211–236

Hinds-Frey KR, Nishikata H, Litman RT, Litman GW (1993) Somatic variation precedes extensive diversification of germline sequences and combinatorial joining in the evolution of immunoglobulin heavy chain diversity. J Exp Med 178:815–824

Hohman VS, Schluter SF, Marchalonis JJ (1992) Complete sequence of a cDNA clone specifying sandbar shark immunoglobulin light chain: gene organization and implications for the evolution of light chains. Proc Natl Acad Sci USA 89:276–280

Hohman VS, Schuchman DB, Schluter SF, Marchalonis JJ (1993) Genomic clone for sandbar shark lambda light chain: generation of diversity in the absence of gene rearrangement. Proc Natl Acad Sci USA 90:9882–9886

Home WA, Ford JE, Gibson DM (1992) L chain isotype regulation in horse. I. Characterization of Ig lambda genes. J Immunol 149:3927–3936

Honjo T, Matsuda F (1995) Immunoglobulin heavy chain loci of mouse and human. In: Honjo T, Alt FW (eds) Immunoglobulin genes. Academic Press, London, pp 145–171

Hordvik I, Voie AM, Glette J, Male R, Endresen C (1992) Cloning and sequence analysis of two isotypic IgM heavy chain genes from Atlantic salmon, *Salmo* salar L. Eur J Immunol 22:2957–2962

Hsu E, Lefkovits I, Flajnik M, Du Pasquier L (1991) Light chain heterogeneity in the amphibian *Xenopus*. Mol Immunol 28:985–994

Hsu E (1994) The variation in immunoglobulin heavy chain constant regions in evolution. Semin Immunol 6:383–391

Kabat EA, Wu TT, Perry HM, Gottesmann KS, Foeller G (1991) Sequences of proteins of Immunological Interest. 5th edn. Department of Health Services, National Institutes Health, Bethesda

Kerr MA (1990) The structure and function of human IgA. Biochem J 271:285–296

Klapper D, Clem LW, Small PA (1971) Proteolytic fragmentation of elasmobranch immunoglobulins. Biochemistry 10:645–652

Klein J (1998) In an immunological twilight zone. Proc Natl Acad Sci USA 95:11504–11505

Kobayashi K, Tomonaga S, Kajii T (1984) A second class of immunoglobulin other than IgM present in the serum of a cartilaginous fish, the skate *Raja kenojei*: isolation and characterization. Mol Immunol 21:397–404

Kobayashi K, Tomonaga S (1988) The second immunoglobulin class is commonly present in cartilaginous fish belonging to the order Rajiformes. Mol Immunol 25:115–120

Kokubu F, Hinds K, Litman R, Shamblott MJ, Litman GW (1988) Complete structure and organization of immunoglobulin heavy chain constant region genes in a phylogenetically primitive vertebrate. EMBO J 7:1979–1988

Kubo RT, Rosenblum IY, Benedict AA (1970) The unblocked N-terminal sequences of chicken IgG λ-like light chain. J Immunol 105:534–536

Lake JA, Moore JE (1998) Phylogenetic analysis and comparative genomics. Trends Guide to Bioinformatics, Trends Suppl, Elsevier Science, Amsterdam, pp 22–23

Lee MA, Bengtén E, Daggfeldt A, Rytting AS, Pilström L (1993) Characterisation of rainbow trout cDNAs encoding a secreted and membrane-bound Ig heavy chain and the genomic intron upstream of the first constant exon. Mol Immunol 30:641–648

Leslie GA, Clem LW (1969) Phylogeny of immunoglobulin structure and function. III. Immunoglobulins of the chicken. J Exp Med 130:1337–1352
Litman GW, Anderson MK, Rast JP (1999) Evolution of antigen binding receptors. Annu Rev Immunol 17:109–147
Lobb CJ, Olson MO, Clem LW (1984) Immunoglobulin light chain classes in a teleost fish. J Immunol 132:1917–1923
Lobb CJ (1987) Secretory immunity induced in catfish, *Ictalurus punctatus*, following bath immunization. Dev Comp Immunol 11:727–738
Lucero JE, Rosenberg GH, Miller RD (1998) Marsupial light chains: complexity and conservation of λ in the opossum *Monodelphis domestica*. J Immunol 161:6724–6732
Lundqvist ML, Strömberg S, Pilström L (1998) Ig heavy chain of the sturgeon *Acipenser baeri*: cDNA sequence and diversity. Immunogenetics 48:372–382
Lundqvist M, Bengtén E, Strömberg S. Pilström L (1996) Immunoglobulin light chain gene in the Siberian sturgeon (*Acipenser baeri*). J Immunol 157:2031–2038
Lutz C, Ledermann B, Kosco-Vilbois MH, Ochsenbein AF, Zinkernagel RM, Kohler G, Brombacher F (1998) IgD can largely substitute for loss of IgM function in B cells. Nature 393:797–801
Magor B, Wilson MR, Miller NW, Clem LW, Middleton DL, Warr GW (1994) An Ig heavy chain enhancer of the channel catfish *Ictalurus punctatus*: evolutionary conservation of function but not structure. J Immunol 153:5556–5563
Magor KE, Warr GW, Middleton D, Wilson MR, Higgins DA (1992) Structural relationship between the two IgY of the duck *Anas platyrhynchos*: molecular genetic evidence. J Immunol 149:2627–2633
Magor KE, Higgins D, Middleton DL, Warr GW (1994) One gene encodes the heavy chain for the three different forms of IgY in the duck. J Immunol 153:5549–5555
Magor KE, Warr GW, Bando Y, Middleton DL, Higgins DA (1998) Secretory immune system of the duck (*Anas platyrhynchos*). Identification and expression of the genes encoding IgA and IgM heavy chains. Eur J Immunol 28:1063–1068
Magor KE, Higgins DA, Middleton DL, Warr GW (1999) Opposite orientation of the alpha and upsilon constant region genes in the immunoglobulin heavy chain of the duck. Immunogenetics 49:692–695
Mansikka A (1992) Chicken IgA H chains: implications concerning the evolution of H chain genes. J Immunol 149:855–861
Marchalonis JJ, Weltman JK (1971) Relatedness among proteins: A new method of estimation and its application to immunoglobulins. Comp. Biochem. Physiol (B) 38:609–625
Marchalonis JJ, Schluter SF, Bernstein RM, Shen S, Edmundson AB (1998) Phylogenetic emergence and molecular evolution of the immunoglobulin family. Adv Immunol 70:417–506
Max EE (1999) Immunoglobulins: Molecular genetics. In: Paul WE (ed) Fundamental Immunology. 4th edn. Lippincott-Raven, Philadelphia, pp 111–182
McCormack WT, Carlson LM, Tjoelker LW, Thompson CB (1989) Evolutionary comparison of the avian IgL locus: combinatorial diversity plays a role in the generation of the antibody repertoire in some avian species. Int Immunol 1:332–341
McCormack WT, Thompson CB (1990) Somatic diversification of the chicken immunoglobulin light chain gene. Adv Immunol 48:41–67
McDougal JS, Harisdangkul V, Christian CL (1975) Naturally-occurring low molecular weight IgM in patients with rheumatoid arthritis, systemic lupus erythematosus, and macroglobulinemia. II. Structural studies and comparison of some physicochemical properties of reduced and alkylated IgM, and low molecular weight IgM. J Immunol 115:223–229
McLaughlin PJ, Dayhoff MO (1972) Evolution of species and proteins: a time scale. In: Dayhoff MO (ed) Atlas of protein sequence and structure, vol 5. National Biomedical Research Foundation, Georgetown University, Washington DC, pp 47–52
Michael N, Accavitti MA, Masteller E, Thompson CB (1998) The antigen binding characteristics of mAbs derived from in vivo priming of avian B cells. Proc Natl Acad Sci USA 95:1166–1171
Mikoryak CA, Steiner LA (1988) Amino acid sequence of the constant region of immunoglobulin light chains from *Rana catesbeiana*. Mol Immunol 25:695–703
Miller NW, Sizemore RC, Clem LW (1985) Phylogeny of lymphocyte heterogeneity: the cellular requirements for in vitro antibody responses of channel catfish leukocytes. J Immunol 134:2884–2888
Miller NW, Rycyzyn MA, Wilson MR, Warr GW, Naftel JP, Clem LW (1994) Development and characterization of channel catfish long term B cell lines. J Immunol 152:2180–2189
Miller N, Wilson M, Bengten E, Stuge T, Warr G, Clem W (1998) Functional and molecular characterization of teleost leukocytes. Immunol Rev 166:187–197

Mußmann R, Wilson M, Marcuz A, Courtet M, Du Pasquier L (1996a) Membrane exon sequences of the three *Xenopus* Ig classes explain the evolutionary origin of mammalian isotypes. Eur J Immunol 26:409–414

Mußmann R, Du Pasquier L, Hsu E (1996b) Is *Xenopus* IgX an analog of IgA? Eur J Immunol 26:2823–2830

Mußmann R, Courtet M, Du Pasquier L (1998) Development of the early B cell population in *Xenopus*. Eur J Immunol 28:2947–2959

Nitschke L, Kosco MH, Kohler G, Lamers MC (1993) Immunoglobulin D (IgD)-deficient mice can mount normal immune responses to thymus-independent and -dependent antigens. Proc Natl Acad Sci USA 90:1887–1891

Ng PLK, Higgins DA (1986) Bile immunoglobulin of the duck (*Anas platyrhynchos*). I. Preliminary characterization and ontogeny. Immunol 58:323–327

Ota T, Rast JP, Litman GW, Amemiya CT (1997) Lungfish immunoglobulins and their evolutionary implications. Dev Comp Immunol 21:159

Parham P (1995) Antibody structure. The ducks dilemma. Nature 374:16–17

Partula S, Schwager J, Timmusk S, Pilström L, Charlemagne J (1996) A second immunoglobulin light chain isotype in the rainbow trout. Immunogenetics 45:44–51

Parvari R, Avivi A, Lenter F (1988) Chicken immunoglobulin gamma-heavy chains: limited gene repertoire, combinatorial diversification by D gene segments and evolution of the heavy chain locus. EMBO J 7:739–744

Perchalski J, Clem LW, Small PA (1968) 7SγM immunoglobulin in normal human cord serum. Amer J Med Sci 256:107–111

Peterson ML (1994) RNA processing and expression of immunoglobulin genes. In: Snow EC (ed) Handbook of B and T lymphocytes. Academic Press, San Diego, pp 321–342

Peterson ML, Perry RP (1989) The regulated production of μm and μs mRNA is dependent on the relative efficiencies of μs poly(A) site usage and the Cμ4-to-M1 splice. Mol Cell Biol 9:726–738

Pilström L, Lundquist ML, Wermenstan NE (1998) The immunoglobulin light chain in poikilothermic vertebrates. Immunol Rev 166:123–132

Pleiman CM, D'Ambrosio D, Cambier JC (1994) The B-cell antigen receptor complex: structure and signal transduction. Immunol Today 15:393–399

Pumphrey R (1986) Computer models of the human immunoglobulins. Shape and segmental flexibility. Immunol Today 7:174–178

Putnam FW, Takahashi N, Tetaert D, Debuire B, Lin LC (1981) Amino acid sequence of the first constant region domain and the hinge region of the delta heavy chain of human IgD. Proc Natl Acad Sci USA 78:6168–6172

Rast JP, Anderson MK, Ota T, Litman RT, Margittai M, Shamblott MJ, Litman GW (1994) Immunoglobulin light chain class multiplicity and alternative organizational forms in early vertebrate phylogeny. Immunogenetics 40:83–99

Reynaud CA, Dahan A, Weill JC (1983) Complete sequence of a chicken lambda light chain immunoglobulin derived from the nucleotide sequence of its mRNA. Proc Natl Acad Sci USA 80:4099–4103

Reynaud CA, Anquez V, Grimal H, Weill JC (1987) A hyperconversion mechanism generates the chicken light chain preimmune repertoire. Cell 48:379–388

Reynaud CA, Dahan A, Anquez V, Weill JC (1989) Somatic hyperconversion diversifies the single gene of the chicken with a high incidence in the D region. Cell 59:171–183

Roes J, Rajewsky K (1993) Immunoglobulin D-deficient mice reveal an auxiliary receptor function for IgD in antigen-mediated recruitment of B cells. J Exp Med 177:45–55

Rombout WM, Joosten EPHM (1998) Mucosal immunity. In: Pastoret PP, Griebel P, Bazin H, Govaerts A (eds) Immunology of fishes, in Handbook of vertebrate immunology, Academic Press, San Diego, pp 39–40

Ross DA, Wilson MR, Miller NW, Clem LW, Warr GW (1998) Evolutionary variation of immunoglobulins μ heavy chain RNA processing pathways: origins, effect and implications. Immunol Rev 166:143–151

Roux KH, Greenberg AS, Greene L, Strelets L, Avila D, McKinney EC, Flajnik MF (1998) Structural analysis of the nurse shark (new) antigen receptor (NAR): Molecular convergence of NAR and unusual mammalian immunoglobulins. Proc Natl Acad Sci USA 95:11804–11809

Rycyzyn MA, Wilson MR, Warr GW, Clem LW, Miller NW (1996) Membrane immunoglobulin-associated molecules on channel catfish B lymphocytes. Dev Comp Immunol 20:341–351

Sanchez C, Alvarez A, Castillo A, Zapata A, Villena A, Dominguez J (1995) Two different subpopulations of Ig bearing cells in lymphoid organs of rainbow trout. Dev Comp Immunol 19:79–86

Saitou N, Nei M (1987) The neighbor joining method: a new method for reconstructing phylogenetic trees. Mol Biol Evol 4:406–425

Schluter SF, Bernstein RM, Marchalonis JJ (1997) Molecular origins and evolution of immunoglobulin heavy-chain genes of jawed vertebrates. Immunol Today 18:543–549

Schumaker VN, Phillips ML, Hansen DC (1991) Dynamic aspects of antibody structure. Mol Immunol 28:1347–1360

Schwager J, Burckert N, Schwager M, Wilson M (1991) Evolution of immunoglobulin light chain genes: analysis of *Xenopus* IgL isotypes and their contribution to antibody diversity. EMBO J 10:505–511

Shamblott MJ, Litman GW (1989) Genomic organization and sequences of immunoglobulin light chain genes in a primitive vertebrate suggest coevolution of immunoglobulin gene organization. EMBO J 8:3733–3739

Shankey TV, Clem LW (1980a) Phylogeny of immunoglobulin structure and function VIII. Intermolecular heterogeneity of shark 19 s IgM antibodies to pneumococcal polysaccaride. Mol Immunol 17:365–375

Shankey TV, Clem LW (1980b) Phylogeny of immunoglobulin structure and function IX. Intramolecular heterogeneity of shark 19 s IgM antibodies to the dinitrophenyl hapten. J Immunol 125:2690–2698

Sitnikova T, Nei M (1998) Evolution of immunoglobulin Kappa chain variable region genes in vertebrates. Mol Biol Evol 15:50–60

Slattery J, Holowka D, Baird B (1985) Segmental flexibility of receptor-bound immunoglobulin E. Biochemistry 24:7810–7820

Small PA, Klapper DG, Clem LW (1970) Half-lives body distribution and lack of interconversion of serum 19 s and 7 s IgM of sharks. J Immunol 105:29–37

Stewart SE, Du Pasquier L, Steiner LA (1993) Diversity of expressed V and J regions of immunoglobulin light chains in *Xenopus lavis*. Eur J Immunol 23:1980–1986

Swofford DL (1993) PAUP: Phylogentic analysis using parsimony (Illinois Nat Hist Survey, Champagne IL), version 3.1.1

Thompson JD, Higgins DG, Gibson TJ (1994) CLUSTALW: improving the sensitivity of progressive multiple sequence alignment through sequence weighting, position-specific gap penalties and weight matrix choice. Nucl Acids Res 11:4673–4680

Thorbecke GJ, Leslie GA (eds) (1982) Immunoglobulin D: structure and function. Annals New York Acad Sciences, New York

Tonegawa S (1983) Somatic generation of antibody diversity. Nature 302:575–581

Tucker PW, Slightom JL, Blattner FR (1981a) Mouse IgA heavy chain gene sequence: implications for evolution of immunoglobulin hinge exons. Proc Natl Acad Sci USA 78:7644–7648

Tucker PW, Cheng H-L, Richards JE, Fitzmaurice L, Mushinski JF, Blattner FR (1981b) Genetic aspects of IgD expression: III. Functional implications of the sequence and organization of the Cδ gene. In: Thorbecke GJ, Leslie GA (eds) Annals New York Acad Sciences, New York, pp 26–40

van Ginkel FW, Pascual DW, Clem LW (1991) Proteolytic fragmentation of channel catfish antibodies. Dev Comp Immunol 15:41–51

van Ginkel FW, Miller NW, Cuchens MA, Clem LW (1994) Activation of channel catfish B cells by membrane immunoglobulin cross-linking. Dev Comp Immunol 18:97–107

Warr GW (1995) The immunoglobulins of fish. Dev Comp Immunol 19:1–12

Warr GW, Magor KE, Higgins DA (1995) IgY: clues to the origins of modern antibodies. Immunol Today 16:392–398

Watanabe H, Kobsyashl K (1974) Peculiar IgA system identified in chicken. J Immunol 113:1405–1409

Wilson MR, Warr GW (1992) Fish immunoglobulins and the genes that encode them. Ann Rev Fish Diseases 2:201–221

Wilson MR, Marcuz A, van Ginkel F, Miller NW, Clem LW, Middleton D, Warr GW (1990) The immunoglobulin M heavy chain constant region gene of the channel catfish, *Ictalurus punctatus*: an unusual mRNA splice pattern produces the membrane form of the molecule. Nucl Acids Res 18:5227–5233

Wilson MR, van Ravenstein E, Miller NW, Clem LW, Middleton DL, Warr GW (1995a) cDNA sequences and organization of IgM heavy chain genes in two holostean fish. Dev Comp Immunol 19:153–164

Wilson MR, Ross DA, Miller NW, Clem LW, Middleton DL, Warr GW (1995b) Alternate pre-mRNA processing pathways in the production of membrane IgM heavy chains in holostean fish. Dev Comp Immunol 19:165–177

Wilson M, Bengtén E, Miller NW, Clem LW, Du Pasquier, Warr GW (1997) A novel chimeric Ig heavy chain from a teleost fish shares similarities to IgD. Proc Natl Acad Sci USA 94:4593–4597

Wu TT, Kabat EA (1970) An analysis of the sequences of the variable regions of Bence-Jones proteins and myeloma light chains and their implications for anti-body complementarity. J Exp Med 132: 211–250

Yamawaki-Kataoka Y, Miyata T, Honjo T (1981) The complete nucleotide sequence of mouse immunoglobulin gamma 2a gene and evolution of heavy chain genes: further evidence for intervening sequence-mediated domain transfer. Nucl Acids Res 9:1365–1381

Zezza DJ, Stewart SE, Steiner LA (1992) Genes encoding *Xenopus laevis* Ig L chains. Implications for the evolution of kappa and lambda chains. J Immunol 149:3968–3977

Evolution of Vertebrate Immunoglobulin Variable Gene Segments

T. Ota[1], T. Sitnikova[2], and M. Nei[3]

1 Introduction

Immunoglobulins (Igs), also known as antibodies, play major roles in the vertebrate humoral immune system. They recognize and bind to foreign antigens, such as viruses, bacteria and parasites, and initiate a series of immunological responses

[1] Center for Human Genetics, Boston University School of Medicine, 715 Albany St., Boston, MA 02118, USA
e-mail: Ota@soken.ac.jp
The Graduate University for Advanced Studies, Hayama, Kanagawa, Japan
[2] Eisai Merrimack Valley Lab Inc., 100 Federal St., Andover, MA 01810, USA
[3] Institute of Molecular Evolutionary Genetics and Department of Biology, The Pennsylvania State University, 328 Mueller, University Park, PA 16802, USA

(effector function). The dual function of the Ig protein is facilitated by its unique structure consisting of two functionally distinct domains, i.e., the variable (V) domain for antigen recognition and the constant (C) domain for effector function (FRAZER and CAPRA 1998). Ig is generally composed of two identical heavy (IgH) and two identical light (IgL) chains, both of which contribute to the formation of V and C domains.

The diversity necessary for recognizing a variety of antigens is generated by genetic and somatic processes (MAX 1998). In general, organisms maintain large multigene families with diversified members to code for numerous V domains and the diversity is further enhanced by somatic mechanisms, such as recombination, imprecise joining of gene segments, hypermutation, and gene conversion. As illustrated in Fig. 1, at least three fundamentally different systems of generating diversity have been found in vertebrates. In cartilaginous fishes clusters of gene segments are repeated many times and chromosomally dispersed in their genome, and each cluster produces either IgL or IgH (LITMAN et al. 1993). In mice and humans two or three multigene families of gene segments are present in separate arrays and one member of each gene segment family is selected to join together to produce a functional Ig V domain (TONEGAWA 1983). In chickens only one "functional" V gene segment is present in the genome, but it is somatically modified by gene conversion after gene rearrangement of the functional gene segments (REYNAUD et al. 1994).

Recent progress in comparative immunology has shown that vertebrates share some basic immunological features, such as the presence of Ig, T-cell receptor and major histocompatibility complex, but that distinct molecular and developmental immune mechanisms have been established in different evolutionary lineages (DU PASQUIER and FLAJINIK 1998). Here we review the humoral immune system of human and other vertebrates and discuss the evolution of Ig diversity generation and its effect on the evolution of V gene segments.

2 Human Ig Genes

2.1 Gene Organization and Germline Diversity of V Gene Segments

Human and mouse are the best characterized vertebrates in terms of the molecular genetics of Ig genes and the developmental biology of B-cells in which Ig genes are expressed. In both species Ig V region is produced by the somatic rearrangement of distinct gene segments (TONEGAWA 1983). The V region of IgH is encoded by variable (V_H), diversity (D), and joining (J_H) gene segments, whereas the V region of IgL is encoded by variable (V_L) and joining (J_L) gene segments. The gene organization of human Ig loci has recently been clarified (FRIPPIAT et al. 1995; KAWASAKI et al. 1997; MATSUDA et al. 1998; TOMLINSON and COOK 1997; ZACAHU 1995). The human IgH locus, located at chromosome 14q, contains 123 V_H

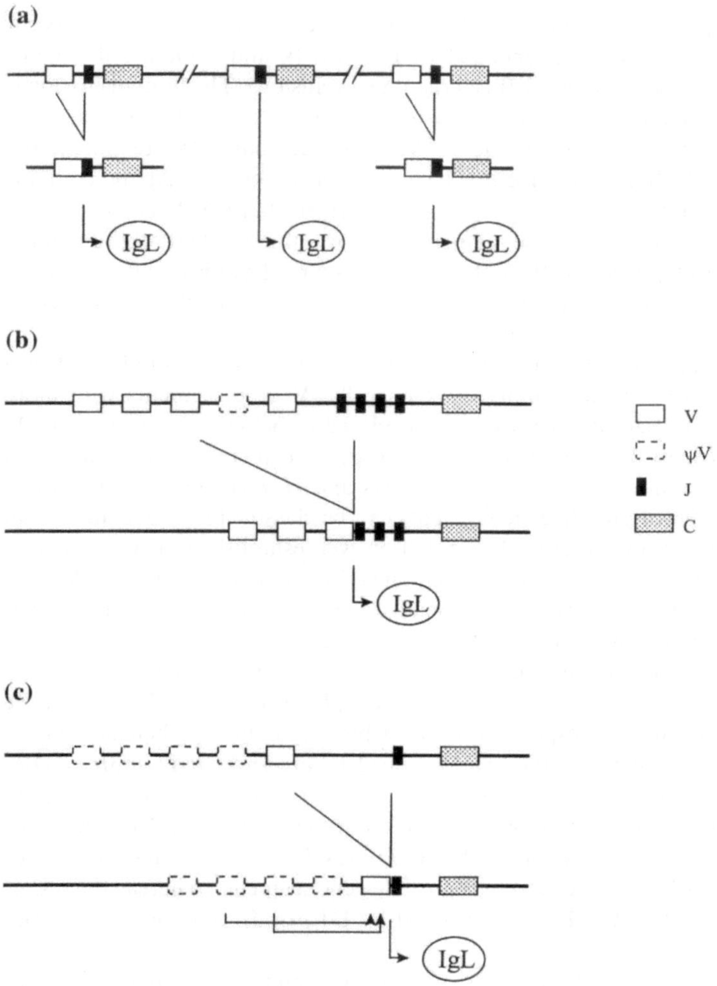

Fig. 1a–c. Immunoglobulin light-chain (IgL) gene organization. **a** Cartilaginous fishes. **b** Mice. **c** Chickens. Similar gene organizations exist at the heavy-chain loci of respective species with presence of D gene segments between V and J gene segments. Cartilaginous fishes have the cluster type gene organization and each cluster is capable of producing a unique IgL chain. Mice have the translocon type gene organization, where one V and one J gene segment are joined by gene rearrangement during B-cell lymphopoiesis. Combination of different gene segments generates the primary diversity. Chickens have a single functional V gene segment and a single J gene segment. The rearranged region is later modified by somatic gene conversion with pseudogenes as donors

(MATSUDA et al. 1998), 27 D (CORBETT et al. 1997), and nine J_H gene segments (RAVETCH et al. 1981). Of the 123 V_H gene segments, 44 are potentially functional, but the remaining genes are likely to be pseudogenes. Additional V_H and D gene segments are present on chromosomes 15 and 16 and are referred to as orphan genes (NAGAOKA et al. 1994; TOMLINSON et al. 1994). The orphan genes do not contribute to Ig production and probably represent evolutionary relics of recent duplicated genes. Human IgL is encoded by two loci (κ and λ loci), only one of which is exclusively used in a given B-cell (isotype exclusion). The human Igκ locus located on chromosome 2p contains 76 V_κ and 5 J_κ gene segments (SCHÄBLE and ZACHAU 1993; ZACHAU 1995). Of the 76 V_κ gene segments, 35 are potentially functional genes. The human Igλ locus is located on chromosome 22q and contains more than 69 V_λ and 7 J_λ gene segments (FRIPPIAT et al. 1995; KAWASAKI et al. 1997). Of the more than 69 V_λ gene segments approx. 30 are potentially functional (WILLIAMS et al. 1996). Orphan V_κ and V_λ gene segments have been also found on chromosomes 1, 2, and 22 (ZACHAU 1995) and on chromosome 8 (FRIPPIAT et al. 1997; QUEIROZ et al. 1997), respectively.

The V_H and V_L gene segments show great diversity and are subdivided into families according to sequence similarity. Generally, V gene segments with more than approx. 80% nucleotide identity are considered to belong to the same family (BRODEUR and RIBLET 1984) and the human V gene segments are classified into seven V_H, five V_κ, and ten V_λ families (MATSUDA and HONJO 1996; ZACHAU 1995; WILLIAMS et al. 1996). Four V_κ gene segments can be classified into two additional V_κ families, but their functionality has not been well established. The number of member genes in a V gene family varies from one (V_H6, $V_\kappa4$, $V_\kappa5$, $V_\lambda6$, $V_\lambda8$, $V_\lambda9$) to 65 (V_H3). Most V gene families contain pseudogenes, and in a few V gene families more than two-thirds of member genes are pseudogenes.

The V domain is divided into the complementarity determining regions (CDRs) and the framework regions (FRs; WU and KABAT 1970). CDRs form the hypervariable loops and correspond to antigen binding regions, whereas FRs encode the two β-pleated sheets of the Ig fold. Four FRs are separated by three CDRs at the primary nucleotide sequence, but CDRs are brought to the C-distal end of Ig domain to form the antigen binding pocket in the tertiary structure. The first three FRs (FR1, FR2, FR3), the first two CDRs (CDR1, CDR2) and a small portion of the third CDR (CDR3) are encoded by V_H or V_L gene segments, whereas the rest of CDR3 and the fourth FR (FR4) are encoded by DJ_H or J_L gene segments (MAX 1998).

CDRs are highly variable in terms of sequence length and amino acid composition. The loop structures formed by CDRs are often classified into a set of main chain structures called canonical structures (CHOTHIA et al. 1987). The number of amino acids involved in the loop formation and the presence of certain amino acids at structurally critical sites in CDR primarily determine canonical structures. The canonical structures formed at the antigen binding pocket provide valuable clues on the antigen specificity of V gene segments (VARGAS-MADRAZO et al. 1995; WEBSTER et al. 1994). For example, the canonical structures with long loop form small pocket at the antigen binding sites and are more often used in the recognition of small molecules such as haptens, whereas the canonical structures with short loop are more adapted for recognition of large molecules such as proteins. In humans, one to six different canonical classes have been identified in CDR1 and CDR2 of V_H and V_L domains, and a structurally related set of canonical structures are often found for a given V gene family (AL-LAZIKANI et al. 1997; VARGAS-MADRAZO et al.

1997). This observation together with the fact that the sequence and the structure in FRs are conserved within a V gene family suggest that V gene families are pre-adapted for the recognition of distinct classes of epitopes (KIRKHAM et al. 1992).

2.2 Polymorphism of V Gene Segments

In the V gene segment region of human Ig loci two types of polymorphism are mainly observed. One is single nucleotide polymorphism and the other is V gene segment insertion/deletion polymorphism. The extensive list of single nucleotide polymorphism is reported in VBASE (TOMLINSON et al. 1996a) and summarized in Table 1. Excluding a few alleles that have deleterious mutations or small nucleotide deletion in CDRs, only 146 nucleotide differences were observed at 105 functional V gene segment loci. This level of polymorphism is much less than those of the classical MHC locus but is more comparable to the low nucleotide diversity found at many nuclear gene loci in human populations (LI and SADLER 1991). Of the 146 nucleotide differences, approximately one-third of the differences were synonymous and the rest were amino acid altering differences. Half the amino acid altering differences were observed in CDRs despite much shorter amino acid lengths in CDRs than in FRs. The higher ratio of amino acid altering/synonymous nucleotide differences is apparent in CDRs of V_H gene segments, and it suggests a role of positive selection to promote the amino acid variation at the antigen binding sites (TANAKA and NEI 1989).

V gene segment insertion/deletion polymorphisms were found at both IgH and IgL loci. In the IgH locus genomic characterization of the entire V gene segment region has revealed both single- and multiple- V gene segment insertion/deletion polymorphism at multiple locations (MATSUDA and HONJO 1996). In the Igκ locus about 5% of individuals do not have a region called "d contig", which contains 33 V_κ gene segments of recent duplicates. The insertion/deletion of V gene segments has

Table 1. The number of codon differences observed among alleles of V gene segments reported in VBASE (from TOMLINSON 1996a)[a]

	$V_H(43)$[b]		$V_\kappa(35)$[b]		$V_\lambda(27)$[b]	
	FR	CDR	FR	CDR	FR	CDR
Synonymous difference[c]	22	5	4	6	4	3
Amino acid altering difference[c]	30	32	5	8	8	7

[a] Only the allelic differences between functional V gene segments were counted. If the same difference was observed between different pairs of alleles for a given locus, it was counted only once for the difference.
[b] The numbers in parentheses are the numbers of V gene segment loci at which more than two nucleotide sequences have been reported independently.
[c] Most codon changes are due to a single nucleotide difference but a few codon changes are due to two or three nucleotide changes. In total, 146 nucleotide differences are observed in 134 codon differences.

little effect on individual's health, indicating that the V gene repertoire is functionally redundant to some degree (PARGENT et al. 1991; SCHAIBLE et al. 1993).

2.3 Somatic Generation of Ig diversity during B-Cell Development

In order to understand the evolution of V gene segments it is important to know how the V gene segment repertoire plays a role in the recognition of foreign antigens. In the following two sections we briefly review how the V gene segments are somatically modified during B-cell development and how the diversity created is employed for antigen recognition.

The development of B-cell can be divided into two discrete stages: (a) the lymphopoietic development stage where B-cells develop from hematopoietic stem cells (see MELCHERS and ROLINK 1998 and references therein); and (b) the B-cell activation stage where Igs with high affinity against antigens are produced and B-cells differentiate into plasma and memory B-cells (DEFRANCO 1998; PICKER and SIEGELMAN 1998). In the first stage Ig loci are subjected to gene rearrangement, and the primary Ig repertoire is generated independent of antigens. It also has a mechanism built in to express only one allele of IgH and of IgL, a process known as allelic exclusion. In the second stage somatic mutation and clonal selection reshape the primary repertoire in the presence of antigens.

During the lymphopoietic development the Ig diversity is furnished by combinatorial and joining diversity (see MAX 1998 and references therein). Recombination signal sequences (RSSs) juxtaposed to the V_H, D, J_H, V_L, and J_L gene segments are recognized by a recombination complex and one gene segment from each gene family is selected to be joined to form the V region of IgL or IgH. Selection of gene segments is biased but the combination of functional gene segments generates substantial diversity. Despite high specificity of RSS mediated joining the recombination process is imprecise, and nucleotide insertion/deletion is commonly observed at the junction region. Nucleotide deletion is presumably due to exonuclease activity of the gene recombination complex, whereas nucleotide insertion is due either to nucleotide addition caused by terminal deoxynucleotidyl transferase (N diversity) or to palindromic nucleotide addition (P diversity) caused by repairs of asymmetrically cleaved intermediate hairpin structures generated during the gene rearrangement. Insertion/deletion of nucleotides also results in the use of the D gene segment in three different reading frames. The combination of gene segments and the joining diversity created during gene rearrangement provide extensive diversity in CDR3.

Somatic mutation observed in the secondary response is highly specific but its molecular mechanism is not yet well understood (BEREK 1998; JOLLY et al. 1996). The somatic mutation targets mostly the V region irrespective to FRs or CDRs but not C region, and it is mainly of point mutation and nucleotide insertion/deletion is rarely observed. However, it is efficient to attain affinity maturation due to the fact that a few critical amino acid changes may increase its affinity tremendously and somatic mutation/selection can undergo repeatedly. It is also noteworthy that the diversity at the main antigen binding sites appears to be encoded in germline and the primary role of somatic mutation is to extend the diversity at the surrounding regions to increase fine specificity (TOMLINSON et al. 1996b). Mutational hot spots are often found in CDRs, partly because the biased pattern of somatic mutation favors the modification of codons frequently found in CDRs. It has been speculated that the biased codon usage in CDRs has been selected during the evolution to increase the mutability at the antigen binding sites (JOLLY et al. 1996; KEPLER 1997).

2.4 Ig Repertoire and Antigen Recognition

Systematic studies to examine the use of the Ig repertoire against specific antigens are important to understand the function of individual V gene segments and the significance of V gene repertoire encoded in germline. Here we describe two cases, one against a T-cell independent (TI) type II antigen and the other against a T-cell dependent (TD) antigen. TI type II antigens are generally poor antigens and typically of repeated polymer such as of bacterial capsular polysaccharides (PS). It often induces primary but not secondary immune response, resulting in no affinity maturation. In contrast, TD antigens are usually protein/peptide antigens. In response to TD antigens, B-cells undergo hypermutation and clonal selection and subsequently, IgGs or IgAs with high affinity against antigens are produced.

Infectious bacteria possess PS capsules on their surface and induce limited B-cell response presumably due to the repeated structure of PS. This type of TI antigen is called TI type II antigen and distinguished from TI type I antigens, such as bacteria lipopolysaccharides, which are mitogenic to B-cells irrespective of antigen specificity (MOND et al. 1995). *Haemophilus influenza* type B (*Hi*b) are PS encapsulated bacteria that cause meningitis among children. In the adult immune response against *Hi*b PS, a limited set of V_H and V_L gene segments are utilized (SCOTT et al. 1989; SCOTT and NAHM 1992). Mostly members of V_H3 gene family are detected in the response, and, interestingly, about half the V_L gene segments are derived from the A2 V_κ gene segment with little modification of the germline sequence. Other V_κ as well as V_λ gene segments are also observed in the response but they are variably expressed among patients and often contain somatic mutations (CHUNG et al. 1993). The importance of A2 V_κ gene segment in *Hi*b PS recognition is highlighted by the evidence that about tenfold-increased incidences of *Hi*b disease have been reported in Navajos population, where a defective allele of A2 V_κ gene segment is found at higher frequency (FEENEY et al. 1996; NADEL et al. 1998). The results indicate that a single germline gene segment can play major role in the TI antigen recognition and may provide significant selectional advantage by its presence.

Studies on V gene segment use against TD antigens are difficult because a prior exposure to antigens can bias Ig repertoire due to memory function of B-cells. IKEMATSU et al. (1993, 1998) have circumvented the problem by using rabies virus vaccine, since most individuals are not experienced with prior infection and no human virus has been known to have crossreacting epitope. The rabies virus is a typical TD antigen and the Ig response is examined after three sequential injections of the virus vaccine. Although only ten Ig variants are reported, no V_H gene segment seems to be used multiple times and variable V_κ and V_λ gene segments were used in the response. Seven of nine V_H gene segments belong to the V_H3 gene family, partly due to the large size of V_H3 gene family and partly due to the preferential expression of V_H3 gene family. High-affinity binding capability was found among the Igs against various component of rabies virus vaccine. Sequence analysis of V_L regions showed the accumulation of 2 to14 somatic nucleotide substitutions from candidate V_L gene segments, where the substitutions favor amino acid replacements in CDRs in a few cases. The results suggest that mutation and clonal selection are indeed important in TD responses, and the diversified germline V gene repertoire can be used to generate high-affinity Igs against various epitopes for an antigen.

It is clear that the germline V gene segment repertoire, combinatorial, and joining diversity provides the primary diversity and that the somatic mutation and clonal selection reshape it to increase affinity and/or avidity. In some cases, however, only a limited number of V gene segments have significant effect on the antigen recognition, as seen in some TI antigen response. Therefore the individual V gene segments and the entire repertoire encoded by V gene families are equally important factors for the immune response.

2.5 Evolution of Human V Gene Segments

As illustrated by previous phylogenetic analyses (Matsuda and Honjo 1996; Matsuda et al. 1998; Ota 1997; Ota and Nei 1994; Sitnikova and Nei 1998), many duplications of V gene segments have occurred during the evolution. The first duplication of gene segments to lead to major lineages of human V_H and V_L gene families likely occurred more than 350 and 470 million years (MY) ago, respectively, and others, such as the duplication of "d contig" in Igκ, occurred about 1–2MY ago. The duplication of gene segments has provided opportunities in which mutation could create new variants that were selected for or against depending on the nature of products. As seen in the case of A2 V_κ gene segment, some V gene segments would have provided immediate selective advantage due to the high affinity against specific antigens in germline configuration or with its little modification. Positive selection has apparently played a role in the fixation of such variants, as indicated by higher amino acid altering nucleotide substitution than one expected from neutral evolution (Tanaka and Nei 1988; Sitnikova and Nei 1998). The extensive diversity in CDRs is thus partly attributable to the positive selection operating on CDRs to enhance the antigen recognition against infectious agents. The great amount of mutation is, on the other hand, deleterious because of its random nature of mutation and has generated pseudogenes. In many instances the V gene segments are functionally redundant, whether it is due to the presence of recent duplicates or due to the presence of V gene segments that can be used against similar or different epitope of same antigens. In this regard, deleterious mutations are likely subject to random genetic drift rather than purifying selection. This explains the abundance of pseudogenes among the human Ig loci. Eventually these pseudogenes were eliminated from the genome by deletion or rapid accumulation of mutations. Considering continuous generation of duplicated genes and the removal of genes from the genome, the evolution of V gene segments is a dynamic process, and their repertoire is constantly subject to change. Theoretically this type of evolutionary process results in expansion and contraction of gene families and eventually homogenization of gene family (concerted evolution). However, in reality there is no indication of concerted evolution among V gene families. In contrast, the various groups of V gene families have been present for long evolutionary time. This implies the significant role of diversifying selection to preserve the germline V gene segment repertoire. Restricted sets of canonical structures are often associated with a V gene family, and it is likely that the maintenance of V gene family results in the preservation of diverse sets of canonical structures, which in turn provides the diversity to recognize distinct classes of epitopes. In summary, the evolution of human V gene segment is characterized by two evolutionary processes: (a) the generation of gene segments by gene duplication and the elimination of gene segments by accumulation of deleterious mutations or by deletion (the evolution by "the birth and death process"), and (b) diversifying selection, which emphasize the importance of purifying selection to maintain the established V gene segment repertoire and positive selection to enhance the repertoire (OTA and NEI 1994; NEI et al. 1997; SITNIKOVA and NEI 1998).

3 Cartilaginous Fish Ig Genes

3.1 Gene Organization and Generation of Diversity

Cartilaginous fishes possess a distinct Ig gene organization (LITMAN et al. 1993; MACHALONIS et al. 1998). Four or three types of gene segments are used as in the case of higher vertebrate but hundreds of clusters of V, (D,) J, and C are present at multiple sites. Somatic gene rearrangement occurs within a cluster, and in many instances gene segments in a cluster are joined at germline level. Overall combinatorial and joining diversity is thus limited. Somatic mutation is known to occur at the Ig loci of cartilaginous fishes, but the significance of somatic mutation remains uncertain, since it is unknown whether allelic exclusion is present in the organism (HSU 1998). Clonal selection without allelic exclusion would be hardly effective to attain affinity maturation and potentially increase the risks of producing autoantibodies. The presence of germline joined functional genes specifically rejects the allelic exclusion models proposed for mammals (ALT et al. 1981; COLECLOUGH et al. 1981).

3.2 Evolution of Cartilaginous Fish V Gene Segments

Variable numbers of V_H and V_L gene families are found among cartilaginous fishes (Table 2). Figure 2 presents a phylogenetic tree of cartilaginous fish V_H gene segments. There are three distinct major clusters of V_H gene segments, one (group E) associated with Cμ gene segment and the others (groups F and G) mostly associated with a second C gene segment (C_ω/IgNARC/C_x), as noted by SHEN et al. (1996). The only exception to this pattern of association of V_H gene segments with C gene segments is the monotypic horned shark V_H gene family represented by Hfr1113. The C gene segment associated with Hfr1113 is partly characterized and is only slightly different from other C_μ gene segments (HINDS-FREY et al. 1993). However, excluding this exception, the coevolution of the V gene family and C gene segments generally exists and the similar association is also found in cartilaginous fish IgL (RAST et al. 1994). These results suggest that the gene segments have been

Table 2. Numbers of V gene families so far detected in the four cartilagenious fish species

Species	V_H gene family	V_L gene family	References
Horned shark	2	3	LITMAN et al. 1993; RAST et al. 1994
Carcharhine shark	7	1	BERSTEIN et al. 1996; HOHMAN et al. 1993; SHEN et al. 1996
Little skate	2	2	HARDING et al. 1990; RAST et al. 1994
Spotted ratfish	4	2	RAST et al. 1994, 1998

Fig. 2. A phylogenetic tree of 69 cartilaginous fish V_H gene segments and 2 outgroup V_L gene segments. The phylogenetic tree was estimated by neighbor-joining method (SAITOU and NEI 1986) with Poisson correction distances. Only amino acid sequences in FRs were used in the analysis. The branch lengths are measured in terms of the number of amino acid substitutions per site, with the scale given below the tree. *Asterisks*, probabilities at which the branch length estimation is different from zero (confidence probabilities): *, >90%; **, >95%; ***, 99%. Most sequences are retrieved from Genbank (accession number): Horned shark (*Heterodontus francisci*): HfrF101 (X13449), HfrHN13 (Z11788), HfrHN21 (Z11791), HfrHN72 (Z11792), HfrHN84 (Z11790), HfrHXIA (M12195), HfrX (Z11789), Hfr1113 (Z11776), Hfr1315 (X13447), Hfr1320 (X13448), Hfr1403 (Z11777), Hfr12214 (Z11783), Hfr12215 (Z11780), Hfr12221 (Z11785), Hfr12241 (Z11784), Hfr12251 (Z11787), Hfr12271 (Z11781), Hfr12272 (Z11782), Hfr12273 (Z11786), Hfr33141 (Z11778). Nurse shark (*Ginglymostoma cirratum*): GciA4-2 (M92851), GciIgNARC (U51450). Bull shark: (*Carchahinus leucas*): Clu2Thy (U50613), Clu5Thy (U50612), Clu6Thy (U50611), Clu8Thy (U50614). Sandbar shark (*Carchahinus plumbeus*): Cpl1RACE (U50606), Cpl1spleen (U50609), Cpl2spleen (U50607), Cpl4RACE (U40560), Cpl6spleen (U50604), Cpl7RACE (U50610), Cpl7spleen (U50608), Cpl8RACE (U50605). Clearnose skate (*Raja eglanteria*): RegVCC2 (U08010). Little skate (*Raja erinacea*): RerRe20 (M29672), RerRe102 (X16146), RerRe107 (X15124), RerRe110 (X15125). Spotted ratfish (*Hydrolagus colliei*): HcoCos5 (AF003841), Hco20182 (AF003946), Hco20441 (AF003861), Hco20442 (AF003839), Hco2659 (AF003860), Hco3430 (AF003853), Hco3431 (AF003854), Hco3432 (AF003855), Hco3433 (AF003856), Hco3434 (AF003857), Hco3435 (AF003858), Hco3436 (AF003859). Other carcharhine shark IgM sequences are taken from SHEN et al. (1996). Two cartilaginous fish V_L sequences, HfrII (L25560), RerII (L25566), were used as outgroup

rarely exchanged between clusters, and that the duplication of entire cluster rather than duplication of individual gene segments have been more important during cartilaginous fish evolution. Duplication of individual gene segments has apparently occurred in the ratfishes, but the V_H gene segments unlinked with C gene segment appear to be pseudogenes (RAST et al. 1998).

In general, the number of V gene families for a given C gene segment isotype is limited. So far only one V gene family has been found for *Raja* type I and type II IgLs, carcharhine shark type II IgL, nurse shark type III IgL, and ratfish type II IgL, with exception of Hco712 (ANDERSON et al. 1995; BARRÉ et al. 1994; HOHMAN et al. 1995; RAST et al. 1994). In horned shark a single major and second monotypic

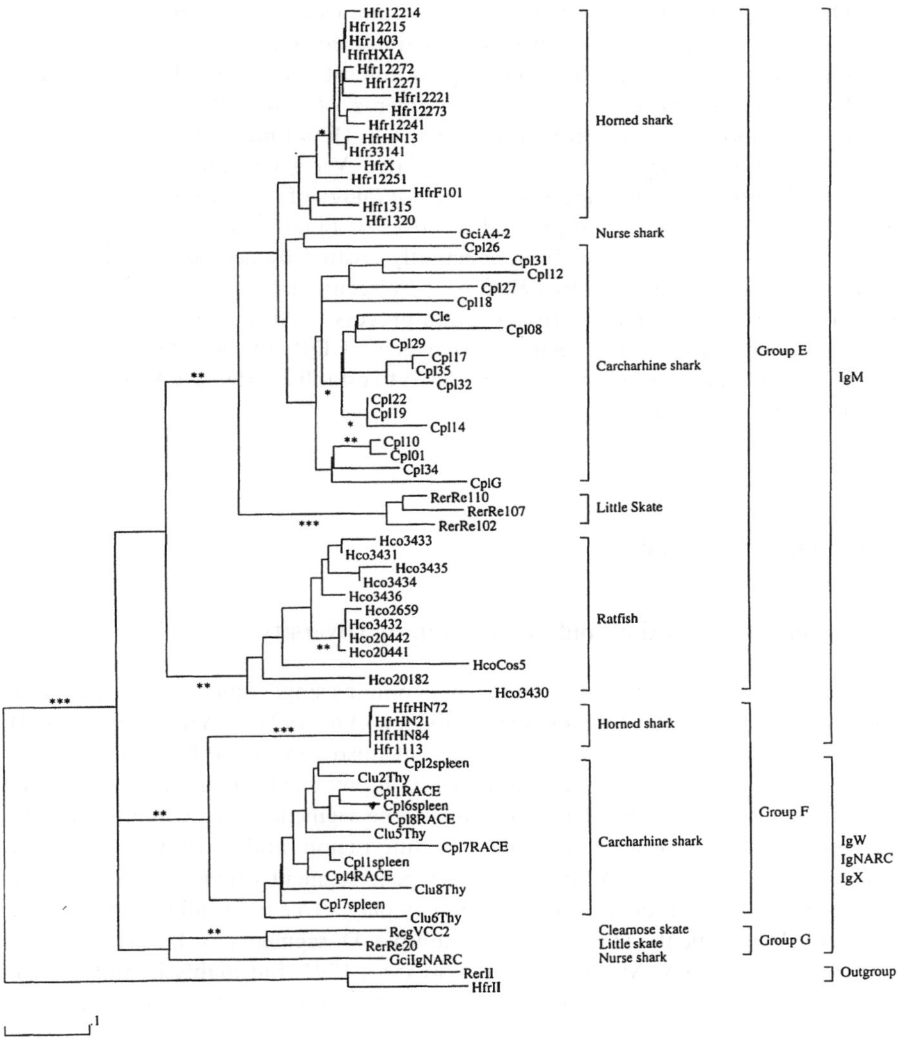

V_H gene families were found in association with C_μ (LITMAN et al. 1993). Even though carcharhine shark V_H gene segments can be classified into seven V_H gene families, they are basically separated into two groups, one group consisting of six closely related C_μ-associated V_H gene families and the other group of one C_ω/IgNARC/C_x-associated V_H gene family (Fig. 2). Despite of the limited diversity among V gene families, however, the amount of variation within a V gene family is not negligible (KOKUBU et al. 1988a; ANDERSON et al. 1995; SHEN et al. 1996). A high frequency of amino acid altering nucleotide differences are found in CDRs, and the positive selection seems to have operated on the diversification of antigen recognition sites in cartilaginous fishes as well (ANDERSON et al. 1995; OTA et al. 1995; MATSUNAGA and ANDERSSON 1997).

One intriguing question in term of cartilaginous fish Ig evolution is how the homology of C gene segments has been maintained among different clusters to sustain its capacity for effective signaling to downstream host defense molecules. Nucleotide sequences among C_μ gene segments of different clusters show an average of approx. 90% identity in horned sharks (KOKUBU et al. 1988b). The fact that a single V_H gene family or closely related V_H gene families are generally associated with an isotype suggests that the homology of C gene segments has been gained by the continuous and gradual turnover of Ig clusters. That is to say, new cluster of gene segments have been constantly created by gene duplication, while other clusters have been lost by deletion or by accumulating by deleterious mutations (the evolution by the "birth and death process"). Although positive selection has been operating at V gene segments to diversify CDRs to some degree, extensive diversity would have been hardly accumulated to establish extensive V gene families due to turnover of clusters.

4 Chicken Ig Genes

4.1 Gene Organization and Generation of Diversity

Chicken Ig gene organization is unique and contains single functional V and J gene segments at the IgL and IgH loci (MCCORMACK et al. 1991; REYNAUD et al. 1994). Many "pseudo"-V gene segments are present upstream of the functional V gene segment, but the lack of proper RSSs makes it impossible for them to be rearranged with D or J gene segments. These pseudo-V gene segments often contain no leader peptide exon, are truncated at 5′ and 3′ coding region, and carry nonsense mutation. The use of a single V and a single J gene segment originally results in the production of almost identical Igs, which may play a role in proliferation of B-cells at early development. The fusion of multiple D segments is known to occur (REYNAUD et al. 1989; MANSIKKA and TOIVANEN 1991) but highly homologous D gene segments, the exclusive use of Igλ, and the absence of N diversity limit the diversity of the primary antibody repertoire. Although the humoral immune system in chicken may be inefficient against TI-2 antigens (JEURISSEN et al. 1998), it seems to be effective against TD antigens. As it turns out, in chicken, the Ig diversity is mainly generated by somatic mechanisms, i.e., somatic gene conversion and somatic point mutation, in the specific organ called the bursa of Fabricius. Somatic gene conversion utilizes pseudogenes located upstream of the functional gene and copies small piece of the pseudogenes to the rearranged gene segment unidirectionally and intrachromosomally. There are 25 or 26 pseudo-V_λ genes (REYNAUD et al. 1987; MCCORMACK et al. 1993) and approximately 80 pseudo-V_H genes (Reynaud et al. 1989). Some pseudo-V_H gene segments also contain D- or J-like sequences that contribute to the diversity in CDR3. Among pseudogenes extensive nucleotide differences as well as length variation are observed in CDRs. Repeated somatic gene conversions at multiple positions by different pseudogenes therefore generate enormous diversity, which can be enhanced later by somatic point mutations (PARVARI et al. 1990; ARAKAWA et al. 1998).

4.2 Evolution of Chicken V Gene Segments

All chicken V_H and V_λ gene segments identified belong to a single V gene family and show little variation in FRs, specifically among V_H gene segments (REYNAUD et al. 1987, 1989). Nucleotide difference in CDRs is, however, extensive and are capable of generating high-affinity Igs (MICHAEL et al. 1998). High homology in FRs indicates that the pseudo-V_H gene segments have been derived from recent gene duplication or are subject to gene conversion. High homology of FRs would increase the efficiency of somatic gene conversion, since V gene segments with high similarity to the functional gene segment undergo somatic gene conversion more frequently. Analyses of pseudo-V gene segments have provided evidence of germline gene conversion (MCCORMACK et al. 1993; BENATAR and RATCLIFFE 1993). Frequent germline gene conversion without selection generally results in homogenization of member genes, but it can create diversity in the presence of positive selection. Indeed, positive selection has been shown to operate in CDRs of pseudo-V_H gene segments (OTA and NEI 1995).

Almost all vertebrates except cartilaginous fishes have human-type gene organization (Vn-Dn-Jn-Cn), or so-called "translocon" gene organization (LITMAN et al. 1993; MARCHALONIS et al. 1998). Therefore the chicken gene organization has likely been derived from the translocon gene organization. It remains to be examined, however, what caused the loss of functional genes in chickens (OTA and NEI 1995). It is possible that the utilization of somatic mutation has reduced the selection pressure to maintain functional genes and all V gene segments except one became nonfunctional owing to accumulation of deleterious mutations in the promoter region or coding region. Alternatively, population size bottlenecks or an extended period of population size reduction would have caused accumulation of deleterious mutations by genetic drift, as suggested by OTA and NEI (1995).

One unique feature of chicken pseudo-V_H gene segments is the presence of germline joined V_HD and V_HDJ_H gene segments. How are D- or DJ_H-like sequences brought to V gene segments, while keeping functional D or V gene segments intact? This might be attributable to another unique feature of chicken Ig gene organization, i.e., the mutually inverted orientation of pseudo-V gene segments (REYNAUD et al. 1987, 1989). Since the gene rearrangement of a D or DJ_H gene segment with a V gene segment in opposite transcriptional polarity would result in only inversion of the fragment but not deletion between them, incidences of gene rearrangement in germline cells by leaky expression of gene rearrangement complex can join D- or DJ_H-like gene segments to V gene segments while maintaining D gene segment diversity and D-proximal V gene segments intact. Alternatively, all features observed in chicken pseudogenes are also consistent with the hypothesis that the reverse-transcribed products of rearranged genes were integrated into genome (OTA and NEI 1995; see also BLANDEN and STEEL (1998) for the possible role of reverse transcription on somatic mutations in higher vertebrates). Further studies are necessary for testing the evolutionary hypotheses on the origin of chicken Ig genes.

5 Evolution of V Gene Families

As mentioned above, variable number of V gene families have been detected among vertebrates (see also Table 3). How did this large number of gene families originate? In order to understand the evolution of V gene families their phylogenetic rela-

Table 3. Number of V gene families in five vertebrate species

Species	V_H gene family	V_κ gene family	$V_{non\text{-}\kappa}$ gene family
Human	7	7	10
Chicken	1	–	1
Xenopus laevis	11	1	8
Rainbow trout	11	1	1
Horned shark	2	1	2

Fig. 3. Neighbor-joining tree of 92 V_H gene segments and 2 outgroup V_L gene segments. The first three letters of genes were designated the first letter of species genus name plus the first two letters of species name: bull shark (*Cle*), cattle (*Bta*), channel catfish (*Ipu*), chicken (*Gdo*), horned shark (*Hfr*), human (*Hsp*), little skate (*Rer*), mexican axolotl (*Ame*), mouse (*Mmu*), nurse shark (*Gci*), pig (*Ssc*), rabbit (*Ocu*), rainbow trout (*Omy*), sandbar shark (*Cpl*), sheep (*Oar*), short-tailed opossum (*Mdo*), spotted ratfish (*Hco*), sturgeon (*Aba*), and *Xenopus laevis* (*Xla*). Please refer to OTA (1997) and ROMAN et al. (1996) as well as the legend of Fig. 1 for the source of most sequences. Additional sequences were retrieved from Genbank (accession number): Aba18.1 (Y13261), Aba205 (AJ222823), Aba3.3 (Y13255), Ame1.182M (AF027252), Ame2.29M (AF027253), Ame3.73M (AF027256), Ame4.43M (AF027257), Ame5.43M (AF027259), Ame6.65Y (AF027260), Ame7.15M (AF027261), Ame8.46M (AF027262), Ame9.103M (AF027267), Ame10.33Y (AF027268), Ame11.5M (AF027269), Ame1.182M (AF027252), Bta8 (U55164), IpuVH7a (U3410), MdoVH2 (AF012124), Mdo257 (AF007070), Mmu15A (U39293), OarV1B (Z49188), and SscVHA (AF064688)

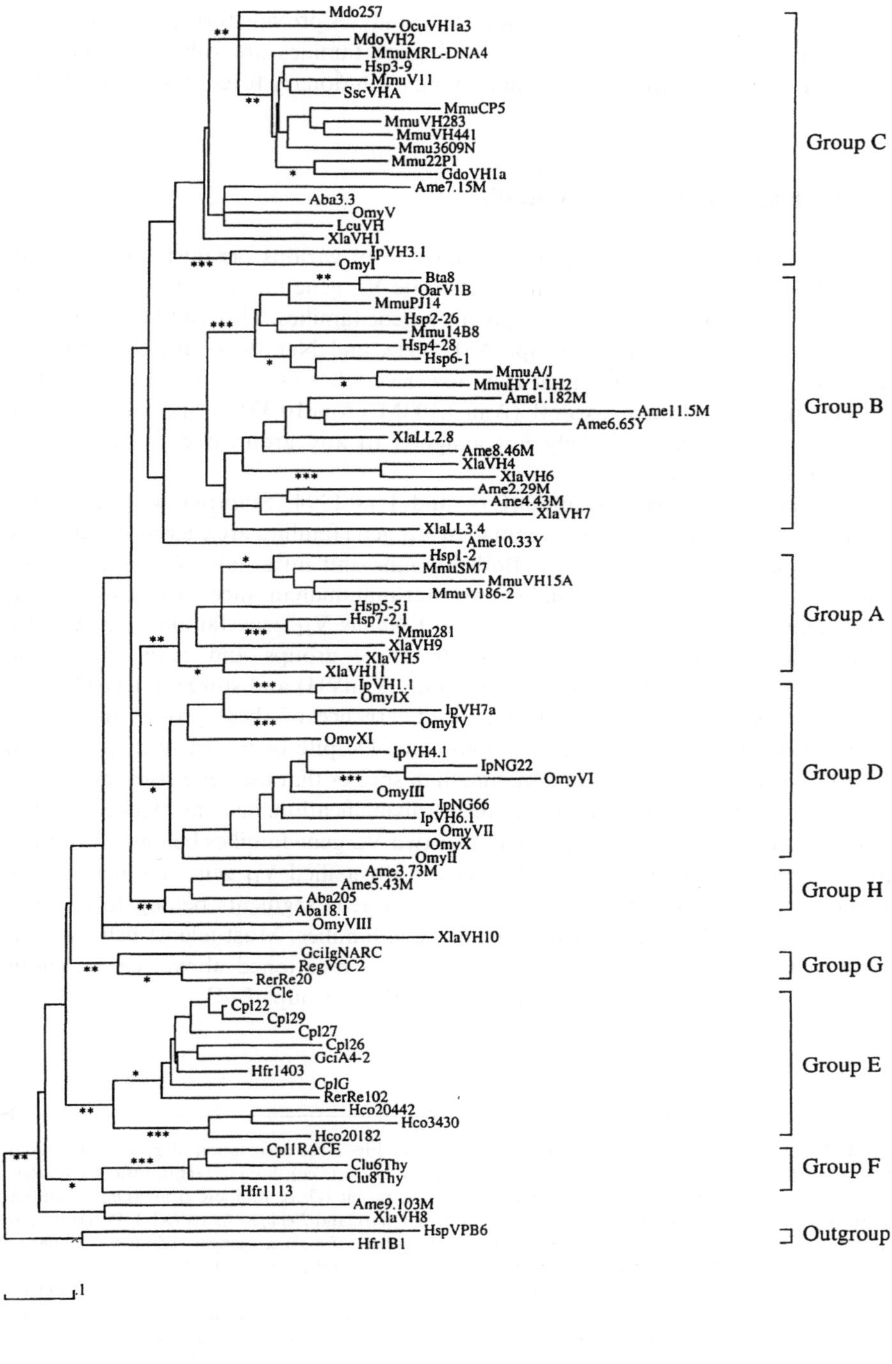
Mdo257
OcuVH1a3
MdoVH2
MmuMRL-DNA4
Hsp3-9
MmuV11
SscVHA
MmuCP5
MmuVH283
MmuVH441
Mmu3609N
Mmu22P1
GdoVH1a
Ame7.15M
Aba3.3
OmyV
LcuVH
XlaVH1
IpVH3.1
OmyI
Bta8
OarV1B
MmuPJ14
Hsp2-26
Mmu14B8
Hsp4-28
Hsp6-1
MmuA/J
MmuHY1-1H2
Ame1.182M
Ame11.5M
Ame6.65Y
XlaLL2.8
Ame8.46M
XlaVH4
XlaVH6
Ame2.29M
Ame4.43M
XlaVH7
XlaLL3.4
Ame10.33Y
Hsp1-2
MmuSM7
MmuVH15A
MmuV186-2
Hsp5-51
Hsp7-2.1
Mmu281
XlaVH9
XlaVH5
XlaVH11
IpVH1.1
OmyIX
IpVH7a
OmyIV
OmyXI
IpVH4.1
IpNG22
OmyVI
OmyIII
IpNG66
IpVH6.1
OmyVII
OmyX
OmyII
Ame3.73M
Ame5.43M
Aba205
Aba18.1
OmyVIII
XlaVH10
GciIgNARC
RegVCC2
RerRe20
Cle
Cpl22
Cpl29
Cpl27
Cpl26
GciA4-2
Hfr1403
CplG
RerRe102
Hco20442
Hco3430
Hco20182
Cpl1RACE
Clu6Thy
Clu8Thy
Hfr1113
Ame9.103M
XlaVH8
HspVPB6
Hfr1B1
Group C
Group B
Group A
Group D
Group H
Group G
Group E
Group F
Outgroup
.1

tionships are investigated in this section. The analyses provide further insight on the evolution of V gene segments, since multigene families are subject to different modes of evolution depending on underlying evolutionary forces (NEI et al. 1997).

5.1 Evolution of V_H Gene Families

A phylogenetic tree for 92 V_H gene families from various organisms is shown in Fig. 3. The V_H gene tree shows that vertebrate V_H gene families are classified into eight groups, A–H, and a few unassigned V_H gene families. Three additional groups were added to five previous groups A–E (OTA and NEI 1994) due to the recent accumulation of V_H gene segment from lower vertebrates. Two *Xenopus* (XlaVH8; XlaVH10), two Mexican axolotl (Ame9.103M, Ame10.33Y), and one rainbow trout (OmyVIII) are tentatively not assigned to any group due to inconsistent results among analyses.

As shown by earlier studies (OTA and NEI 1994; SCHROEDER et al. 1990; TUTTER and RIBLET 1989), the mammalian V_H gene families are classified into three groups A–C or three clans I–III. Both humans and mice have V_H gene families belonging to all three groups, indicating that mammalian ancestors possessed V_H gene segments of the three groups. Since *Xenopus* V_H gene families cluster with mammalian V_H gene families within the three groups and even teleost fish (IpuVH1.1, OmyVHI, OmyVHV), coelacanth (LcuVH) and sturgeon (Aba3.3) V_H gene families are found in group C, the divergence of these groups must have occurred before the emergence of amphibians. Despite of the early divergence of these three groups, the avian and mammalian species that use gene conversion and/or hypermutation as the primary source of diversification, such as chicken, rabbit, sheep, cattle, and pig, have only a single or two V_H gene families belong to group B and/or group C. Meanwhile, most extensive diversified V_H gene families are observed for amphibian species, where many V_H gene segments belong to the four groups A–C and H and a few unassigned gene families. Most teleost fish V_H gene families are classified into groups C and D. As mentioned above, cartilaginous fishes V_H gene families are classified into the three groups E–G.

Fig. 4. Neighbor-joining tree of 54 V_L gene segments. See Fig. 3 for designation of genes: cattle (*Bta*), channel catfish (*Ipu*), chicken (*Gga*), horned shark (*Hfr*), horse (*Eca*), human (*Hsp*), little skate (*Rer*), mouse (*Mmu*), muscoby duck (*Cmo*), nurse shark (*Gci*), rabbit (*Ocu*), rainbow trout (*Omy*), sandbar shark (*Cpl*), sheep (*Oar*), spotted ratfish (*Hco*), sturgeon (*Aba*), and *Xenopus laevis* (Xla). Please refer to HAIRE et al. (1996) and SITNIKOVA and SU (1998) for the source of most sequences. Additional sequences were retrieved from Genbank (accession number): Aba106.1(X90552), Btaλ2 (U32258), Ecak (X75611), IpuLF(U25705), and OmyL2(U69987). Cartilaginous fish type I V_L gene segments were used as outgroup instead of V_H gene segments to increase the number of amino acids used in the analysis. Note that the phylogenetic analysis including V_H gene segments (data not shown) and previous phylogenetic analyses (HAIRE et al. 1996; RAST et al. 1994) have shown that the cartilaginous fish type I V_L gene segments were the first groups diverged from the rest of V_L gene segments

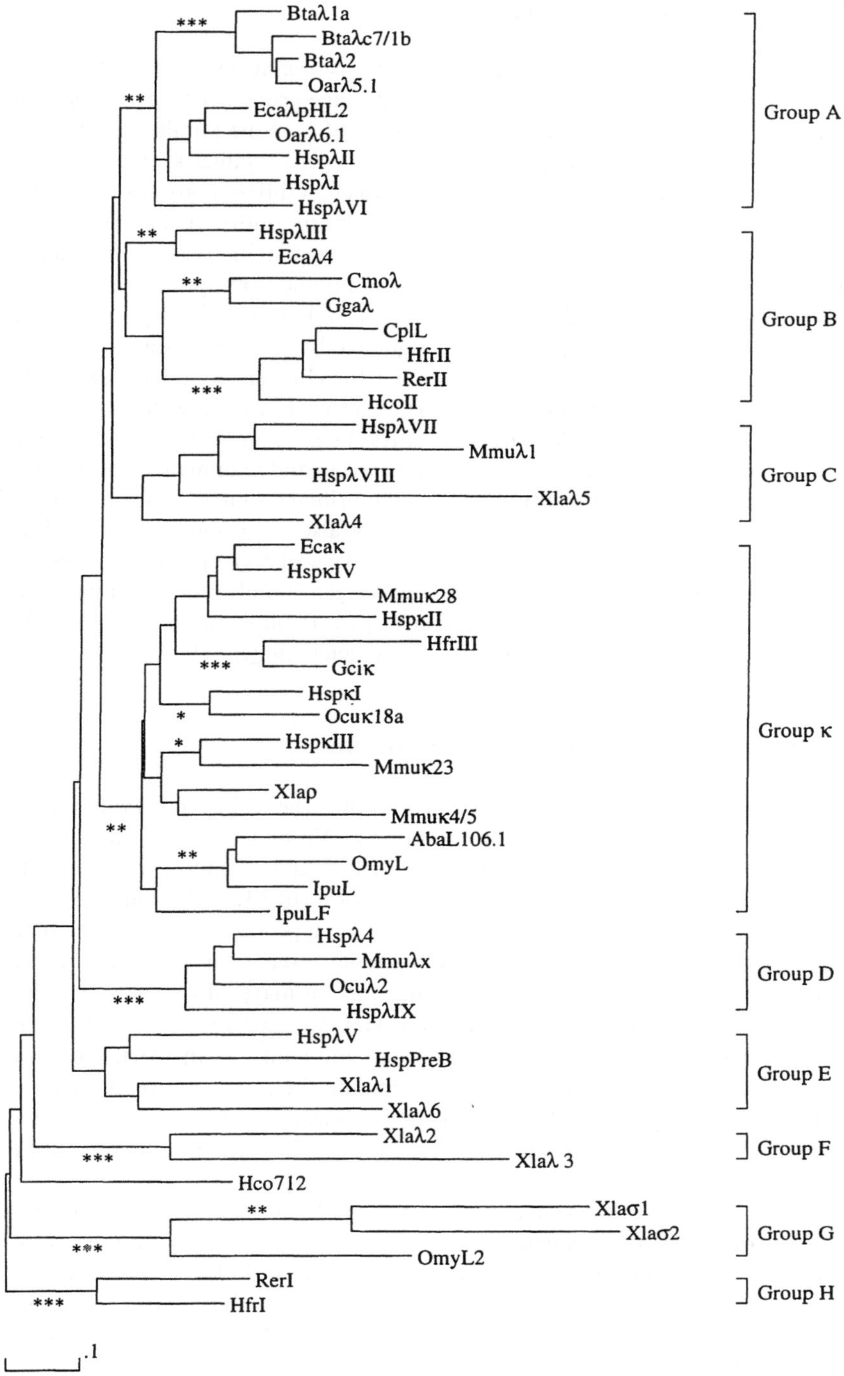

Btaλ1a
Btaλc7/1b
Btaλ2
Oarλ5.1
EcaλpHL2
Oarλ6.1
HspλII
HspλI
HspλVI
Group A
HspλIII
Ecaλ4
Cmoλ
Ggaλ
CplL
HfrII
RerII
HcoII
Group B
HspλVII
Mmuλ1
HspλVIII
Xlaλ5
Xlaλ4
Group C
Ecaκ
HspκIV
Mmuκ28
HspκII
HfrIII
Gciκ
HspκI
Ocuκ18a
HspκIII
Mmuκ23
Xlaρ
Mmuκ4/5
AbaL106.1
OmyL
IpuL
IpuLF
Group κ
Hspλ4
Mmuλx
Ocuλ2
HspλIX
Group D
HspλV
HspPreB
Xlaλ1
Xlaλ6
Group E
Xlaλ2
Xlaλ 3
Group F
Hco712
Xlaσ1
Xlaσ2
OmyL2
Group G
RerI
HfrI
Group H
.1

5.2 Evolution of V_L Gene Families

A phylogenetic tree for V_L gene families from various organisms is shown in Fig. 4. Here vertebrate V_L gene families are classified into nine groups and one unassigned V_L gene family. One monophyletic group consists of V_κ gene families from cartilaginous fishes to mammals. On the other hand, V_λ gene families are polyphyletic with respect to V_κ gene families and separated into eight different groups. These V_L groups diverged at an early stage of vertebrate evolution (more than 470MY ago), since both groups B and κ have V_L gene families from cartilaginous fish to mammals.

Among species studied here, human V_L gene families are by far the most extensively diverged and are classified into six groups, A–E and κ. One V_L gene family HspλX is not included because it may have originated by recombination (Sitnikova and Su 1998). Only one V_λ gene family known for chicken or rabbit belongs to group B or D, respectively. Rabbits are also known to use Igκ and so far a single V_κ gene family has been detected. All sheep and cattle V_λ gene families belong to group A. As in the case of V_H gene family evolution, the avian and mammalian species that use gene conversion and/or hypermutation as the primary source of diversification show limited diversity of V_L gene families (see below also). *Xenopus* V_L gene families are classified into the four groups C, F, G, and κ. Teleost fish V_L gene families are mostly of the V_k group with the exception of one V_L gene family reported in rainbow trout (Partula et al. 1996). Cartilaginous fish V_L gene families are classified into the three groups, B, H and κ, and the unassigned V_L gene segment Hco712.

5.3 Coevolution of V_H and V_L Gene Segments

From the phylogenetic analyses conducted above the following evolutionary patterns are commonly observed in the evolution of V_H and V_L gene segments. (a) One dominant V gene family or closely related V gene families are generally associated with each C isotype in cartilaginous fishes. (b) More diversified V gene families have been maintained under "translocon" type gene organization than "cluster" type gene organization for a given species, except for those that utilize gene conversion/hypermutation to generate the primary diversity (see also Table 3). (c) Amphibians, both *Xenopus* and Mexican axolotl, show one of the most extensive V gene family diversity. (d) Avian and mammalian species that use somatic gene conversion/hypermutations to generate the primary diversity have a reduced number of V gene families. (e) The major lineages of V gene families have been separated for more than 350MY for V_H gene families and more than 470MY for V_L gene families.

The above features of the coevolution of V_H and V_L gene segments are striking and likely attributable to some common evolutionary factors acting on the V gene family but not to random genetic events (Sitnikova and Su 1998). The next section reviews the change in diversity generation that occurred during the vertebrate evolution and examine the underlying evolutionary factors affecting the evolution of V gene family.

6 Evolution of Immune System and Ig Diversity

In terms of the generation of Ig diversity, three main changes have occurred during the vertebrate evolution: (a) the establishment of translocon gene organization, (b) the establishment of an organismal system for efficient clonal selection, and (c) the use of somatic gene conversion/hypermutation to generate the primary diversity.

The establishment of translocon gene organization likely occurred before the emergence of teleost fishes at IgH (LITMAN et al. 1993; MARCHALONIS et al. 1993). The establishment of translocon gene organization was very important since it would enable the combinatorial diversity and increase the diversity in CDR3 tremendously. Furthermore, it also has facilitated the accumulation of V gene family, since the V gene segments have evolved independently from the evolution of C gene segments and diversified V gene segments are more productively utilized to create Ig diversity by combinatorial diversity. It is still questionable if it occurred at IgL loci around the same time, since teleost fish Igκ locus has cluster type gene organization ($(V_{1,2}\text{-J-C})_n$), similar to cartilaginous fishes (MARCHALONIS et al. 1998 and references therein). The recent report by LUNDQVIST et al. (1998), however, suggests that the sturgeon has the translocon type gene organization at the IgL locus and raises a possibility that the teleost fish cluster type gene organization might have derived secondarily.

The lower vertebrates are known for relatively "poor" humoral immune response (DU PASQUIER 1980) despite the diversified V gene families seen among them. A high rate of somatic mutation is present in *Xenopus* and the poor immune response has been attributed to the lack of system to select B-cells that generate higher affinity Igs (WILSON et al. 1992). Germinal centers, where the clonal selection of B-cells occurs during secondary response in avian and mammalian species, are not found among lower vertebrates and the invention of such apparatus would have had significant impact on the generation of secondary Ig diversity. Specifically, it has solved one problem which "anticipatory" immune system encounters. In most vertebrates each B-cell expresses only one kind of Ig (allelic exclusion and isotypic exclusion). Consequently, the antigen recognition capacity is dictated not only by the ability to generate diversity but also by the number of cells to hold the diversity. By integrating efficient evolutionary systems, i.e., mutation and selection, into its somatic processes, B-cells can expand Ig repertoire temporarily, select B-cells producing high-affinity Igs and eliminate unnecessary B-cells under the limited resources in its secondary response.

Lastly, among many avian and mammalian species the mechanism to generate the primary repertoire by hypermutation and/or somatic gene conversion has been found (BUTLER 1997). In chickens, rabbits, cattle, and pigs somatic gene conversion and hypermutation play major roles in the primary diversification (KNIGHT 1992; LUCIER et al. 1998; McCORMACK 1993; PARNG et al. 1996; REYNAUD et al. 1994; SUN et al. 1996). In sheep hypermutation is the primary source of diversity (REYNAUD et al. 1991). In these organisms a few closely related V gene families are expressed (KNIGHT 1992; LOPEZ et al. 1998; REYNAUD et al. 1987, 1989). This is partly due to the fact that the gene rearrangement occurs only at the early stage of organismal development in some species (McCormack et al. 1993; Knight and Winstead 1997). However, genetic data provided so far have indicated the absence of a diverse array of V gene family in their genome. Since the various lineages of V gene families were present at the emergence of tetrapods (see Figs. 3, 4), a reduction in germline V gene repertoire must have occurred during their evolution. The loss of germline diversity have apparently occurred independently in several lineages of evolution (Nei et al. 1997; Sitnikova and Su 1998). For example, swine and sheep are phylogentically closely related to each other but swine have V_H gene families belonging to group C (Sun et al. 1994), whereas sheep have V_H gene families belonging to group B (Dufour et al. 1996). This can be explained only if ancestors of these organisms had V_H gene families belonging to groups B and C and sheep and swine lost or are losing one of them (Sitnikova and Su 1998). Consistent with this hypothesis, cattle, another artiodactyl species, have two gene families similar to mouse Q52 (group B) and X81 (group C) but express only the V_H gene family belonging to group B (Lopez et al. 1998). In these artiodactyl species, the utilization of somatic gene conversion or hypermutation for the generation of primary diversity has likely reduced the selection pressure to maintain diversified V gene families in germline. In this respect, it is interesting to note that only two V_H gene families belonging to group C are expressed and no other V_H gene family seems to be present in short-tailed opossum (Miller et al. 1998; see also Fig. 3). Since the marsupials are the most closely related animals to placental mammals, further characterization of their immune system will shed light on the evolution of somatic gene conversion and hypermutation system in placental mammals.

7 Summary

Evolution of Ig V gene segments are generally characterized by (a) evolution by "the birth and death process" and (b) diversifying selection. However, the detailed evolutionary pattern of V gene segments varies among species due to the fact that the humoral immune system itself has changed during vertebrate evolution. The change in somatic diversification system coupled with the change in lymphocyte development has imposed a significant impact on the evolution of Ig genes. In order to understand the evolution of immunological genes it is important to view it in the context of the evolution of the entire immune system itself.

Acknowledgements. We thank Chris T. Amemiya, David Garrity, and Su Chen for critical reading of the manuscript.

References

Al-Lazikani B, Lesk AM, Chothia C (1997) Standard conformations for the canonical structures and immunoglobulins. J Mol Biol 273:927–948

Alt F, Rosenberg N, Lewis S, Tomas E, Baltimore D (1981) Organization and reorganization of immunoglobulin genes in A-MuLV-transformed cells: rearrangement of heavy but not light chains genes. Cell 27:381–390

Anderson MK, Schamblott MJ, Litman RT, Litman GW (1995) Generation of immunoglobulin light chain gene diversity in *Raja erinacea* is not associated with somatic rearrangement, an exception to a central paradigm of B-cell immunity. J Exp Med 182:109–119

Arakawa H, Kuma K, Yasuda M, Furusawa S, Ekino S, Yamagishi H (1998) Oligoclonal development of B cells bearing discrete Ig chains in chicken single germinal centers. J Immunol 160:4232–4241

Barré S, Greenberg AS, Flajnik MF, Chothia C (1994) Structural conservation of hypervariable regions in immunoglobulin evolution. Nat Struct Biol 1:915–920

Benatar T, Ratcliffe MJ (1993) Polymorphism of the functional immunoglobulin variable region genes in the chicken by exchange of sequence with donor pseudogenes. Eur J Immunol 23:2448–2453

Berek C (1998) Affinity Maturation. In: Paul WE (ed) Fundamental immunology, 4th edn. Lippincott-Raven, Philadelphia, pp 862–878

Bernstein R, Schluter SF, Shen S, Marchalonis JJ (1996) A new high molecular weight immunoglobulin class from the carcharhine shark: implications for the properties of the primordial immunoglobulin. Proc Natl Acad Sci USA 93:3289–3293

Blanden RJ, Steele EJ (1998) A unifying hypothesis for the molecular mechanism of somatic mutation and gene conversion in rearranged immunoglobulin variable genes. Immunol Cell Bio 76:288–293

Brodeur PH, Riblet R (1984) The immunoglobulin heavy chain variable region (Igh-V) locus in the mouse. I. One hundred IgH-V gene comprise seven families of homologous genes. Eur J Immunol 14:922–930

Butler JE (1997) Immunoglobulin gene organization and the mechanism of repertoire development. Scand J Immunol 45:455–462

Chothia C, Lesk AM (1987) Canonical structures for the hypervariable regions of immunoglobulins. J Mol Biol 196:901–917

Chung GH, Scott MG, Kim KH, Kearnery J, Siber GR, Anbrosino DM, Nahm MH (1993) Clonal characterization of the human IgG antibody repertoire to Haemophilus influenzae type b polysaccharide. V. In vivo expression of individual antibody clones is independent on Ig CH haplotypes and the categories of antigen. J Immunol 151:4352–4361

Coleclough C, Perry RP, Karjalaninen K, Weigert M (1981) Aberrant rearrangements contribute significantly to the allelic exclusion of immunoglobulin gene expression. Nature 290:372–378

Corbett S, Tomlinson I, Sonnhammer E, Buck D, Winter G (1997) Sequence of the human immunoglobulin diversity (D) segment locus: a systematic analysis provides no evidence for the use of DIR segments, inverted D segments, "mirror" D segments or D-D recombination. J Mol Biol 271:587–597

DeFranco AL (1998) B-lymphocyte activation. In: Paul WE (ed) Fundamental Immunology, 4th edn. Lippincott-Raven, Philadelphia, pp 225–261

Dufour LV, Malinge S, Nau F (1996) The sheep Ig variable region repertoire consists of a single V_H family. J Immunol 156:2163–2170

Du Pasquier L (1982) Antibody diversity in lower vertebrates – why is it so restricted? Nature 296: 311–313

Du Pasquier L, Flajnik M (1998) Origin and evolution of the vertebrate immune system. In: Paul WE (ed) Fundamental immunology, 4th edn. Lippincott-Raven, Philadelphia, pp 605–650

Feeney AJ, Atkinson MJ, Cowan MJ, Escuro G, Lugo G (1996) A defective V_κ A2 allele in Navajos which may play a role in increased susceptibility to *Haemophilus influenzae* type b disease. J Clin Invest 97:2277–2282

Frazer JK, Capra JD (1998) Immunoglobulins: Structure and Function. In: Paul WE (ed) Fundamental immunology, 4th edn. Lippincott-Raven, Philadelphia, pp 37–74

Frippiat JP, Williams SC, Tomlinson IM, Cook GP, Cherif D, Le-Paslier D, Colliins JE, Dunham I, Winter G, Lefranc MP (1995) Organization of the human immunoglobulin lambda light-chain locus on chromosome 22q11.2. Hum Mol Genet 4:983–991

Frippiat JP, Dard P, Marah S, Winter G, Lefranc MP (1997) Immunoglobulin lambda light chain orphans on human chromosome 8q11.2. Eur J Immunol 27:1260–1265

Harding F, Amemiya CT, Litman RT, Cohen N, Litman GW (1990) Two distinct immunoglobulin heavy chain isotypes in a primitive, cartilaginous fish, *Raja erinacea*. Nuc Acids Res 18:6369–6376
Hinds-Frey K, Nishikata H, Litman RT, Litman GW (1993) Somatic variation precedes extensive diversification of germline sequences and combinatorial joining in the evolution of immunoglobulin heavy chain diversity. J Exp Med 178:815–824
Hohman VS, Schuchman DB, Schluter SF, Marchalonis JJ (1993) Genomic clone for sandbar shark λ light chain: generation of diversity in the absence of gene rearrangement. Proc Natl Acad Sci USA 90:9882–9886
Hohman VS, Schlter SF, Machalonis JJ (1995) Diversity of Ig light chain clusters in the sandbar shark (*Carcharhinus plumbeus*). J Immunol 155:3922–3928
Hsu E (1998) Mutation, selection, and memory in B lymphocytes of exothermic vertebrates. Immunol Rev 162:25–36
Ikematsu W, Harindranath N, Ueki Y, Notkins AL, Casali P (1993) Clonal analysis of human antibody response. II. Sequence of the V_H genes of human IgM, IgG, and IgA to rabies virus reveal preferential utilization of V_H3 segments and somatic hypermutation. J Immunol 150:1325–1337
Ikematsu W, Kobarg J, Ikematsu H, Ichiyoshi Y, Casali P (1998) Clonal analysis of a human antibody response. III. Nucleotide sequences of monoclonal IgM, IgG, and IgA to rabies virus reveal restricted V_κ gene utilization, junctional $V_\kappa J_\kappa$ and $V_\lambda J_\lambda$ diversity, and somatic hypermutation. J Immunol 161:2895–2905
Jeurissen SH, Janse EM, van Rooijen N, Claassen E (1998) Inadequate anti-polysaccharide antibody responses in the chicken. Immunobiol 198:385–395
Jolly C, Wagner SD, Rada C, Klix N, Milstein C, Neuberger MS (1996) The target of somatic mutation. Seminar Immunol 8:159–168
Kawasaki K, Minoshima S, Nakato E, Shibuya K, Shintani A, Schmeits JL, Wang J, Shimizu N (1997) One-megabase sequence analysis of the human immunoglobulin lambda gene locus. Genome Res 7:250–261
Kepler TB (1997) Codon bias and plasticity in immunoglobulins. Mol Biol Evol 14:637–643
Kirkham PM, Mortani F, Newton JA, Schroeder HW Jr (1992) Immunoglobulin V_H clan and family identity predicts variable domain structure and may influences antigen binding. EMBO J 11:603–609
Knight LK (1992) Restricted V_H gene usage and generation of antibody diversity in rabbit. Ann Rev Immunol 10:593–616
Knight LK, Winstead CR (1997) Generation of antibody diversity in rabbits. Curr Opin Immunol 9: 228–232
Kokubu F, Litman R, Shamblott MJ, Hinds K, Litman GW (1988a) Diverse organization of immunoglobulin V_H gene loci in a primitive vertebrate. EMBO J 7:3413–3422
Kokubu F, Hinds K, Litman R, Shamblott MJ, Litman GW (1988b) Complete structure and organization of immunoglobulin heavy chain constant region genes in a phylogenetically primitive vertebrate. EMBO J 7:1979–1988
Li WH, Sadler LA (1991) Low nucleotide diversity in man. Genetics 129:513–523
Litman GW, Rast JP, Schamblott MJ, Haire RN, Hulst M, Roess W, Litman RT, Hinds-Frey KR, Zilch A, Amemiya CT (1993) Phylogenetic diversification of immunoglobulin genes and the antibody repertoire. Mol Biol Evol 10:60–72
Lopez O, Perez C, Wylie D (1998) A single VH family and long CDRs are the target for hypermutation in bovine immunoglobulin heavy chains. Immunol Rev 162:55–66
Lucier MR, Thompson RE, Waire J, Lin AW, Osborne BA, Goldsby RA (1998) Multiple sites of Vλ diversification in cattle. J Immunol 161:5438–5444
Lundqvist M, Bengtén E, Strömberg S, Pilström L (1996) Ig light chain gene in the Siberian sturgeon (*Acipenser barei*): implications for the evolution of the immune system. J Immunol 157:2031–2038
Machalonis JJ, Schluter SF, Bernstein RM, Shen S, Edmundson AB (1998) Phylogenetic emergence and molecular evolution of the immunoglobulin family. Adv Immunol 70:417–506
Mansikka A, Toivanen P (1991) D-D recombination diversifies the CDR 3 region of chicken immunoglobulin heavy chains. Scand J Immunol 33:543–548
Matsuda F, Honjo T (1996) Organization of the human immunoglobulin heavy chain locus. Adv Immunol 62:1–29
Matsuda F, Ishii K, Bourvagnet P, Kuma K, Hayashida H, Miyata T, Honjo T (1998) The complete nucleotide sequence of the human immunoglobulin heavy chain variable region locus. J Exp Med 188:2151–2162

Matsunaga T, Andersson E (1997) Analysis of VH gene diversity in rainbow trout (*Oncorhynchus mykiss*): both nonsynonymous and synonymous nucleotide changes are more frequent in CDRs than in FRs. Immunogenetics 45:201–208

Max EE (1998) Immunoglobulins: Molecular Genetics. In: Paul WE (ed) Fundamental immunology, 4th edn. Lippincott-Raven, Philadelphia, pp 111–182

McCormack WT, Tjoelker LW, Thompson CB (1991) Avian B-cell development: generation of an immunoglobulin repertoire by gene conversion. Annu Rev Immunol 9:219–241

McCormack WT, Hurley EA, Thompson CB (1993) Germ line maintenance of the pseudogene donor pool for somatic immunoglobulin gene conversion in chickens. Mol Cell Biol 13:821–830

Melcher F, Rolink A (1998) B-lymphocyte development and biology. In: Paul WE (ed) Fundamental immunology, 4th edn. Lippincott-Raven, Philadelphia, pp 183–224

Michael N, Accavitti MA, Masteller E, Thompson CB (1998) The antigen-binding characteristics of mAbs derived from *in vivo* priming of avian B-cells. Proc Natl Acad Sci USA 95:1166–1171

Miller RD, Grade H, Resenberg GH (1998) V_H repertoire of a marsupial (*Mondelphis domestica*). J Immunol 160:259–265

Mond JJ, Lees A, Snapper CM (1995) T-cell independent antigens type 2. Ann Rev Immunol 13:655–692

Nadel B, Tang A, Lugo G, Love V, Escuro G, Feeney AJ (1998) Decreased frequency of rearrangement due to the synergistic effect of nucleotide changes in the heptamer and nonamer of the recombination signal sequence of the V_κ gene A2b, which is associated with increased susceptibility of Navajos to *Haemophilus influenza* Type b disease. J Immunol 161:6068–6073

Nagaoka H, Ozawa K, Matsuda F, Hayashida H, Matsumura R, Haino M, Shin EK, Fukita Y, Imai T, Anand R, Yokoyama K, Eki T, Soeda E, Honjo T (1994) Recent translocation of variable and diversity segments of the human immunoglobulin heavy chain from chromosome 14 to chromosome 15 and 16. Genomics 22:189–197

Nei M, Gu X, Sitnikova T (1997) Evolution by the birth-and-death process in multigene families of the vertebrate immune system. Proc Natl Acad Sci USA 94:7799–7806

Ota T (1997) Molecular evolution of immunoglobulin V_H segments. In: Herzenberg LA, Weir DM, Herzenberg LA, Blackwell C (eds) Weir's handbook of experimental immunology, 5th edn., vol. 1: immunochemistry and molecular immunology. Blackwell Scientific Press, Oxford, pp 8.1–8.8

Ota T, Nei M (1994) Divergent evolution and evolution by the birth and death process in the immunoglobulin VH gene family. Mol Biol Evol 11:469–482

Ota T, Nei M (1995) Evolution of immunoglobulin VH pseudogenes in chickens. Mol Biol Evol 12: 94–102

Ota T, Margittai M, Rast JP, Litman RT, Litman GW, Amemiya CT (1995) Studies of the early evolution of immunoglobulin genes. In: Nei M, Takahata N (eds) Current topics of molecular evolution. The Pennsylvania State University, University Park

Pargent W, Schäble KF, Zachau HG (1991) Polymorphism and haplotypes in the human immunoglobulin κ locus. Eur J Immunol 21:1829–1835

Parng CL, Hansal S, Goldsby RA, Osborne BA (1996) Gene conversion contributes to Ig light chain diversity in cattle. J Immunol 157:5478–5486

Parvari R, Ziv E, Lantner F, Heller D, Schechter I (1990) Somatic diversification of chicken immunoglobulin light chains by point mutations. Proc Natl Acad Sci USA 87:3072–3076

Pastula S, Schwager J, Timmusk S, Pilström L, Charlemagne J (1996) A second immunoglobulin light chain isotype in the rainbow trout. Immunogenetics 45:44–51

Picker LJ, Siegelman MH (1998) Lymphoid tissue and organ. In: Paul WE (ed) Fundamental immunology, 4th edn. Lippincott-Raven, Philadelphia, pp 479–531

Queiroz RG, Carrier A, Victorero G, Jordan B, Passos GA Jr (1997) Chromosomal location of the human immunoglobulin lambda variable 8 (IGLV8) gene family outside the major lambda locus on chromosome 22q11.2. Immunol Lett 59:177–180

Rast JP, Amemiya CT, Litman RT, Strong SJ, Litman GW (1998) Distinct patterns of IgH structure and organization on a divergent lineage of chondrichthyan fishes. Immunogenetics 47:234–245

Rast JP, Anderson MK, Ota T, Litman R, Margittai M, Schamblott MJ, Litman GW (1994) Immunoglobulin light chain class multiplicity and alternative organizational forms in early vertebrate phylogeny. Immunogenetics. 40:83–99

Ravetch JV, Siebenlist U, Korsmeyer SJ, Waldmann T, Leder P (1981) Structure of the human immunoglobulin mu locus: characterization of embryonic and rearranged J and D genes. Cell 27: 583–591

Reynaud CA, Anquez V, Grimal H, Weill JC (1987) A hyperconversion mechanism generates the chicken light chain pre-immune repertoire. Cell 48:379–388

Reynaud CA, Dahan A, Anquez V, Weill JC (1989) Somatic hyperconversion diversifies the single VH gene of the chicken with a high incidence in the D region. Cell 59:171–183
Reynaud CA, Mackay CR, Müller RG, Weill JC (1991) Somatic generation of diversity in mammalian primary lymphoid organ: the sheep Ileal Peyer's patches. Cell 64:995–1005
Reynaud CA, Bertocci B, Dahan A, Weill JC (1994) Formation of the chicken B-cell repertoire: ontogenesis, regulation of Ig gene rearrangement, and diversification by gene conversion. Adv Immunol 57:353–378
Roman T, Andersson E, Bengtén E, Hansen J, Kaattari S, Pilström L, Charlemage J, Matsunaga T (1996) Unified nomenclature of Ig VH genes in rainbow trout (*Oncorhynchus mykiss*): definition of eleven VH families. Immunogenetics 43:325–326
Saitou N, Nei M (1986) The neighbor-joining method: a new method for reconstructing phylogenetic trees. Mol Biol Evol 4:406–425
Schäble KF, Zachau HG (1993) The variable genes of the human immunoglobulin k locus. Biol Chem Hoppe Seyer 374:1001–1022
Schaible G, Rappold GA, Pargent W, Zachau HG (1993) The immunoglobulin kappa locus: polymorphism and haplotypes of Caucasoid and non-Caucasoid individuals. Hum Genet 91:261–267
Schroeder HW Jr, Hillson JL, Permutter RM (1990) Structure and evolution of mammalian V_H families. Int Immunol 20:41–50
Scott MG, Nahm MH (1992) Characterization of the human IgG antibody VL repertoire to *Haemophilus influenza* type b polysaccharide. J Infect Dis 65:S53–S56
Scott MG, Tarrand JJ, Crimmins DL, McCourt DW, Siegel NR, Smith CE, Nahm MH (1989) Clonal characterization of the human IgG antibody repertoire to *H. influenza* type b polysaccharide. II. IgG antibodies contain V_H genes from a single V_H gene family and V_L genes from at least four V_L gene families. J Immunol 143:292–298
Shen SX, Bernstein RM, Schluter SF, Marchalonis JJ (1996) Heavy-chain variable regions in carcharhine sharks: development of a comprehensive model for the evolution of VH domains among the gnathanstomes. Immunol Cell Biol 74:357–364
Sitnikova T, Nei M (1998) Evolution of immunoglobulin kappa chain variable region genes in vertebrates. Mol Biol Evol 15:50–60
Sitnikova T, Su C (1998) Coevolution of immunoglobulin heavy- and light-chain variable region gene families. Mol Biol Evol 15:617–626
Sun J, Kackovics I, Brown WR, Butler JE (1994) Expressed swine V_H genes belong to a small V_H gene family homologous to human V_HIII. J Immunol 153:5618–5627
Tanaka T, Nei M (1989) Positive Darwinian selection observed at the variable-region genes of immunoglobulins. Mol Biol Evol 6:447–459
Tonegawa S (1983) Somatic generation of antibody diversity. Nature 302:575–581
Tomlinson IM, Cook GP (1997) Human immunoglobulin variable region genes. In: Herzenberg LA, Weir DM, Herzenberg LA, Blackwell C (eds) Weir's handbook of experimental immunology, 5th edn., vol. 1. Immunochemistry and molecular immunology. Blackwell Scientific Press, Oxford, pp 4.1–4.8
Tomlinson IM, Cook GP, Carter NP, Elaswarapu R, Smith S, Walter G, Buluwela L, Rabbitts TH, Winter G (1994) Human immunoglobulin V_H and D segments on chromosomes 15q11.2 and 16p11.2. Hum Mol Genet 3:853–860
Tomlinson IM, Williams SC, Ignatovich O, Corbett SJ, Winter G (1996a) VBASE sequence directory. MRC Center for Protein Engineering, Cambridge
Tomlinson IM, Waler G, Jones PT, Dear PH, Sonnhammer ELL, Winter G (1996b) The imprint of somatic hypermutation on the repertoire of human germline V genes. J Mol Biol 256:813–817
Tutter A, Riblet R (1989) Conservation of an immunoglobulin variable-region gene family indicates a specific, noncoding function. Proc Natl Acad Sci USA 86:7460–7464
Vargas-Madrazro E, Lara-Ochoa F, Almagro JC (1995) Canonical structure repertoire of the antigen-binding site of immunoglobulins suggests strong geometrical restrictions associated to the mechanism of immune recognition. J Mol Biol 254:497–504
Vargas-Madrazo E, Lara-Ochoa F, Ramirez-benites M, Almagro JC (1997) Evolution of the structural repertoire of the human V_H and V_k germline genes. Int Immunol 9:1801–1815
Webster DM, Henry AH, Rees AR (1994) Antibody-antigen interactions. Curr Opin Struct Biol 4: 123–129
Williams SC, Frippiat JP, Tomlinson IM, Ignatovich O, Lefranc MP, Winter G (1996) Sequence and evolution of the human germline V_λ repertoire. J Mol Biol 264:220–232

Wilson M, Hsu E, Marcuz A, Courtet M, Du Pascuier L, Steinberg C (1992) What limits affinity maturation of antibodies in Xenopus – the rate of somatic mutation or the ability to select mutants? EMBO J 11:4337–4347

Wu TT, Kabat EA (1970) An analysis of the sequences of the variable regions of Bence-Jones proteins and myeloma light chains and their implications for antibody complementarity. J Exp Med 132: 211–250

Zachau HG (1995) The human immunoglobulin κ genes. In: Honjo T, Alt F (eds) Immunoglobulin genes. 2nd Edn. Academic Press, San Diego

Elasmobranchs

The Immune System of Cartilaginous Fish

M.F. FLAJNIK and L.L. RUMFELT

1 Introduction

Chondrichthyes, or cartilaginous fish, are a highly successful group of vertebrates. Most extant species are long-lived, display an amazing diversity, and fill all predatory oceanic niches (and a few other niches as well; WILSON 1992). For our purposes, they are the oldest group of vertebrates shown to possess an adaptive immune system grounded on immunoglobulins (Ig), T cell receptors (TCR), and major histocompatibility complex (MHC) class I and class II molecules (LITMAN et al. 1999). The examination of these molecules and genes, the mechanisms of

Department of Microbiology and Immunology, University of Maryland School of Medicine, 655 West Baltimore Street, Baltimore, MD 21201, USA

recombination and the somatic mutation of antigen receptor genes, as well as the manner by which immune responses occur in secondary (and other?) lymphoid tissues, in comparison to vertebrates of other classes, allow us to develop paradigms to illustrate the immune system in the common ancestor of all jawed vertebrates.

Prior to the discovery of these genes, molecules, and mechanisms it was believed by many investigators that the building blocks of the cartilaginous fish adaptive immune system were somehow simple; this is clearly *not* true and we must reexamine immune responses in light of the molecular diversity of this system. There have been several recent comprehensive reviews of the cartilaginous fish immune system (Schluter et al. 1997; Marchalonis et al. 1998a,b; Litman et al. 1999); here we concentrate and call on those aspects that are relatively new and/or not well understood for future studies.

2 The Creatures

The cartilaginous fish arose about 450 million years ago from a jawed vertebrate ancestor related to the placoderms (Carroll 1988). Only one group, the holocephalins (e.g., the ratfish), survived from the early emergence of that class. A second evolutionary wave occurred about 250 million years ago, giving rise to the "modern" sharks, skates, and rays (Fig. 1).

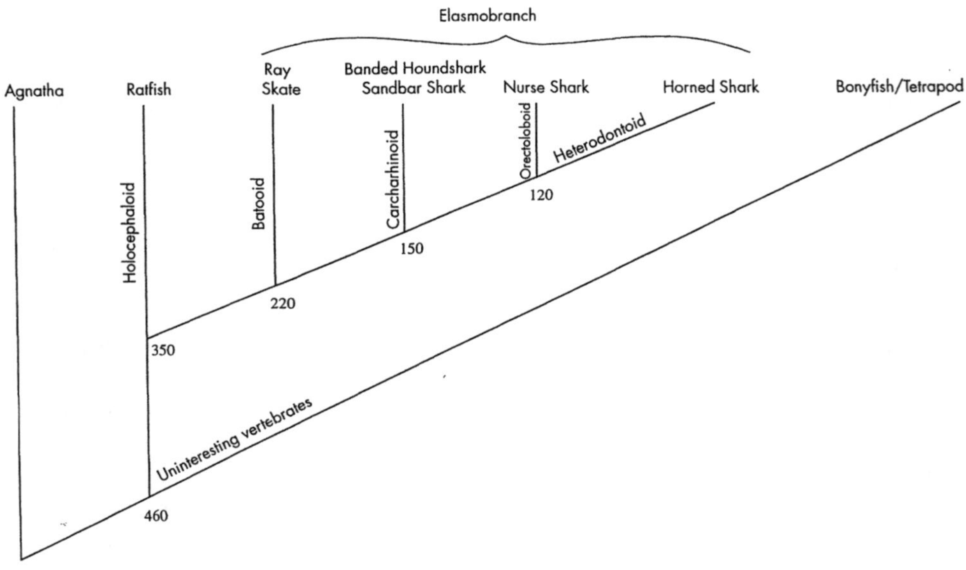

Fig. 1. Phylogenetic relationships among the major taxa of cartilaginous fish for which the immune system has been examined at the molecular and/or functional levels

Reproductively, all cartilaginous fish produce relatively few embryos (Carroll 1988; Matthews 1996). In some species egg cases are laid on the ocean floor from which the embryos hatch; in other cases the egg cases are maintained internally (ovoviviparous) where the embryos hatch and live for several weeks before birth; and finally some species show true viviparity with "placental" attachments. There are, in addition, bizarre manners of reproduction: the tiger shark mother produces only two living young, these two having consumed all of their brothers and sisters in utero! Generally, then, the selection is for the production of few young by each mother, a general rule for long-lived predators. Such a reproductive strategy has probably selected for efficient immune responses within this taxon, and may even have been one of the forces behind the emergence of the adaptive immune system (Flajnik 1998).

3 The Lymphoid Tissues

All examined species have a thymus and spleen, as those found in most other jawed vertebrates (Good et al. 1966; Zapata et al. 1996). In some species the thymus can be maintained into old age, perhaps not surprising for animals that continue to grow for many years. At one time it was thought that the thymus might not be fully formed (and by inference, functional) in this group, but this notion was dismissed by the finding of Luer et al. (1995) that typical cortices and medullae are found in the thymi of every species studied. In addition, expression of terminal deoxynucleotidyl transferase (TdT) and the peanut agglutinin receptor by cortical lymphocytes is also consistent with studies of thymus in other vertebrates. Considering that polymorphic class I and class II molecules are expressed by sharks (and by inference all cartilaginous fish), it is expected that positive and negative selection of T cells must occur in the thymus as is found in all other vertebrates.

The spleen of cartilaginous fish is highly vascularized and can be extremely large. It is composed of white and red pulp, as in other vertebrates, but no distinct marginal zones or periarteriolar lymphatic sheaths with central arterioles are apparent (reviewed in Zapata et al. 1996; see also Zapata and Amemiya, this volume). Follicular dendritic cells appear to be absent, although it is presumed (with no functional experiments to date) that antigen presenting cells of some type (melanomacrophages?) are definitely present. This organ is described in somewhat more detail in the section on immune responses. The intestine contains cells that secrete Ig, and accumulations of ill-defined lymphocytes are found in the intestinal lining (Tomonaga et al. 1986).

There are, in addition, immune tissues without a prototypic organization specific to the cartilaginous fish. Physically associated with the gonad is the epigonal organ, filled mainly with granulocytes but also with some IgM-positive lymphocytes. The Leydig organ is found attached to the esophagus in some species, and its function may be similar to the epigonal since some species may have only one tissue or the other (reviewed in ZAPATA et al. 1996). There are also accumulations of lymphocytes and other leukocytes in the meninges of the brain that have been proposed to be important in the protection of nervous tissue (CHIBA et al. 1989). In addition to the morphology and presence of Ig-positive cells in these tissues, we know next to nothing of the functions of cells residing there.

One study examined the ontogenic appearance of Ig-positive cells (with an antiserum, not a monoclonal antibody) in the dogfish (LLOYD-EVANS 1993). The first cells to express Ig occur in the liver, followed by the kidney and then the spleen and in other tissues mentioned above such as the Leydig and epigonal organs. Development of the peripheralized B cell system in the spleen is correlated with the time of hatching when the shark pups become exposed to seawater in the uterus. There are no published studies of recombination-activating gene expression to support these findings; thus it is also not known whether cells continue to rearrange their Ig genes throughout life or do so only in early development.

4 The Molecules and Genes

4.1 IgM

Ig was first discovered as bona fide IgM by MARCHALONIS and EDELMAN (1965, 1966) in the dogfish shark in two forms, 19-S (pentamer?) and a 7-S (monomer, Fig. 2). Short-term immunizations suggested that IgM is the predominant, if not the only, type of Ig in sharks. Further work by Clem in the lemon (CLEM and SMALL 1967) and nurse (CLEM et al. 1967) sharks replicated the dogfish data and carried the research forward with long-term immunizations to several different classes of antigens (see below). Clem's studies also demonstrated that IgM makes up at least half of the serum protein in the blood, perhaps a complicating factor when specific immune responses are examined.

In contrast to mammals, in which monomeric IgM is present in very low amounts in the serum, the 7-S form in sharks is a major serum protein in most species. Injection of radiolabeled 19-S IgM into the blood of sharks did not result in the subsequent appearance of labeled monomeric forms (and vice versa), suggesting that 7-S and 19-S IgM are made by different cells or at least are the result of different processing pathways (SMALL et al. 1970). However, the difference in polymerization is not likely to be caused by differences in the secretory tail since both forms seem to have this region (KLAPPER and CLEM 1977; Rumfelt and Flajnik, unpublished; Fig. 2). Thus it remains a major problem to understand how this polymerization is regulated. CLEM and colleagues suggested that 19S is the predominant intravascular form, with 7-S IgM being the form of Ig that equilibrates between intravascular and intervascular sites. Their idea may need to be broadened with the finding of at least two other monomeric, secreted Igs in elasmobranchs (see below).

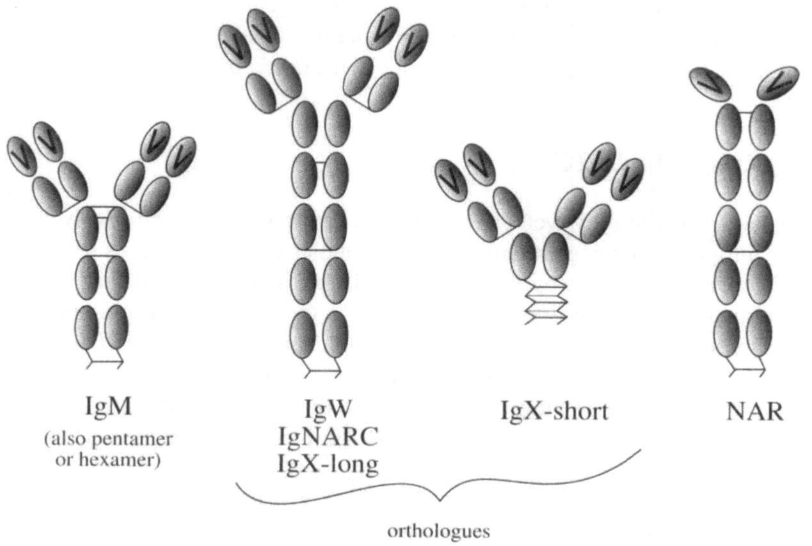

Fig. 2. The major Ig isotypes detected so far in cartilaginous fish. Interchain disulfide bonds are noted. The five bonds shown for the secretory tail of IgX-short are only proposed; it is not at all clear which cysteine residues form inter- vs. intrachain disulfide bridges in this region

Ig detected in shark pups is produced by the pups themselves, not transferred from the yolk (GITLIN et al. 1973). Plasma 19-S IgM is present in low amounts at the "birth" of nurse sharks, with 7-S IgM appearing approximately 1 week later (FIDLER et al. 1969). Similarly, in most cases (but not all, see below), antigen-specific 7-S IgM appears later than 19-S IgM during an antigen-specific response, suggesting regulation of polymerization in the course of a response, or selection of B cells that only make one type or another at different stages of the response (CLEM et al. 1967). One report suggested that antigen-specific 7-S IgM had a higher affinity than 19-S IgM after long-term immunizations (VOSS and SIGEL 1972).

IgM H chain genes of the horned shark were isolated by cross-hybridization with a mammalian V_H probe (LITMAN et al. 1985). These genes are not organized in a similar fashion to the mouse/human/*Xenopus*/teleost "translocon" prototype but instead in the well-known "cluster-type" organization with each cluster composed of one V, two D, one J, and one set of C exons (HINDS and LITMAN 1986). Rearrangement with accompanying N and P addition does not appear to occur between clusters (HINDS-FREY et al. 1993), and there may be as many as 200 clusters in the horned shark (KOKUBU et al. 1987, 1988a) and skate (not quantified in this case; HARDING et al. 1990a). The V genes do not make up families as in all other vertebrate groups examined but form only a single group (KOKUBU et al. 1988b; except for one unusual V_H group that allowed a study of somatic mutation, see below). All cartilaginous fish, including the ratfish (RAST et al. 1998) have all of their Ig genes in such a configuration. Finally, there are varying degrees (e.g.,

VD-DJ, VDD-J, V-DDJ, VDDJ) of "germline-joined" genes in all species examined (reviewed in LITMAN et al. 1999). This is a prevalent feature of cartilaginous fish Ig genes, the functional significance of which (if any) is unknown.

4.2 IgR/IgX/IgW/IgNARC

While IgM has been the only well-characterized isotype shown to be involved in responses to antigenic challenge, there have been reports of other isotypes found in the blood. The first report was by FULLER et al. (1978), who suggested that a light (L) chain-binding molecule of approximately 50kDa (coincident with mammalian IgG H chain by sodium dodecyl sulfate–polyacrylamide gel electrophoresis) was the predominant Ig found in the sera of nurse sharks immunized to bovine γ-globulin. This Ig was shown not to be simply a proteolytic product of IgM based on the different antigenic properties as compared to 7-S IgM. There have been no other reports of such an antigen-reactive molecule in any species.

In both the skate (several species) and the frill shark, believed by most investigators to be the most primitive extant shark, a 2nd light chain-binding molecule clearly has been found and is called IgR (for Ig*Raja*; KOBAYASHI et al. 1984, 1988, 1992). As with Fuller's molecule, the IgR heavy chain is also approximately 50kDa by sodium dodecyl sulfate–polyacrylamide gel electrophoresis and makes monomers (2H, 2L) and dimers. IgR-specific antisera were used to demonstrate that IgR is not expressed in the same lymphocytes that produce IgM in adult skates, but double-staining cells exist in the cells of embryos (KOBAYASHI et al. 1985). The response to antigen has not been described for this molecule. It was proposed that IgR might not be present in the majority of shark species (KOBAYASHI et al. 1988).

Another isotype, IgX, was molecularly cloned from the skate (HARDING et al. 1990b; Figs. 2, 3). Its deduced amino acid sequence suggests a three domain molecule, as with the short form of IgY in the duck (MAGOR et al. 1992). The sequence of the IgX secretory tail is not at all similar to the tails of IgM (or any other isotype), and it contains multiple cysteine residues (Fig. 2) as well. It was proposed, and is generally accepted (HARDING et al. 1990b), that the IgX gene encodes IgR, but this hypothesis still awaits verification with biochemical analyses. Indeed, as described above the molecular weight of the IgR heavy chain is suggestive of a four-domain molecule, as mammalian IgG, rather than the deduced three domain molecule encoded by IgX genes. Protein sequence is required to determine whether these indeed are the same molecules.

In the original northern blot with an IgX variable region probe in the skate, a high molecular weight, uncharacterized species also was detected (HARDING et al. 1990b). This species (ANDERSON et al. 1999) turned out to be the orthologue of another cDNA cloned from the sandbar shark, called IgW (BERNSTEIN et al. 1996), and the nurse shark, called Ig new antigen receptor (IgNARC) from cartilaginous fish (GREENBERG et al. 1996). This molecule, in the three divergent cartilaginous fish species, has one V followed by six C domains (Figs. 2, 3). The two deduced forms in

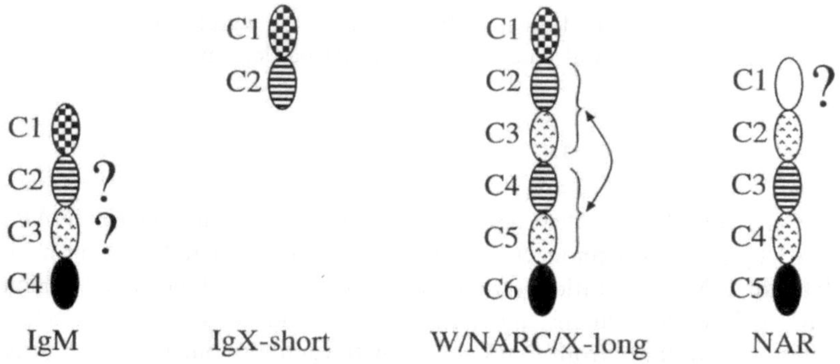

Fig. 3. Evolutionary relationships among the various constant domains in IgM, IgX, IgW, and NAR. The IgW C4/C5 are homologous to C2/C3 and were derived in one or two duplication events. IgM C2 and C3 may be homologous to the same domains of IgM (Anderson et al. 1999), but support for this in phylogenetic trees is controversial. The NAR C1 domain is difficult to place into trees, although one study suggests an IgX/W/NARC C2 homology (Schluter et al. 1997)

the skate are so similar in their first three domains, and even in noncoding regions within the genes, that they are either derived from very similar clusters, and the sequences are homogenized by gene conversion, or they are the products of alternative splicing (Fig. 3 and see below; Anderson et al. 1999). The secretory tail of the long form in all three species is similar to the IgM tail in size, sequence, and features (glycosylation site, the penultimate residue is a cysteine, Fig. 2). In the nurse shark we have isolated a putative NARC protein from metabolically labeled spleen cells both with a monoclonal antibody (mAb) specific for one of the L chains and with an antiserum specific for the NARC secretory tail; immunoprecipitations with another L chain-specific mAb did not immunoprecipitate the putative NARC, suggesting an L chain preference for this heavy chain (Greenberg et al. 1996). The protein seems to be found only at very low levels in nurse shark serum (Flajnik and Rumfelt, unpublished).

The NARC/W/X/R genes have been examined only in skates, and only for the short form (Harding et al. 1990b; Anderson et al. 1994). As with IgM, the IgX clusters have one V, two D, one J, and one set of C exons. Southern blotting analysis suggests a much lower complexity of genes in comparison to IgM. In nurse sharks there are only a few NARC (X) genes, with the same number of V- and C-hybridizing fragments detected by genomic Southern blotting (Greenberg et al. 1996). Transmembrane forms, either at the genomic or cDNA levels for the short and long forms, have not been described. Finally, the short form has not been identified in sharks, suggesting that insertion of a genetic element encoding the strange secretory tail took place after skates/rays diverged from sharks.

In summary, then, the short and long forms of these molecules in the skate (IgX-short and IgX-long, respectively), and the long forms in two shark species (IgW or IgNARC), are orthologues (Figs. 2, 3 and see below). It is not clear whether the IgX short form encodes the antibody described in frill shark and skates called IgR – we suggest that it does not. In any case, the function(s) and tissue distribution of these deduced and true molecules must be determined.

4.3 NAR

In the nurse shark a molecule called new, or nurse, shark antigen receptor (NAR) was discovered by immunoprecipitation with one of a panel of mAbs from mice immunized to an IgM preparation (Greenberg et al. 1995). One mAb, which apparently recognized Ig L chains under denaturing conditions, immunoprecipitated a disulfide-linked dimer from shark serum, with each monomer being slightly larger than IgH chains. Peptides sequenced from this protein fortuitously matched a cDNA clone obtained with a degenerate oligonucleotide primer that amplified an Ig/TCR-like cDNA. NAR monomers are composed of an N-terminal V domain and five C domains. The V domains do not form dimers but instead are free and flexible (Fig. 2). A pronounced "swing" can occur between the third and fourth constant domains in the vicinity of an interchain disulfide bond, which may have implications for the function (Roux et al. 1998). In addition, this could be a proteolytically sensitive site, explaining why the original mAb recognized a band in the L chain region on western blots (both the L chain and the proposed NAR proteolytic product have two Ig superfamily domains). Experiments must be set up to test this possibility.

NAR genes are also in the cluster-type organization, with very few genes being present in each species examined (Greenberg et al. 1995; Greenberg and Flajnik, unpublished). Each cluster contains one V, three D, one J and one set of C exons; thus four rearrangement events must occur within each cluster, with attendant N- and P-region additions and oligonucleotide-capture events, to produce a functional gene (Roux et al. 1998). One of the four identified clusters/haploid genome contains two germline-joined D regions, but this gene has not been shown to be expressed (Greenberg and Flajnik, unpublished).

The biggest surprise to us when we began experiments with NAR was the high level of somatic mutation detected in cDNA clones (Greenberg et al. 1995; Du Pasquier et al. 1998; Diaz et al. 1998). The mutation frequency, especially from clones derived from the spleen, is very high, reaching levels found at the upper limit for mammalian B cells (Betz et al. 1993). Hallmarks of somatic mutation including hotspots, strand preference, and complementarity-determining region (CDR) clustering are all similar in NAR and in mammalian Ig genes (Diaz et al. 1999); these features are in contrast to what has been found for mutation of shark (and frog; Wilson et al. 1992) IgM H (Hinds-Frey et al. 1993) and skate L (Anderson et al. 1995) chain genes, where there is a GC bias (Du Pasquier et al. 1998), rather low levels of mutation, and poor CDR clustering. One prominent characteristic specific to mutation of the NAR V gene is the apparent insertion of tandem nucleotides during the mutation process, often associated with palindromic sequences, oligonucleotide runs, and other secondary structures (Diaz et al. 1999). The mutations do not appear in the *primary* NAR repertoire but are likely to be the result of an antigen-driven process (DIAZ et al. 1998). Thus we believe that the entire primary repertoire is generated in NAR CDR3 via the rearrangement events, and somatic mutation occurs only when cells come into contact with antigen. Indeed, at face value NAR seems to offer a stripped down approach to adaptive immunity.

4.4 Relationships Among the Genes

IgM and IgW (NARC, IgX) both have V regions with canonical L chain-interacting residues and thus are predicted to associate with L chain V regions. In addition, the canonical framework (FR)2 sequences found in all V_H genes are present in both genes. Nevertheless, the two V families are divergent enough in sequence to form their own clusters in phylogenetic trees (MARCHALONIS et al. 1998b). The C1 domains in both IgM and IgW have a cysteine in position to make the disulfide bridge with the invariant cysteine in L chains. Indeed, as mentioned above, we have obtained biochemical evidence that the IgW H chain associates with L chains (GREENBERG et al. 1996). The common ancestor of IgM and IgW predates (at least) the common ancestor of all of the cartilaginous fish (Fig. 3; SCHLUTER et al. 1997; ANDERSON et al. 1999).

The NAR V region is unusually small, stemming from a tiny CDR2/FR2 region. The sequence of the entire V, especially strands A–D, is more similar to L chain and TCR V sequences than to V_H genes, a situation suggesting that either (a) the NAR V domain is very old, perhaps derived from a V domain in a molecule predating the split between TCR and Ig or between IgH and L chains (GREENBERG et al. 1995), or (b) that the NAR V is in a "runaway" IgW(NARC) cluster that changed rapidly once the protein lost its association with L chains (SCHLUTER et al. 1997; KLEIN 1998). The three D regions are included in almost all NAR rearrangements, with extensive N- and P-region additions. In the two types of NAR so far described different combinations of noncanonical cysteine residues are found that likely form bridges stabilizing the single domain (ROUX et al. 1998); two of these disulfide bridges are also found (convergently) in the single domain camel IgG V_H (DESMYTER et al. 1996). The cysteines in CDR3 are encoded by preferred reading frames in codons of the D segments (ROUX et al. 1998), the first time such a clear purpose for such preferred reading frames has been uncovered. Such stabilizing disulfide bonds must be searched for in the Ig of other species.

As mentioned above, the first three domains of IgW are orthologous to skate IgX and the last four domains are homologous to those of NAR (Fig. 3). The split between the NAR and IgW preceded the shark/skate divergence 220 million years ago; indeed, we have found NAR in all extant cartilaginous fish carefully examined so far (Rumfelt, Guo, and Flajnik, unpublished). NAR is either derived from a NARC/W/X cluster, with the V domain being modified to a much greater extent than the C domains over time, or there was a recombination between an ancestral NAR cluster and a NARC/W/X cluster sometime during the evolution of cartilaginous fish. The ancient sharing of these last four domains (the split occurring at about the time that ancient birds appeared on earth), suggests that NAR and IgW may have similar effector functions, distinct from those of IgM (see below).

4.5 Light Chains

Cartilaginous fish L chains are described in detail in two other contributions to this volume (LEE et al. and BENGTÉN et al.) and are touched on only briefly here. As with the H chain genes, L chains were also originally expected to be rather simple, but this is not the case either. Peptide sequencing studies suggested that nurse sharks had an L chain somewhat similar to mammalian κ (STANTON et al. 1974), and that tiger sharks had a variety of light chains (MARCHALONIS et al. 1988). Molecular genetic data have now shown that most, if not all, cartilaginous fish have at least three L chains, one with a V region orthologous to κ (GREENBERG et al. 1993; SITNIKOVA and NEI 1998), and the other two somewhat more similar to λ (RAST et al. 1994). Expression of the three L chains seems to vary according to the species, with nurse shark expressing predominantly κ-type III (STANTON et al. 1974; GREENBERG et al. 1993), the horned shark λ-type I (SHAMBLOTT and LITMAN 1989a) and the sandbar shark λ-type II (HOHMAN et al. 1992, 1993, 1995). It is not known whether expression is correlated with gene complexity for each isotype, but the number of genes can vary greatly between species (Greenberg and Flajnik, unpublished).

L chain genes are also arranged in clusters, with one V, one J and one C gene in each cluster (SHAMBLOTT and LITMAN 1989b; HOHMAN et al. 1993; GREENBERG et al. 1993). Each of the isotypes can be germline-joined: in some cases the L chains are all joined (ANDERSON et al. 1995), e.g., type I in the skate but not the horned shark, and type II for all cartilaginous fish examined including the ratfish (RAST et al. 1994; HOHMAN et al. 1993; E. Hsu, personal communication). As mentioned in the section on H chains, the significance of this phenomenon is not known, but it may be perpetuated for binding to conserved epitopes on common pathogens, as in innate immune responses, or clusters that are in small numbers and may be selected for somatic mutation purposes (E. Hsu, personal communication).

Considering there are at least three very divergent L chain isotypes, exclusion experiments can now be performed at the RNA or protein level to begin to understand the regulation of rearrangement and expression of joined and unjoined genes. In addition, further studies of the L chain preference for IgM vs. IgW, should be performed as was suggested from the one previously described biochemical analysis (GREENBERG et al. 1996).

4.6 Class II Molecules

The class II α chain genes, *DAA* and *DBA*, were cloned from nurse sharks with degenerate PCR primers (KASAHARA et al. 1992) and later shown to be polymorphic (Kasahara et al. 1993). Two distinct isotypes were found, with one (*DBA*) found only in some nurse sharks and not others. Polymorphism of the alleles fits the pattern found in all other MHC genes with a ratio of replacement to silent substitutions that is higher than 1 in codons specifying amino acids that contact the peptide (Hughes and Nei 1988). This suggests that the types of selection of T cells by MHC are the same in almost all jawed vertebrates.

The β chain genes were also cloned from nurse sharks and found to be quite similar to homologues in other vertebrates (Bartl and Weissman 1994). These genes have also been analyzed in the horned shark (Ohta and Flajnik, unpublished). For neither the α nor β chain gene does there seem to have been an expansion, as seen in some teleost species (Klein and Sato 1998); rather the genes seem quite stable (as with class I, see below).

4.7 Class I Molecules

Two groups cloned MHC classical class I genes from three shark species. Studies of polymorphism in the carcharhinoid *Triakis* (banded houndshark) showed a great diversity of class I alleles, on par with that found in mammals (Okamura et al. 1997; Hashimoto et al. 1999). Furthermore, a high amount of recombinational diversity is generated among the alleles. Some haplotypes bear two class I genes, while others bear only one, a situation reminiscent of the aforementioned nurse shark class II α chain genes.

In nurse and horned shark, both nonclassical and classical class I genes were identified (Bartl et al. 1997). The class Ia genes seem to form old orthologous groups distinct from the nonclassical genes, which also might form old groupings (Bartl 1998). Where examined, the MHC genes have an exon/intron structure typical of genes in all other vertebrates. In class I, there is a tiny intron between the α1 and α2 domains (Okamura et al. 1997; Ohta and Flajnik, unpublished). Expression of the class I genes mirrors that of classical genes in all other species, with highest levels in the spleen and intestine (Bartl et al. 1997).

We have performed linkage studies in the nurse shark between class I and class II genes. Such studies became crucial when it was found that all teleost fish examined show no linkage of their class I and class II genes (reviewed in Klein and Sato 1998). In two nurse shark families (17 and 39 individuals) we found that class I and class II RFLP cosegregate, strongly suggesting that this was the primordial organization of the genes, i.e., that class I and class II genes were originally derived from cis-duplication and exon transfer events (Ohta, Bartl, Hashimoto, McKinney, and Flajnik, unpublished). Our results with nurse sharks are being confirmed in *Triakis* families. Other genes including *TAP1*, *TAP2*, *lmp7*, *lmp2*, *ring3*, *HSP70* (Ohta and Flajnik, unpublished), and *factorB* (Smith, unpublished) have been cloned and are now being examined for their MHC linkage.

4.8 T Cell Receptors

Litman's group has found that all four of the TCR genes defined in mammals also exist in the cartilaginous fish in studies of skate and the horned shark (RAST and LITMAN 1994; HAWKE et al. 1996; RAST et al. 1997). To date no biochemical or cellular analyses have been performed, nor have good tissue distribution studies been carried out.

The TCR β genes appear to be in the "translocon" organization (mouse and human type) in cartilaginous fish (although the horned shark may have multiple Cβ genes), suggesting that either shark Ig genes are secondarily in the cluster organization, or the primitive character was maintained only in Ig. As in all other vertebrates examined, D regions appear to be lacking in TCR α and γ genes and present in β and δ (two D regions inferred from δ cDNAs). The α and δ loci seem to be linked, based on pulsed field gel electrophoresis; thus this feature is ancient and reveals the ancestral character for TCR genes (RAST et al. 1997; CHIEN et al. 1987). The diversity generated by rearrangement for all of the TCR loci is on par with all other species examined, and there are multiple Vβ families. The robust potential TCR repertoire is perhaps surprising considering difficulties in demonstrating full-fledged T-dependent functions in the cartilaginous fish (see below); such diversity is not unexpected, however, considering the aforementioned mammalianlike shark MHC.

5 Functional Studies

Sharks are notorious for their poor experimental response to foreign antigens (reviewed in HSU 1998). Graft rejection has been reported to be "chronic" with rather lackluster memory responses (BORYSENKO and HILDEMANN 1970; HILDEMANN 1970). Review of these data, however, suggests that rejection times are not so different from those of other vertebrates. Furthermore, all of the experiments were performed with animals maintained at 22°C, and it is well known that T-dependent responses in poikilotherms are highly temperature dependent. Another study suggested that memory to allografts indeed occurred in cartilaginous fish (PEREY et al. 1968).

Extensive experiments that examined affinity maturation of antibody responses were performed by Clem and colleagues (e.g., CLEM and SIGEL 1963; SIGEL and CLEM 1966; CLEM and LESLIE 1971; SHANKEY and CLEM 1980a,b), Voss and SIGEL (1972), and MÄKELÄ and LITMAN (1980). There is no question that long-term immunizations yield specific responses, but by and large neither a typical "mammalianlike" memory reaction nor rise in affinity of the antibodies have been detected. Thus far only 19-S and 7-S IgM responses have been studied, with no hint of a role for either NAR or IgW.

Clem and colleagues examined antibody responses to carbohydrates (CLEM and LESLIE 1971; SHANKEY and CLEM 1980a), viruses, heterologous erythrocytes

(Sigel and Clem 1966), globular proteins (bovine serum albumin; Clem and Small 1967; Clem et al. 1967), and a hapten (dinitrophenol, DNP; Shankey and Clem 1980b). Mäkelä and Litman (1980) also studied the response to a hapten (2-furyloxazalone), measuring affinity with the phage-neutralization procedure. Only in the case of the virus immunizations did there appear to be a possible memory response, with a subsequent higher titer upon secondary immunizations (Sigel and Clem 1966); it was suggested that the dose of antigen used in this case was instrumental in generating memory. In all other cases responses peaked not long after immunization, and often remained at the same levels for as long as the animals were examined. With some antigens, e.g., bovine serum albumin, the response shifted from 19-S to 7-S IgM long (6 months) after the primary immunization (Clem et al. 1967), but for some carbohydrate antigens there was no shift to 7-S (Shankey and Clem 1980a), and for the viral antigens 7-S was present from the earliest test bleeds (Sigel and Clem 1966). Indeed, in one experiment there was a shift from 19-S to 7-S back to 19-S (Voss and Sigel 1971)! Importantly, in most cases when the specific antibody titer was allowed to drop, a secondary immunization produced levels of circulating Ig equivalent to those obtained after the primary immunization. In the experiments of Mäkelä and Litman (1980) there was no increase in affinity over the 25-days in which the response was examined, but only the 19-S IgM specific antibody was examined (see Blank et al. 1972). In contrast to all of the other experiments, Voss and Sigel (1972) suggested that 7-S antibodies elicited to DNP are of a higher affinity than the 19-S.

The general absence of an increase in affinity in most studies is surprising, considering the high levels of somatic mutation found in NAR, unless NAR is performing some other type of adaptive function. The poor responses are further enigmatic since we have found evidence for selection on the NAR binding site, based on replacement/silent mutation ratios in CDR codons (Diaz et al. 1998). We also have obtained evidence that the IgW (NARC) in nurse sharks is undergoing somatic mutation (unpublished data): the low number of genes in this species has allowed such an analysis. It is possible that NAR and/or IgW, being present in much lower amounts than IgM, were not detected in most of the experiments described above, or perhaps were only evident in the studies of Voss and Sigel (1972), i.e., the high-affinity 7-S fraction may have contained NAR as well as monomeric IgM.

Sharks, and in fact all ectothermic vertebrates studied, do not appear to produce germinal centers in the typical mammalian fashion (Good and Finstad 1967; Wilson et al. 1992; Zapata et al. 1996; Hsu 1998). We have now reinvestigated this problem by immunizing animals to biotinylated-bovine serum albumin in order to follow the antigen and the specific response of IgM and NAR (Rumfelt and Flajnik, unpublished). Preliminary findings have tracked the antigen into the splenic white pulp 4 weeks after the immunization. At this point we do not know whether the antigen is on the surface of "antigen-presenting cells" in a whole form or as peptides, or perhaps it is present only on the surface of lymphocytes in the white pulp. Furthermore, we now examine whether IgM- and NAR-positive cells in the white pulp are specific for the antigen and undergo any morphological changes (do they proliferate or become secretory cells?).

6 Predictions and Future Experiments

6.1 H and L Isotypes

One of the very interesting features of the cartilaginous fish is that despite their long history, almost all of the immune system components described above are found in all species examined. This is exemplified by the presence of all three L chain isotypes, IgM, NAR, and IgW/NARC/X/R in all rigorously examined shark, skate, and ray species. The ratfish is an important touchstone in determining whether these genes are found in all cartilaginous fish. However, it should be taken into account that the ratfish is just one species from the first expansion of this class of vertebrates, and may not have the full complement of immune genes found in the common ancestor. Furthermore, the divergence of orthologous genes appears to be much slower in the cartilaginous fish, i.e., we can easily obtain cross-hybridization of all of the immune genes in all elasmobranch species (220 million years divergence time, Fig. 1), while the same cannot be for cross-hybridizations between mammalian and avian genes, which have had approximately the same divergence times. For example, when L chain genes were cloned from the sandbar shark (Hohman et al. 1992), the level of sequence identity to the horned shark L chain seemed to be indicative of the divergence of these two shark species 150 million years ago. Instead, these L chain genes were found not to be orthologues, but two different isotypes of high sequence identity and found in all species (Greenberg et al. 1993; Rast et al. 1994). It has been suggested from studies of mitochondrial DNA that the basic mutation rate may be up to one order of magnitude less in sharks than in other groups (Martin et al. 1992). Since this slower evolutionary mutational rate also appears to be true of the nuclear genes, it provides an opportunity for comprehensive comparative studies in elasmobranchs, and perhaps in all cartilaginous fish. Despite the high sequence similarity of the various genes, how conserved are the numbers, expression and function of immune genes over very long periods of time? One would predict, because of the diversity of niches encountered by elasmobranchs, that the genes could be utilized in different ways.

It is our opinion that other Ig isotypes (H and L) will found in addition to the ones described here. We have detected what appears to be other H chains that are immunoprecipitated with our anti-L chain mAbs. Thus, although we are close to an understanding of all adaptive immune genes (at least the recognition molecules) in elasmobranchs, more work remains to be carried out.

The nomenclature is very messy for NARC/X/R/W and should be resolved. IgNARC is a cumbersome name and there is no Greek letter corresponding to "NARC." More importantly, this name was given for historical reasons, since NAR was found first, but if NAR indeed is derived from an old "NARC" cluster, the "new antigen receptor from cartilaginous fish" is not only cumbersome, it is confusing. By rights, IgX was the first member of this family to be cloned, but IgX was already taken (even at that time) as an isotype specific for *Xenopus* (Hsu et al. 1985) and has always caused confusion. If IgX/NARC/W really proves to be the

IgR protein, IgR has priority for the name. However, the isotype seems to be present in other species in addition to skates and thus the *Raja* appellation becomes inappropriate; further the *Xenopus* κ L chain was formerly called ρ (Schwager et al. 1992). IgW seems to be the choice with the fewest failings, and we vote to standardize the nomenclature with this name. Our previous suggestion to call this isotype IgSlash (Ig/or IgKordellStewart) should not be taken seriously, unlike every other morsel in this scholarly work!

6.2 Isotype Functions

As described, it has been shown time and again that IgM is induced to bind to antigen, but most experiments have shown neither an increase in affinity nor a memory component for either the 19-S or 7-S form. Thus the appearance of antigen-specific Ig resembles more that which one finds in an induced nonadaptive response. There are, however, other considerations. First, the concentration of IgM is very high in the blood, and the antibodies (at least the 19-S) are quite "sticky" (see Marchalonis et al. 1993). How these features affect the response is not known, i.e., how do preexisting "sticky" antibodies impact antigen presentation? One idea is that some B cells do not regulate their Ig clusters well and simply pour IgM into the blood, and/or at least some clusters (the germline-joined ones?) encode polyreactive antibodies. The 19-S and 7-S antibodies specific for DNP show intramolecular heterogeneity (Shankey and Clem 1980b), suggestive that one cell may indeed express more than one Ig (this, however, is not the case for antibodies raised to carbohydrate antigens; Shankey and Clem 1980a). Another complicating factor in most of the immunization studies was the high amount of antigen used; it was recognized early on by Sigel and Clem (1966) that the experiments in sharks were not performed, for the most part, in the optimal manner to examine affinity maturation, i.e., classically, antibody titers must drop and small amounts of antigen must be given subsequently. Comprehensive kinetic and antigen-dose studies have never been carried out, mainly because of the difficulty in maintaining large numbers of sharks and skates for long periods of time. We must take this into account. Finally, the route of immunization is very important to consider. There is no true lymphatic system in the cartilaginous fish (Zapata et al. 1996), and therefore subcutaneous or intramuscular injections may force the initiation of immune responses in ectopic sites, leading to inefficient collaboration between lymphocytes and APC.

There are appreciable levels of NAR in the serum, and we have preliminary evidence for at least some level of antigen-specific NAR after immunization (unpublished). The C regions, being so different from IgM, suggest a different function for NAR than IgM – better complement fixation, induction of inflammation, targeting of killer cells? Furthermore, the high levels of mutation in NAR, bearing the footprints of the mammalian mechanism and apparently resulting in selection on the binding site, strongly suggest that cells expressing NAR are under different constraints than IgM (Diaz et al. 1998). How and why in this case? Why select for a single V antigen-combining site? Is there a certain class of antigen for which a narrower binding site is optimal? Is there something specific to shark physiology that selects for a single binding site? For example, do the high urea levels in the blood select for a molecule that does not make dimers, but rather for an unassociated V stabilized by disulfide bonds? It was suggested that bona fide shark IgM H and L chains are more tightly associated by noncovalent bonds than IgM from other species, perhaps counteracting the effects of the urea (FROMMEL et al. 1971). Finally, was a single V found in the primordial secreted antigen receptor?

IgW and IgM (but not NAR) are expressed in the same tissues as detected by northern blotting (Flajnik et al. unpublished), but there seem to be only low amounts of IgW protein in the serum. Since the C domains of IgW are homologous to NAR (although they have been separated from one another for at least 250 million years!), IgW and NAR may have similar effector functions. The IgX-short form in skates, as has been proposed for the short form of duck IgY (MAGOR et al. 1992; PARHAM 1995), may be used in cartilaginous fish to prevent inflammation and only perform a neutralizing function.

For the L chains, there are two competing theories for the evolution of several forms: (a) the CDR of different L chains make different conformations in combinations with the same H chains, perhaps allowing responses to different classes of antigens, and (b) different L chains may have a preference for different H chains, or some may be more promiscuous than others (GREENBERG et al. 1993; RAST et al. 1994). Considering that the IgM and IgX H chain vs. form different V families (MARCHALONIS et al. 1998b), it is possible that they have coevolved with their own L chains.

Finally, how excluded are the different loci? There are at least three types of heavy chains and three types of L chains and multiple versions of each family. Expression cannot simply be regulated via rearrangement, as has been proposed as an original driving force for rearrangement (JANEWAY 1992), because there are germline-joined versions of most of the loci. Single cell PCR analyses and studies associated with isotype-specific mAbs, as they become available, will begin to answer these questions. Such studies will also begin to examine whether there is sequential rearrangement of the L chain gene families, as has been found in mammals (DURDIK et al. 1984).

For the germline-joined genes, most evidence suggests that this is a derived characteristic in cartilaginous fish, and not indicative of the primordial rearranging gene (LITMAN et al. 1999; Lohr and Bartl, personal communication). However, is there a significance to such genes; i.e., selected for binding to particular epitopes such as JANEWAY's (1972) pattern-associated recognition molecules, or for mutation purposes (LEE et al., this volume), or perhaps for expression early in ontogeny?

6.3 Lymphoid Tissue Organization and Function

Very few studies have tracked antigen into the spleen and followed the course of a response. One study showed that injection of allogeneic spleen cells resulted in their migration into the white pulp of the spleen (MORROW et al. 1982), but this might not have been indicative of a response but rather to a homing mechanism. Obviously much more needs to be accomplished. Considering the high mutation rate of NAR, our initial hypothesis was that IgM provided a first line of defense and NAR provided the specificity, in line with the studies of Klinman and colleagues, proposing that B cell populations are present in animals that provide primary and secondary immune responses (LINTON et al. 1989; GREENBERG et al. 1995). If so, we should be able to detect NAR-positive clusters inside the spleen (in white pulp?) after immunization. Such studies, in combination with affinity measurements, will determine whether this hypothesis has any merit. How is antigen retained in the spleen? Are cells, at the least, analogous to follicular dendritic cells present in sharks? How large is the burst size in antigen-specific clones and what limits the production of specific antibodies? Even if it is proven that no memory exists for humoral responses, there still *must be* some sort of selection for antigen-reactive cells.

Then there are other shark-specific things. What is the function of the Leydig organ and epigonal organ: are they primary or secondary lymphoid tissues, or are there specialized immune reactions that occur in these tissues, or are they repositories of cells with particular functions? What is the function of the lymphocytes found in the brain; are there specialized lymphocytes in nervous tissue that secrete particular types of antibodies? What is the origin of B (and T) cells? Again, how do cells "know" whether to express NAR vs. IgM vs. IgW? In contrast to representatives of all other vertebrate classes, each isotype has its own set of V regions in elasmobranchs; thus one would predict that there would be lineages of each type of cell. Do they all become stimulated in the spleen, and are T cells required for stimulation of each cell type?

6.4 MHC and T Cells

A priori, it appears that T cell function is not optimal in cartilaginous fish. The lack of affinity maturation, a suggestion of chronic graft rejection (HILDEMANN 1970), and difficulty to demonstrate MLR and other in vitro T cell functions all imply a "defect" in these cells. However, the "defect" certainly is not at the level of repertoire diversity, nor at the level of MHC expression or polymorphism. So, what accounts for the "nonmammalian"-type responses? Is there "poor" antigen presentation because dendritic cells do not function in a mammalian fashion? Are certain cytokines or "second signals" lacking that result in less than robust responses? Or are we blinded by our analyses of immune functions in mammals and have neglected the basic system in cartilaginous fish?

And, what of the γ/δ TCR? In which tissues does one find such cells in sharks, and could their function shed some light on γ/δ function in mammals, and more importantly, on the evolution of the different subclasses of T cells? Cellular cytotoxic reactions of various types have been demonstrated in cartilaginous fish (PETTEY and MCKINNEY 1983; HAYNES and MCKINNEY 1991); what is the molecular bases for the recognition? Dive in, the water's fine!

Are the *TAP* and proteasome genes in close linkage with the polymorphic class I genes in cartilaginous fish, as seems to be a rule in nonmammalian vertebrates (Flajnik et al. 1999)? It is also of interest that in several species of sharks two forms of lmp7 were found that seem to match sequences in the active site that are found in two major allelic forms in frogs (Kandil et al. 1996; Namikawa et al. 1995). Are these two lmp7 genes in close linkage to different types of class I genes, and how do they work together as a team? Are the nonclassical class I genes in clusters far from the MHC, as is the case in frogs (Flajnik et al. 1993) and chickens (Briles et al. 1993), capable of rapid expansions and contractions, while the classical class I is relatively stable? Are there more surprises in the class I family, i.e., genes encoding proteins with very different functions? Finally, is there a possibility that some type of "attractive" function can be linked to MHC and easier to uncover in an aquatic species? Stay tuned.

7 Conclusions

The molecular diversity of the immune system in cartilaginous fish, combined with their very interesting physiology and phylogenetic position, mark these awesome animals as research subjects extraordinaire. We must piece together the molecular diversity with immune function, possibly resulting in novel concepts of a prototypic or primordial immune system.

Acknowledgements. Our work described here was supported by NIH grants RR06603 (antigen receptors) and AI 27877 (MHC). We dedicate this chapter to our colleagues Churchill McKinney, Masdnori Kasahara, Andrew Greenberg, and Marilyn Diaz.

References

Anderson MK, Amemiya CT, Luer CA, Litman RJ, Rast JP, Niimura Y, Litman GW (1994) Complete genomic sequence and patterns of transcription of a member of an unusual family of closely related, chromosomally dispersed Ig gene clusters in *Raja*. Int Immunol 6:1661–1670

Anderson MK, Shamblott MJ, Litman RT, Litman GW (1995) Generation of immunoglobulin light chain diversity in *Raja erinacea* is not associated with somatic rearrangement, an exception to a central paradigm of B cell immunity. J Exp Med 182:109–119

Anderson MK, Strong SJ, Litman RT, Luer CA, Amemiya CT, Rast JP, Litman GW (1999) A long form of the skate IgX gene exhibits a striking resemblance to the new shark IgW and Ig NARC genes. Immunogenetics 49:56–67

Bartl S, Weissman IL (1994) Isolation and characterization of the major histocompatibility complex class IIB genes from the nurse shark. Proc Natl Acad Sci 91:262–266

Bartl S, Bash MA, Flajnik MF, Ohta Y (1997) Identification of class I genes in cartilaginous fish, the most ancient group of vertebrates displaying an adaptive immune response. J Immunol 159:6097–6104

Bartl S (1998) What sharks can tell us about the evolution of MHC genes. Immunol Rev 166:317–331

Bernstein RM, Schluter SF, Shen S, Marchalonis JJ (1996) A new high molecular weight immunoglobulin class from the carcharhine shark: implications for the properties of the primordial immunoglobulin. Proc Natl Acad Sci USA 93:3289–3293

Betz AG, Neuberger MS, Milstein C (1993) Passenger transgenes reveal intrinsic specificity of the antibody hypermutation mechanism: clustering, polarity, and specific hotspots. Proc Natl Acad Sci 90:2385–2388

Blank SE, Leslie GA, Clem LW (1972) Antibody affinity and valence in viral neutralization. J Immunol 108:665–673

Borysenko M, Hildemann WH (1970) Reaction to skin allografts in the horn shark, *Heterodontus francisci*. Transplantation 10:545–551

Briles WE, Goto R, Auffray C, Miller M (1993) A polymorphic system related to but genetically independent of the chicken major histocompatibility complex. Immunogenetics 37:408–414

Carroll RL (1988) Vertebrate Paleontology and Evolution. Freeman, New York

Chiba A, Torroba M, Honma Y, Zapata AG (1989) Occurrence of lymphohaematopoietic tissue in the meninges of the stingray *Dasyatis akajei* (Elasmobranchii, Chondricthyes). Amer J Anat 183:268–276

Chien Y-H, Iwashima M, Kaplan KB, Elliott JF, Davis MM (1987) A new T-cell receptor gene located within the α locus and expressed early in T-cell differentiation. Nature 327:677–682

Clem LW, Sigel MM (1963) Comparative immunochemical and immunological reactions in marine fishes with soluble, viral, and bacterial antigens. Fed Proc 22:1138–1144

Clem LW, Small PA (1967) Phylogeny of immunoglobulin structure and function I. Immunoglobulins of the lemon shark. J Exp Med 125:893–920

Clem LW, DeBoutaud F, Sigel MM (1967) Phylogeny of immunoglobulin structure and function II Immunoglobulins of the nurse shark. J Immunol 99:1226–1235

Clem LW, Leslie GA (1971) Production of 19S IgM antibodies with restricted heterogeneity from sharks. Proc Natl Acad Sci 68:139–141

Desmyter A, Transue TR, Ghahroudi MA, Thi MHD, Poortmans F, Hamers R, Muyldermans S, Wyns L (1996) Crystal structure of a camel single-domain V_H antibody fragment in complex with lysozyme. Nature Struc Biol 3:803–811

Diaz M, Greenberg AS, Flajnik MF (1998) Somatic hypermutation of the new antigen receptor (NAR) gene in the nurse shark does not generate the repertoire: possible role in antigen-driven reactions in the absence of germinal centers. Proc Natl Acad Sci 95:14343–14348

Diaz M, Velez J, Singh M, Cerny J, Flajnik MF (1999) Mutational pattern of the nurse shark antigen receptor gene (NAR) is similar to that of mammalian Ig genes and to spontaneous mutations in evolution: the translesion synthesis model of somatic hypermutation. Int Immunol 11: in press

Du Pasquier L, Wilson M, Greenberg A, Flajnik MJ (1998) Somatic mutation in ectothermic vertebrates. Curr Top Microbiol Immunol 299:199–216

Durdik J, Moore MW, Selsing E (1984) Novel kappa light-chain gene rearrangements in mouse lambda light chain-producing B lymphocytes. Nature 307:749–752

Fidler JE, Clem LW, Small Jr PA (1969) Immunoglobulin synthesis in neonatal nurse sharks (*Ginglymostoma cirratum*) Comp Biochem Physiol 31:365–371

Flajnik MF, Kasahara M, Shum BP, Salter-Cid L, Taylor E, Du Pasquier L (1993) A novel type of class I gene organization in vertebrates: a large family of non-MHC linked class I genes is expressed at the RNA level in the amphibian *Xenopus*. EMBO J 12:4385–4396

Flajnik MF (1998) Churchill and the immune system of ectothermic vertebrates. Immunol Rev 166:5–14

Flajnik MF, Ohta Y, Namikawa-Yamada C, Nonaka M (1999) Insight into the primordial MHC from studies in ectothermic vertebrates. Immunol Rev 167:59–67

Frommel D, Litman GW, Finstad J, Good RA (1971) The evolution of the immune response. XI. The immunoglobulin of the horned shark, *Heterodontus francisci*: purification, characterization and structural requirement for antibody activity. J Immunol 106:1234–1243

Fuller L, Murray J, Jensen JA (1978) Isolation from nurse shark serum of immune 7S antibodies with two different molecular weight H-chains. Immunochemistry 15:251–259

Gitlin D, Perricelli A, Gitlin JD (1973) Immunoglobulin synthesis in fetal sharks. Comp Biochem Physiol 45:247–256

Good RA, Finstad J, Pollara B, Gabrielsen AE (1966) Morphologic studies on the lymphoid tissues among the lower vertebrates. In: Smith RT, Miescher PA, Good RA (eds) Phylogeny of immunity. University of Florida Press, Gainesville

Good RA, Finstad J (1967) The phylogenetic development of immune responses and the germinal center system. In: Cottier H, Odartchenko N, Schindler R, Congdon CC (eds) Germinal Centers in Immune Responses. Springer-Verlag, New York

Greenberg AS, Steiner LA, Kasahara M, Flajnik MF (1993) Isolation of a shark immunoglobulin light chain cDNA clone encoding a protein resembling mammalian kappa light chains: implications for the evolution of light chains. Proc Natl Acad Sci 10603–10607

Greenberg AS, Avila D, Hughes M, Hughes A, McKinney EC, Flajnik MF (1995) A new antigen receptor gene family that undergoes rearrangement and extensive somatic diversification in sharks. Nature 374:168–173

Greenberg AS, Hughes AL, Guo J, Avila D, McKinney EC, Flajnik MF (1996) A novel "chimeric" antibody class in cartilaginous fish: IgM may not be the primordial immunoglobulin. Eur J Immunol 26:1123–1129

Harding FA, Cohen N, Litman GW (1990a) Immunoglobulin heavy chain gene organization and complexity in the skate, *Raja ernacea*. Nucl Acids Res 18:1015–1020

Harding FA, Amemiya CT, Litman RT, Cohen N, Litman GW (1990b) Two distinct immunoglobulin heavy chain isotypes in a primitive cartilaginous fish, *Raja ernacea*. Nucl Acids Res 18:6369–6376

Hashimoto K, Okamura K, Yamaguchi H, Ototake M, Nakanishi T, Kurosawa Y (1999) Conservation and diversification of MHC class I and its related molecules in vertebrates. Immunol Rev 167:81–100

Hawke NA, Rast JP, Litman GW (1996) Extensive diversity of transcribed TCR-β in a phylogenetically primitive vertebrate. J Immunol 156:2458–2464

Haynes L, McKinney EC (1991) Shark spontaneous cytotoxicity: characterization of the regulatory cell. Devel Comp Immunol 15:123–134

Hildemann WH (1970) Transplantation immunity in fishes: Agnatha, Chondrichthyes and Osteichthyes. Transpl Proc 2:253–259

Hinds KR, Litman GW (1986) Major reorganization of immunoglobulin V_H segmental elements during vertebrate evolution. Nature 320:546–549

Hinds-Frey KR, Nishikata H, Litman RT, Litman GW (1993) Somatic variation precedes extensive diversification of germline sequences and combinatorial joining in the evolution of immunoglobulin heavy chain diversity. J Exp Med 178:815–824

Hohman VS, Schluter SF, Marchalonis JJ (1992) Complete sequence of a cDNA clone specifying sandbar shark immunoglobulin light chain: gene organization and implications for the evolution of light chains. Proc Natl Acad Sci 89:276–280

Hohman VS, Schuchman DB, Schluter SF, Marchalonis JJ (1993) Genomic clone for sandbar shark lambda light chain: generation of diversity in the absence of rearrangement. Proc Natl Acad Sci 90:9882–9886

Hohman VS, Schluter SF, Marchalonis JJ (1995) Diversity of Ig light chain clusters in the sandbar shark (*Carcharhinus plumbeus*). J Immunol 155:3922–3928

Hsu E, Flajnik MF, Du Pasquier L (1985) A third immunoglobulin class in amphibians. J Immunol 135:1998–2004

Hsu E (1998) Mutation, selection, and memory in B lymphocytes of exothermic vertebrates. Immunol Rev 162:25–36

Hughes AL and Nei M (1988) Pattern of nucleotide substitution at major histocompatibility complex class I loci reveals overdominant selection. Nature 335:167–170

Janeway CA Jr (1992) The immune system evolved to discriminate infectious nonself from noninfectious self. Immunol Today 13:11–16

Kandil E, Namikawa C, Nonaka M, Greenberg AS, Flajnik MF, Ishibashi T, Kasahara M (1996) Isolation of low molecular mass polypeptide complementary cDNA clones from primitive vertebrates. Implications for the origin of MHC class I-restricted antigen presentation. J Immunol 159: 734–740

Kasahara M, Vasquez M, Sato K, McKinney EC, Flajnik MF (1992) Evolution of major histocompatibility complex: isolation of class II A cDNA clones from the cartilaginous fish. Proc Natl Acad Sci 89:6688–6692

Kasahara M, McKinney EC, Flajnik MF, Ishibashi T (1993) The evolutionary origin of major histocompatibility complex: polymorphism of class II α chain genes in the cartilaginous fish. Eur J Immunol 23:2160–2165

Klapper DG, Clem LW (1972) Studies on the mild reduction of shark polymeric and monomeric IgM. Comp Biochem Physiol 42:241–247

Klapper DG, Clem LW (1977) Phylogeny of immunoglobulin structure and function: Characterization of the cysteine-containing peptide involved in the pentamerization of shark IgM. Devel Comp Immunol 1:81–92

Klein J (1998) In an immunological twilight zone. Proc Natl Acad Sci 95:11504–11505

Klein J and Sato A (1998) Birth of the major histocompatibility complex. Scand J Immunol 47:199–209

Kobayashi K, Tomonaga S, Kajii T (1984) A second class of immunoglobulin other than IgM present in the serum of a cartilaginous fish, the skate *Raja kenojei*: isolation and characterization. Mol Immunol 21:397–404

Kobayashi S, Tomonaga S, Teshima K, Kajii T (1985) Ontogenic studies on the appearance of two classes of immunoglobulin-forming cells in the spleen of the Aleutian skate, *Bathyraja aleutica*, a cartilaginous fish. Eur J Immunol 15:952–956

Kobayashi K, Tomonaga S (1988) The second immunoglobulin class is commonly present in cartilaginous fish belonging to the order Rajiformes. Mol Immunol 2:115–120

Kobayashi K, Tomonaga S, Tanaka S (1992) Identification of a second immunoglobulin in the most primitive shark, the frill shark, *Chlamydoselachus anguineus*. Devel Comp Immunol 16:295–299

Kokubu F, Hinds K, Litman R, Shamblott MJ, Litman GW (1987) Extensive families of constant region genes in phylogenetically primitive vertebrate indicate an additional level of immunoglobulin complexity. Proc Natl Acad Sci 84:5868–5872

Kokubu F, Hinds K, Litman RT, Shamblott MJ, Litman GW (1988a) Complete structure and organization of immunoglobulin heavy chain constant region genes in a phylogenetically primitive vertebrate. EMBO J 7:1979–1988

Kokubu F, Litman RT, Shamblott MJ, Hinds K, Litman GW (1988b) Diverse organization of immunoglobulin V_H loci in a primitive vertebrate. EMBO J 7:3413–3422

Linton PJ, Decker DJ, Klinman NR (1989) Primary antibody-forming cells and secondary B cells are generated from separate precursor cell populations. Cell 59:1049–1059

Litman GW, Berger L, Murphy K, Litman RT, Hinds KR, Erickson BW (1985) Immunoglobulin V_H structure and diversity in Heterodontus, a phylogenetically primitive shark. Proc Natl Acad Sci 82:2082–2086

Litman GW, Anderson MK, Rast JP (1999) Evolution of antigen-binding receptors. Annu Rev Immunol 17:109–147

Lloyd-Evans P (1993) Development of the lymphomyeloid system in the dogfish, *Scyliorhinus canicula*. Devel Comp Immunol 17:501–514

Luer CA, Walsh CJ, Bodine AB, Wyffels JT, Scott TR (1995) The elasmobranch thymus: anatomical, histological, and preliminary functional characterization. J Exp Zool 273:342–254

Magor KE, Warr GW, Middleton DL, Wilson MR, Higgins DA (1992) Structural relationship between the two IgY of the duck *Anas platyrhynchos*: molecular genetic evidence. J Immunol 149: 2627–2633

Mäkelä O, Litman GW (1980) Lack of heterogeneity in anti-hapten antibodies of a phylogenetically primitive shark. Nature 287:639–640

Marchalonis J, Edelman GM (1965) Phylogentic origins of antibody structure I. Multichain structure of immunoglobulins in the smooth dogfish (*Mustelus canis*). J Exp Med 122:601–618

Marchalonis J, Edelman GM (1966) Polypeptide chains of the immunoglobulins from the smooth dogfish (*Mustelus canis*). Science 154:1567–1568

Marchalonis JJ, Schluter SF, Rosenshein IL, Wang A-C (1988) Partial characterization of immunoglobulin light chains of carcharhine sharks: evidence for phylogenetic conservation of variable region and divergence of constant region structure. Devel Comp Immunol 12:65–74

Marchalonis JJ, Hohman VS, Thomas C, Schluter SF (1993) Antibody production in sharks and humans: a role for natural antibodies. Devel Comp Immunol 17:41–53

Marchalonis JJ, Schluter SF, Bernstein RM, Shen S, Edmundson AB (1998a) Phylogenetic emergence and molecular evolution of the immunoglobulin family. Adv Immunol 70:417–506

Marchalonis JJ, Schluter SF, Bernstein RM, Hohman VS (1998b) Antibodies of sharks: revolution and evolution. Immunol Rev 166:103–122

Martin AP, Naylor GJP, Palumbi SR (1992) Rates of mitochondrial DNA evolution are slow compared with mammals. Nature 357:153–155

Matthews D (1996) Sharks! The mysterious killers. Random House, New York

Morrow WJW, Harris JE, Pulsford A (1982) Immunological responses of the dogfish (*Scyliorhinus canicula* L.) to cellular antigens. Acta Zool 63:153–159

Okamura K, Ototake M, Nakanishi T, Kurosawa Y, Hashimoto K (1997) The most primitive vertebrates with jaws possess highly polymorphic MHC class I genes comparable to those of humans. Immunity 7:777–790

Parham P (1995) The duck's dilemma. Nature 374:16–17

Perey DYE, Finstad J, Pollara B, Good RA (1968) Evolution of the immune response. VI. First and second set skin homograft reactions in primitive fishes. Lab Invest 19:591–597

Pettey CL, McKinney EC (1983) Temperature and cellular regulation of spontaneous cytotoxicity in the shark. Eur J Immunol 13:133–138

Rast JP, Anderson MK, Ota T, Litman RT, Margittai M, Shamblott MJ, Litman GW (1994) Immunoglobulin light chain class multiplicity and alternative organizational forms in early vertebrate phylogeny. Immunogenetics 40:83–99

Rast JP, Litman GW (1994) T cell receptor homologs are present in the most primitive jawed vertebrates. Proc Natl Acad Sci 91:9248–9252

Rast JP, Anderson MK, Strong SJ, Luer C, Litman RT, Litman GW (1997) α, β, γ, and δT cell antigen receptor genes arose early in vertebrate phylogeny. Immunity 6:1–11

Rast JP, Amemiya CT, Litman RT, Strong SJ, Litman GW (1998) Distinct patterns of *IgH* structure and organization in a divergent line of cartilaginous fishes. Immunogenetics 47:234–245

Roux KH, Greenberg AS, Greene L, Strelets L, Avila D, McKinney EC, Flajnik MF (1998) Structural analysis of the nurse shark (new) antigen receptor (NAR): molecular convergence of NAR and unusual mammalian immunoglobulins. Proc Natl Acad Sci 96:11804–11809

Schluter SF, Bernstein RM, Marchalonis JJ (1997) Molecular origins and evolution of immunoglobulin heavy-chain genes of jawed vertebrates. Immunol Today 81:543–549

Schwager J, Burckert N, Schwager M, Wilson M (1992) Evolution of immunoglobulin light chain genes: analysis of *Xenopus* IgL isotypes and their contribution to antibody diversity. EMBO J 10:505–511

Shamblott MJ, Litman GW (1989a) Complete nucleotide sequence of primitive vertebrate immunoglobulin light chain genes. Proc Natl Acad Sci 86:4684–4688

Shamblott MJ, Litman GW (1989b) Genomic organization and sequence of immunoglobulin light chain genes in a primitive vertebrate suggest coevolution of immunoglobulin gene organization. EMBO J 8:3733–3739

Shankey TV, Clem LW (1980a) Phylogeny of immunoglobulin structure and function VIII. Intermolecular heterogeneity of shark 19S IgM antibodies to pneumococcal polysaccharide. Mol Immunol 17:365–375

Shankey TV, Clem LW (1980b) Phylogeny of immunoglobulin structure and function IX. Intramolecular heterogeneity of shark 19 S IgM antibodies to the dinitrophenol hapten. J Immunol 125:2690–2698

Sigel MM, Clem LW (1966) Immunological anamnesis in elasmobranchs. In: Smith RT, Miescher PA, Good RA (eds) Phylogeny of Immunity. University of Florida Press, Gainesville

Sitnikova T, Nei M (1998) Evolution of immunoglobulin kappa chain variable region genes in vertebrates. Mol Biol Evol 15:50–60

Small PA, Klapper DG, Clem LW (1970) Half-lives, body distribution, and lack of interconversion of serum 19S and 7S IgM of sharks. J Immunol 105:29–37

Stanton T, Sledge C, Capra JD, Woods R, Clem LW, Hood L (1974) A sequence restriction in the variable region of immunoglobulin light chains from sharks, birds, and mammals. J Immunol 112:633–640

Tomonaga S, Kobayashi K, Hagiwara K, Yamaguchi K, Awaya K (1986) Gut-associated lymphoid tissue in elasmobranchs. Zool Sci 3:453–458

Voss Jr EW, Sigel MM (1971) Distribution of 19S and 7S IgM antibodies during the immune response in the nurse shark. J Immunol 106:1323–1329

Voss Jr EW, Sigel MM (1972) Valence and temporal change in affinity of purified 7S and 19S nurse shark andti-2,4 dinitrophenyl antibodies. J Immunol 109:665–673

Wilson EO (1993) The diversity of life. WW Norton & Company, New York, London

Wilson M, Hsu E, Marcuz A, Courtet M, Du Pasquier L, Steinberg C (1992) What limits affinity maturation in *Xenopus* – the rate of somatic mutation or the ability to select mutants? EMBO J 11:4337–4347

Zapata AG, Torroba M, Sacedon R, Varas A, Vicente A (1996) Structure of the lymphoid organs of elasmobranchs. J Exp Zool 275:125–143

Immune-Type Diversity in the Absence of Somatic Rearrangement

J.A. YODER and G.W. LITMAN

1 Introduction

The rearrangement and joining of segmental elements during the differentiation of B and T lymphocytes in apparently all jawed vertebrate species results in the primary diversification of the immune receptor genes encoding both immunoglobulin (Ig) and T cell antigen receptor (TCR) genes. Although an extraordinary degree of interspecific variation in both numbers of recombining elements and mechanisms of secondary diversification of rearranged Ig genes has been described throughout vertebrate phylogeny, the basic mechanisms involved in the somatic rearrangement of individual gene segments have been intensely conserved. Recent studies of the recombination process (reviewed in GRAWUNDER et al. 1998) and the enzymatic machinery that catalyzes DNA-mediated genetic rearrangement events have resulted in an increasingly clearer understanding of not only this highly complex process but also of the probable origin of the rearrangement mechanism itself (AGRAWAL et al. 1998; HIOM et al. 1998).

Moffitt Cancer Center, Department of Pediatrics, University of South Florida, All Children's Hospital, St. Petersburg, FL 33701, USA

Rearrangement of Ig and TCR genes occurs in a relatively small percentage of the genome. However, the coupling of these de novo rearrangement events with other mechanisms of somatic diversification (e.g., junctional diversity, somatic mutation, gene conversion) results in essentially limitless variation in antigen-binding receptors. Cells expressing such receptors are of enormous selective advantage in terms of both genetic "economy" and the capacity to respond to "unanticipated" challenge. Immune receptor function has been optimized through coevolution of receptor signaling mechanisms, cell–cell interactions, and transcriptional control mechanisms as well as through mechanisms that allow the selective expansion of individual, somatically varied forms of receptors. The capacity of the immune system to recall and expand previously created (and selected) antigen-specific receptor variants affords an additional selective advantage. The various mechanisms that are employed to create secondary variation in divergent species often are associated with major differences in the organization of Ig genes. In contrast, the overall structure and organization of TCR genes has remained stable throughout the evolution of jawed vertebrates. Presumably the variation in Ig structure reflects an evolutionary selection process that optimizes antibody response capabilities in individual species (LITMAN et al. 1999).

Of all of the forms of Ig gene variation that have been described to date, none is as unusual as the germline joining of segmental elements, which has been identified only in the various radiations of chondrichthyans (cartilaginous fish). As the germline joined genes do not undergo segmental rearrangement, they lack junctional variation; however, genetic diversity is maintained in the absence of gene rearrangement. Germline joining presumably affords a unique selective advantage in that certain specificities become prefixed and are maintained in the germline (KOKUBU et al. 1988; LITMAN et al. 1999). Equivalent joining of TCR genes has not been described, presumably reflecting a major difference between the organization of these genes and Ig genes. However, another multigene family of receptors, which possess diversified Ig/TCR-like variable (V) and joining (J) regions, but otherwise are distinct from Ig and TCR, has been identified recently (RAST et al. 1995; RAST and LITMAN 1998; STRONG et al. 1999). These novel immune-type receptor (NITR) genes, such as germline-joined Ig genes, are predicted to encode different specificities and likewise do not undergo rearrangement. A variety of other genes, which contain V regions and exhibit varying degrees of relatedness to higher vertebrate V sequences but are not necessarily implicated in direct immunological recognition processes, are described elsewhere (see Du Pasquier, this volume). This review summarizes that which is known about germline joined Ig genes, describes the general features of the newly identified family of immune-type receptors, and speculates as to the evolution and functional significance of germline committed immune-type diversity.

2 Genomic Organization of Ig Genes in Cartilaginous Fish

In order to understand the phenomena of germline joining of Ig genes it first is necessary to understand the unique manner in which Ig genes are organized in the various species of cartilaginous fishes that have been examined to date. Specifically, in the three major radiations of the cartilaginous fish, represented by contemporary sharks, skates, and ratfish, both Ig heavy and light-chain gene loci are organized in multiple clusters. The cluster-type gene organization consists of a single V as well as a J element and, in the case of heavy chains, two to three diversity (D) elements (Fig. 1). Presently, three distinct classes of heavy-chain genes (IgM, IgX, and NAR) and three classes of light-chain genes (type I, type II, and Igκ) have been identified in various species of cartilaginous fish (Litman et al. 1999).

IgM heavy-chain genes, which are structurally equivalent to mammalian IgM genes, have been characterized in various species of sharks, skates and in ratfish. The IgX class of genes, which is unrelated to any known class of higher vertebrate Ig heavy-chain gene, has been described in skate (*Raja erinacea* and *Raja eglanteria*; Litman et al. 1999) and presumably is equivalent to the IgR molecule that was identified previously in skate (Kobayashi and Tomonaga 1988). IgX produces two transcript isoforms that vary in length. The short form of IgX in the skate consists of a single variable region (V_X), two constant domains (C_X1 and C_X2), and a secretory exon (Harding et al. 1990a; Anderson et al. 1994). Subsequent char-

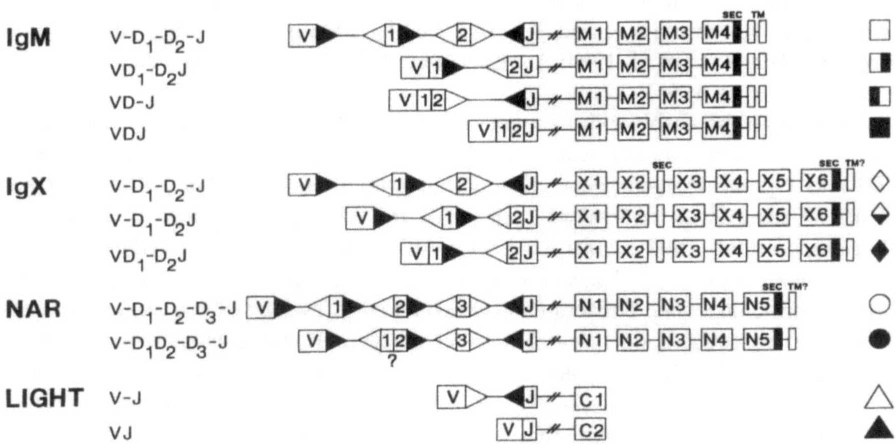

Fig. 1. Schematic depiction of joined and unjoined gene clusters representing the three classes of Ig heavy-chain genes and two of the three classes of light-chain genes that have been identified in cartilaginous fish. *Blackened triangles* (► and ◄), recombination signal sequences with 22 or 23 nucleotide spacers; *unblackened triangles* (▷ and ◁), recombination signal sequences with 12 nucleotide spacers. Each of the different forms of joining shown has been described with the exception of the partially joined NAR locus designed with a question mark. Joined vs. unjoined status of light-chain genes is not class dependent as both joined and unjoined forms of type I light-chain genes have been described; however, all type II light-chain genes described thus far are germline-joined. *Symbols to the right*, the various forms of joined and unjoined genes depicted in Fig. 2

acterization of a long IgX cDNA transcript showed it to encode V_X, C_X1, and C_X2 domains as well as four additional constant region exons (Anderson et al. 1994, 1999). IgW (Bernstein et al. 1996) and NARC (Greenberg et al. 1996) were identified independently in sandbar (*Carcharhinus plumbeus*) and nurse (*Ginglymostoma cirratum*) sharks, respectively, and subsequently were shown to be orthologous with the long transcript form of IgX that was identified in skate (Anderson et al. 1999). All evidence points to the long and short forms of IgX deriving from the same gene locus by differential processing, although there is no unequivocal evidence to support this hypothesis. NAR, which has been described in nurse shark, represents a third class of Ig heavy-chain gene that is distinct from both IgM and IgX (Greenberg et al. 1995; Roux et al. 1998; Diaz et al. 1998) but shares significant sequence identity with the 3′ exons of the long form of IgX (IgW and NARC). Based on phylogenetic analyses, it is likely that NAR represents a derived form of IgX (Anderson et al. 1999). NAR genes are organized in the same cluster-type manner as seen in both IgM and IgX genes but contain three rather than two D segments (Fig. 1). The absence of association with light chains and the high degree of hypermutability in the V region distinguishes NAR from all other cluster-type Ig genes (Greenberg et al. 1995; Diaz et al. 1998; Kokubu et al. 1988).

Chrondrichthyan light-chain genes consist of single V_L and C_L regions that bear an overall structural resemblance to light-chain genes found in other jawed vertebrates. One class of chondrichthyan light-chain genes identified in both nurse and horned (*Heterdontus franciscii*) shark resembles mammalian Igκ (Greenberg et al. 1993; Rast et al. 1994); these genes are not joined in the germline and are not described further. The other two classes of light-chain genes lack significant identity with higher vertebrate Igκ and Igλ and are classified as type I (Shamblott and Litman 1989; Rast et al. 1994; Anderson et al. 1994) and type II (Hohman et al. 1992; Rast et al. 1994).

Although the absolute linkage relationships of the various Ig heavy- and light-chain gene clusters has not been defined, it is estimated that there are approx. 100–200 IgM-type heavy-chain gene clusters (Kokubu et al. 1987, 1988; Marchalonis et al. 1998a) and an equivalent number of IgX clusters (Harding et al. 1990a; Anderson et al. 1994), contrasted with only four clusters of NAR-type genes (Greenberg et al. 1995; Roux et al. 1998). Present estimates are that more than 75 type I light-chain gene clusters (Shamblott and Litman 1989; Anderson et al. 1995) and an equal if not greater number of type II gene clusters are present in cartilaginous fish (Marchalonis et al. 1998b). Relatively few pseudogene clusters have been identified (Anderson et al. 1995). In situ hybridization studies with interphase nuclei have shown that IgM- and IgX-type heavy-chain gene loci are present at multiple locations on different chromosomes in skate (Fig. 2; Anderson et al. 1994). Efforts to establish physical linkage between individual clusters have been unsuccessful although the prominent fluorescence signals achieved with in situ DNA hybridization are consistent with a relatively close physical proximity of at least some clusters (Fig. 2; Anderson et al. 1994). Efforts to establish linkage between individual Ig heavy-chain gene clusters in P1 artificial chromosome inserts from horned shark have been unsuccessful, consistent with linkage distances of

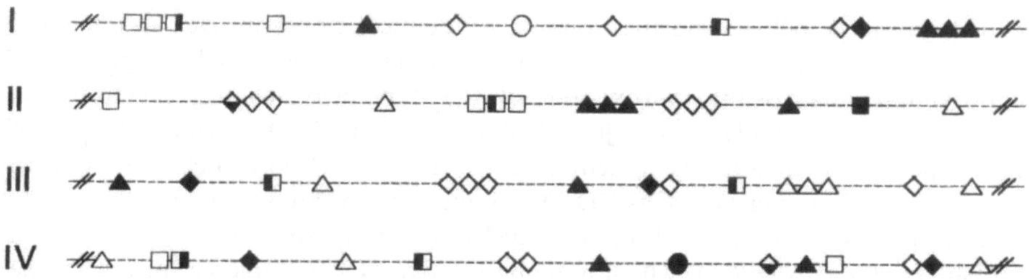

Fig. 2. Hypothetical chondrichthyan genomic regions of unspecified lengths on different chromosomes (designated by Roman numerals) containing various types of joined and unjoined Ig gene clusters. Symbols are as defined in Fig. 1. Distances between clusters are presumed to exceed 100kb (*dashed lines*). Joined and unjoined IgM and IgX heavy-chain genes are found at various locations on different chromosomes. Based on in situ chromosomal hybridization, the number of IgX clusters most likely exceeds the number of IgM clusters, and/or IgX clusters may be linked more closely than IgM clusters; relatively few NAR clusters exist. Immunoglobulin light-chain gene clusters, which are either joined or unjoined within a given type, also are interspersed and possibly are present in as many different clusters as IgM and IgX. The relatively close proximity of IgM, IgX, and light-chain gene clusters shown at several positions is hypothetical since physical linkage has not been established. Relatively close proximity of similar clusters may be obligatory to achieve sufficient signal densities for in situ chromosomal hybridization. The purpose of this diagram, based largely on and combining results from both shark and skate, is to indicate potential relationships that are consistent with present observations and is not intended to represent actual linkage relationships

more than approx. 100kb (T. Ota and C. Amemiya, unpublished observations). Furthermore, type I Ig light-chain genes have been shown to be present at various chromosomal locations in the shark (Fig. 2; C. Amemiya, unpublished observations). No equivalent overall organization of immunoglobulin genes has been observed in any species (other than cartilaginous fish) studied to date. The possibility exists that the adaptive immune system in cartilaginous fish functions through independently regulated gene loci, a marked difference from the Ig genes present in other vertebrates. Furthermore, germline joining, in which two or more of the segmental elements within a cluster are contiguous in the germline occurs in the heavy- and/or light-chain genes of various species of cartilaginous fish.

3 Joined Heavy-Chain Genes

Germline joining was reported first for the IgM heavy-chain genes of the shark (Kokubu et al. 1988). Previous reports had shown that the IgM genes in this species were organized in clusters consisting of a V, two D, and a J region that are closely linked (300–400bp) upstream of the constant region exons (Hinds and Litman 1986). Twelve genomic clones were characterized, and more than half of the clusters were shown to be germline-joined; the VD-J form of joining in IgM occurs more frequently than full VDJ joining (Fig. 1; Kokubu et al. 1988). The possibility that these unusual gene clusters arose by RAG-mediated segmental rearrangement in somatic tissues was discounted both by the high frequency of identification of joined genes (see below) and the isolation of identical clones in liver and gonadal tissue. Although this sampling is statistically limited, these results suggest that approximately half of the approx. 100–200 IgM-type heavy-chain gene clusters in shark are either partially or fully joined (Kokubu et al. 1988). Germline joining of IgM-type clusters also has been demonstrated in skate (Harding et al. 1990b). In this case, VD-J joining as well as VD-DJ joining patterns are seen in approximately half of the clusters (Fig. 1). VDJ joined IgM-type genes also have been identified in ratfish (*Hydrolagus colliei*), which is representative of the chimerae lineage of the cartilaginous fish (Rast et al. 1998).

The same phenomenon of germline joining has been described for both short and long transcripts of IgX (Rast et al. 1998). Specifically, an IgX cDNA has been described in which D_X2J_X joining has occurred but D_X1 remains in the germline configuration (Fig. 1; Anderson et al. 1994). A cDNA representing the long form of IgX has been shown to contain recombination signal sequences (RSSs) and an intervening sequence (IVS) between D_X1 and D_X2; the IVS is related closely to the corresponding region of a genomic IgX gene (Anderson et al. 1999). It has not been possible to identify cDNA transcripts corresponding to joined IgM-type or IgX-type genes in either shark or ratfish; however, efforts are under way to examine the expression patterns of joined genes in the skate owing to its well characterized developmental staging and unique sites of Ig gene expression (A. Miracle, unpublished observations). Evidence has been found that is suggestive of germline joining of D elements in a single NAR gene (M. Flajnik, personal communication). Thus, germline joining has been shown to occur in each of the three major representative lineages of cartilaginous fish as well as in each of the three different classes of heavy-chain gene clusters.

4 Joined Light-Chain Genes

Type I light-chain genes in cartilaginous fish are present in both joined and unjoined forms (Fig. 1). Specifically, all type I light-chain genes in horned shark are not joined (V-J; Fig. 1; Shamblott and Litman 1989); however, type I genes that were identified in a skate genomic DNA library possess contiguous, in-frame V and J segments (VJ-joined; Fig. 1; Litman et al. 1993). Using both genomic screening and direct amplification of skate genomic DNA, 89 unique gene clusters were characterized and grouped into four major families. Only seven of these type I genes were shown to be pseudogenes, of which five are members of the same gene family. It was not possible to identify any type I gene cluster that was not joined in skate (Anderson et al. 1995). The type I light-chain genes in skate are the most comprehensively characterized of all germline joined genes described to date (Litman et al. 1999).

A comparison of the complementarity-determining regions (CDR) and framework regions of the germline joined type I light-chain genes reveals a preferential distribution of substitutions in CDR positions. Furthermore, the findings of greater numbers of replacement than neutral substitutions are consistent with these genes being functional. CDR3 lengths in type I genes are identical, unlike type II joined genes found in skate and other cartilaginous fish species (see below). Finally, several cDNA transcripts have been characterized and shown to represent somatically varied (mutated) forms of the germline joined light-chain genes (ANDERSON et al. 1995). Type I light-chain gene clusters in skate lack the typical upstream octamer motif, which was identified in the type 1 (unjoined) clusters found in horned shark (SHAMBLOTT and LITMAN 1989), suggesting that rearranging and nonrearranging clusters are regulated differently. Taken together, the structural characteristics of joined type I genes strongly suggest that these genes are expressed and, as a family, exhibit patterns of sequence differences that are consistent with biological function.

Ig light-chain genes that differ significantly at the nucleotide and predicted amino acid sequence levels from the type I light-chain genes have been characterized in the sandbar shark (HOHMAN et al. 1992) and skate (RAST et al. 1994). These genes, which are classified as type II, are found in multiple clusters consisting of joined V and J elements as well as a single C_L (Fig. 1). Two different groups of sandbar shark type II light-chain genes have been recognized on the basis of noncoding sequences, which include the upstream octamer and TATA box motifs (HOHMAN et al. 1995). Although the joining phenomenon has been confirmed definitively in only four of what are estimated to be approx. 180–200 type II clusters, PCR analyses of other genomic clones are consistent with the germline joining in the majority of if not all type II light-chain genes in sandbar shark (HOHMAN et al. 1993; MARCHALONIS et al. 1998a).

5 The Novel Immune-Type Receptor Genes

An additional Ig/TCR-like, nonrearranging multigene family termed NITR has been identified recently in Southern pufferfish (*Spheroides nephelus*; STRONG et al. 1999), catfish (*Ictalurus punctatus*; N. Hawke and R. Haire, unpublished observations), and zebrafish (*Danio rerio*; J. Yoder, unpublished observations). Initially, NITR genes were isolated from pufferfish using a PCR strategy that had been employed previously to identify V regions of TCR genes (RAST et al. 1995; LITMAN et al. 1999). The prototypic NITR gene (SN193) consists of a V domain, and a C2 domain as well as transmembrane and cytoplasmic regions. The NITR V region exhibits remarkable similarity to TCR and Ig V regions, beyond that which is displayed by other nonrearranging, V region-containing proteins (see Du Pasquier, this volume). A J-like region is located 3′ of V and the NITR genes are VJ joined. P1 artificial chromosomes that hybridize with probes specific for SN193 and several other candidate NITR genes were identified in tandem linkage over a region greater than 200kb. NITR genes exhibit Ig/TCR-like diversified V regions, most of which contain immunoreceptor tyrosine-based inhibitory motifs (ITIMs) in their cytoplasmic regions (Strong et al. 1999). NITR genes may represent a link between "conventional" adaptive immune recognition, as found in Ig and TCR, and innate functions that are mediated by inhibitory recognition molecules (such as certain natural killer receptors that are members of a newly defined superfamily of ITIM-containing genes, the inhibitory receptor superfamily, IRS; Lanier 1998). NITR orthologs have not yet been identified in any species other than in teleost fish. Efforts are underway to determine the function of NITRs and to address whether these genes are an evolutionary specialization of teleost fish or have a broader phylogenetic distribution.

6 Evolution of Unjoined Genes

Since the initial description of the phenomena of germline joining, there has been considerable speculation as to its phylogenetic origin. In order to address this issue it is necessary to note that the current prevailing consensus is that all Ig genes are derived from a single progenitor gene which has undergone multiple gene and chromosomal duplication and recombination events, thereby increasing the numbers of the genes (segments). The question then arises as to whether the joined genes characterized in modern representatives of phylogenetically ancient species is a precursor form of the rearranging Ig system or reflects a derived character. Both hypotheses are plausible; however, it is instructive to further consider the origins of the somatic rearrangement itself.

Many possible scenarios could explain how a primordial "Ig" gene evolved to its "modern" segmented, rearranging form. The transposase activity of the RAG proteins (Agrawal et al. 1998; Hiom et al. 1998) and the observation that RSSs share limited sequence homology to the ends of certain invertebrate transposons (Dreyfus 1992) support the hypothesis that the primordial Ig gene was segmented by a transposition event (Sakano et al. 1979; Thompson 1995) which introduced inverted repeats into the gene. This event may have involved an ancient RAG-like transposon inserting into the primordial Ig gene and subsequently transposing out leaving behind inverted repeats as its "footprint"; these repeats could then have evolved through genetic drift to become the modern RSS. Alternatively, the RAG proteins and the RSS may have arisen independently. Modern RSSs may be remnants of a non-RAG related transposon which since has lost the ability to transpose; the RSS could have evolved to be recognized by the RAG proteins. It also is possible that vertebrate RSSs have evolved through genetic drift of intronic segments of the primordial Ig gene. Support for this last possibility is seen in the form and function of centromere protein-B (CENP-B), which shares sequence similarity with the pogo superfamily of transposons and binds to "CENP-B boxes" in centromeric satellite DNA. CENP-B is thought to promote the nicking of centromeric DNA and thereby promote recombination (Kipling and Warburton 1997). However, in the African green monkey the CENP-B protein is present, but the centromeric CENP-B boxes are only weakly homologous to prototypes (Goldberg et al. 1996). This observation demonstrates that the CENP-B boxes are not essential for centromere function and suggests that RSS could have resulted from genetic drift.

7 Evolution of Joined Genes

The most direct explanation for the evolution of the joined genes observed in cartilaginous fish is that they escaped the control of the RAG genes and thus, are representative of the primordial Ig receptor genes. However, an alternative hypothesis, which is supported by a number of convincing arguments, is that the joined genes are derived from unjoined genes. First, joined genes are seen only in cartilaginous fish in which the segmental elements of unjoined clusters are typically separated by approx. 200–350bp (Shamblott and Litman 1989); such close linkage presumably would promote recombination. The demonstration that RAG expression is not limited to early ontogeny, but rather occurs in germinal centers, offers support for such a possibility (Zheng et al. 1998). Joined heavy-chain genes bear hallmarks of the junctional variation observed after segmental recombination of unjoined Ig gene clusters in sharks and skates (Kokubu et al. 1988; G. Litman, unpublished observations). Specifically, D core sequences can be recognized in the regions of joined IgM genes that correspond to CDR3, which is entirely typical of that seen in segmental rearrangement (Hinds-Frey et al. 1993). Finally, evidence for the duplication and diversification of a multigene family of germline joined type I light-chain genes in skate has been presented (Anderson et al. 1995).

The phenomenon of germline joining appears to be restricted to Ig heavy and light-chain genes as no evidence has been obtained to date suggesting that TCR genes exhibit equivalent joining. However, in catfish, a duplicated IgH locus has been described in which a VDJ segment is found downstream of multiple V_H elements and upstream of a single J segment. The unjoined V_H and J segments that are adjacent to the joined VDJ segment contain the expected 3′ and 5′ RSSs, respectively (Ghaffari and Lobb 1999). Duplication of gene loci occurs frequently in the genomes of ray-finned fish; however, the occurrence of a joined VDJ segment suggests that the second cluster underwent gene rearrangement after its duplication. This IgH duplication event in catfish is somewhat reminiscent of the IgH loci in chicken which contains one functional V_H and approx. 80 upstream "pseudo" V_H segments. All of the pseudo V_H segments analyzed were VD-joined, some of which also contained J-like sequences and might be considered to be VDJ-joined (Reynaud et al. 1989). It also is noteworthy that the mammalian CD8β and VpreB genes, which are nonrearranging and encode Ig-like V domains, are located close to rearranging genes in the mouse and have functions that are consistent with an ancient association with these antigen recognition systems (Litman et al. 1999). As with most phenomenon that are considered in an evolutionary context, it is difficult to establish an unequivocal proof for the evolution of germline joining; however, evidence accumulated to date supports our working hypothesis that the joining process is derived rather than being representative of an ancestral character.

8 Summary

Immunoglobulin gene diversity has been characterized to varying degrees in modern representatives of all of the major radiations of cartilaginous fish. A pattern of overall chromosomal relationships of the various types of joined and unjoined Ig gene clusters is suggested in which the essential features are: (a) both Ig heavy and light-chain gene clusters occur on multiple chromosomes, (b) various classes of Ig are interspersed, (c) not all individual gene loci appear to be closely linked (Fig. 2). The cluster-type Ig gene system appears to be a series of (potentially) individually regulated loci analogous in part to the olfactory receptor gene system (Buck and Axel 1991) and markedly distinct from Ig loci in other vertebrate groups and TCR genes. Such a system would be ideal for the creation of variation in both form and function in a large number of clusters while preserving or partially preserving specificity in a number of other gene clusters. The full range of joined genes and the relative number of joined genes (as relates to unjoined genes), have yet to be determined. Nevertheless, a number of conclusions can be drawn: (a) four distinct forms of heavy-chain joining have been identified (VDD-J, VD-DJ, V-D-DJ, and VDJ; Fig. 1); (b) light-chain genes, which possess only two recombining elements, can be found in either unjoined (V-J) or joined (VJ) forms (Fig. 1); (c) physical linkage between individual joined and unjoined genes has not been established, although such investigations have not been pursued in a significantly rigorous manner as to rule out this possibility; (d) joined light-chain genes are expressed and can be somatically mutated.

Can germline joining be viewed as an ancestral character? The answer to this needs to be considered in the context of an overall system in which the level of structural and functional redundancy is extremely high. Joining is an adaptation that is unique to multicluster gene families. The phenomenon overcomes the possibility of not generating a specific form of a receptor, a major shortcoming of conventional rearranging Ig and TCR gene systems. The limitation of encoding specific receptors is compensated through large numbers of additional gene clusters that retain the capacity to rearrange and generate new specificities. Commitment of a V region to diverse, fixed specificity also is a property of the NITR genes, which although not related closely to Ig in a structural sense, may reflect an analogous phenomena. The possibility that immune-type diversity is achieved in the absence of somatic rearrangement and that remnants of such systems could be operative in immune recognition in contemporary vertebrates is of extraordinary significance in terms of our overall understanding of the relationships between adaptive and innate immune recognition.

Acknowledgements. We thank Barbara Pryor for editorial assistance and Dr. Tim Bestor for helpful discussions. This work was supported by grants to G.W.L. by the National Institutes of Health, the Valerie Fund, and Pediatric Cancer Foundation. J.A.Y. is a Moffitt Fellow of the H. Lee Moffitt Cancer Center, Tampa, Florida.

References

Agrawal A, Eastman QM, Schatz DG (1998) Transposition mediated by RAG1 and RAG2 and its implications for the evolution of the immune system. Nature 394:744–751

Anderson M, Amemiya C, Luer C, Litman R, Rast J, Niimura Y, Litman G (1994) Complete genomic sequence and patterns of transcription of a member of an unusual family of closely related, chromosomally dispersed immunoglobulin gene clusters in *Raja*. Int Immunol 6:1661–1670

Anderson MK, Shamblott MJ, Litman RT, Litman GW (1995) The generation of immunoglobulin light chain gene diversity in *Raja erinacea* is not associated with somatic rearrangement, an exception to a central paradigm of B cell immunity. J Exp Med 181:109–119

Anderson MK, Strong SJ, Litman RT, Luer CA, Amemiya CT, Rast JP, Litman GW (1999) A long form of the IgX gene in the skate exhibits a striking resemblance to the novel IgW and IgNARC genes in the shark. Immunogenetics 49:56–67

Bernstein RM, Schluter SF, Shen S, Marchalonis JJ (1996) A new high molecular weight immunoglobulin class from the carcharhine shark: implications for the properties of the primordial immunoglobulin. Proc Natl Acad Sci USA 93:3289–3293

Buck L, Axel R (1991) A novel multigene family may encode odorant receptors: a molecular basis for odor recognition. Cell 65:175–197

Diaz M, Greenberg AS, Flajnik MF (1998) Somatic hypermutation of the new antigen receptor gene (NAR) in the nurse shark does not generate the repertoire: possible role in antigen-driven reactions in the absence of germinal centers. Proc Natl Acad Sci USA 95:14343–14348

Dreyfus DH (1992) Evidence suggesting an evolutionary relationship between transposable elements and immune system recombination sequences. Mol Immunol 29:807–810

Ghaffari SH, Lobb CJ (1999) Structure and genomic organization of a second cluster of immunoglobulin heavy chain gene segments in the channel catfish. J Immunol 162:1519–1529

Goldberg IG, Sawhney H, Pluta AF, Warburton PE, Earnshaw WC (1996) Surprising deficiency of CENP-B binding sites in African green monkey alpha-satellite DNA: implications for CENP-B function at centromeres. Mol Cell Biol 16:5156–5168

Grawunder U, West RB, Lieber MR (1998) Antigen receptor gene rearrangement. Cur Op Immunol 10:172–180

Greenberg AS, Steiner L, Kasahara M, Flajnik MF (1993) Isolation of a shark immunoglobulin light chain cDNA clone encoding a protein resembling mammalian type kappa light chains: implications for the evolution of light chains. Proc Natl Acad Sci USA 90:10603–10607

Greenberg AS, Avila D, Hughes M, Hughes A, McKinney EC, Flajnik MF (1995) A new antigen receptor gene family that undergoes rearrangement and extensive somatic diversification in sharks. Nature 374:168–173

Greenberg AS, Hughes AL, Guo J, Avila D, McKinney EC, Flajnik MF (1996) A novel "chimeric" antibody class in cartilaginous fish: IgM may not be the primordial immunoglobulin. Eur J Immunol 26:1123–1129

Harding FA, Amemiya CT, Litman RT, Cohen N, Litman GW (1990a) Two distinct immunoglobulin heavy chain isotypes in a primitive, cartilaginous fish, *Raja erinacea*. Nucleic Acids Res 18:6369–6376

Harding FA, Cohen N, Litman GW (1990b) Immunoglobulin heavy chain gene organization and complexity in the skate, *Raja erinacea*. Nucleic Acids Res 18:1015–1020

Hinds KR, Litman GW (1986) Major reorganization of immunoglobulin V_H segmental elements during vertebrate evolution. Nature 320:546–549

Hinds-Frey KR, Nishikata H, Litman RT, Litman GW (1993) Somatic variation precedes extensive diversification of germline sequences and combinatorial joining in the evolution of immunoglobulin heavy chain diversity. J Exp Med 178:825–834

Hiom K, Melek M, Gellert M (1998) DNA transposition by the RAG1 and RAG2 proteins: a possible source of oncogenic translocations. Cell 94:463–470

Hohman VS, Schluter SF, Marchalonis JJ (1992) Complete sequence of a cDNA clone specifying sandbar shark immunoglobulin light chain: gene organization and implications for the evolution of light chains. Proc Natl Acad Sci USA 89:276–280

Hohman VS, Schuchman DB, Schluter SF, Marchalonis JJ (1993) Genomic clone for the sandbar shark lambda light chain: generation of diversity in the absence of gene rearrangement. Proc Natl Acad Sci USA 90:9882–9886

Hohman VS, Schluter SF, Marchalonis JJ (1995) Diversity of Ig light chain clusters in the sandbar shark (*Carcharhinus plumbeus*). J Immunol 155:3922–3928

Kipling D, Warburton PE (1997) Centromeres, CENP-B and Tigger too. TIGS 13:141–145

Kobayashi K, Tomonaga S (1988) The second immunoglobulin class is commonly present in cartilaginous fish belonging to the order rajiformes. Mol Immunol 25:115–120

Kokubu F, Hinds K, Litman R, Shamblott MJ, Litman GW (1987) Extensive families of constant region genes in a phylogenetically primitive vertebrate indicate an additional level of immunoglobulin complexity. Proc Natl Acad Sci USA 84:5868–5872

Kokubu F, Litman R, Shamblott MJ, Hinds K, Litman GW (1988) Diverse organization of immunoglobulin V_H gene loci in a primitive vertebrate. EMBO J 7:3413–3422

Lanier LL (1998) NK cell receptors. Ann Rev Immunol 16:359–393

Litman GW, Rast JP, Shamblott MJ, Haire RN, Hulst M, Roess W, Litman RT et al. (1993) Phylogenetic diversification of immunoglobulin genes and the antibody repertoire. Mol Biol Evol 10:60–72

Litman GW, Anderson MK, Rast JP (1999) Evolution of antigen binding receptors. Ann Rev Immunol 17:109–147

Marchalonis JJ, Schluter SF, Bernstein RM, Hohman VS (1998a) Antibodies of sharks: revolution and evolution. Immunol Rev 166:103–122

Marchalonis JJ, Schluter SF, Bernstein RM, Shen S, Edmundson AB (1998b) Phylogenetic emergence and molecular evolution of the immunoglobulin family. Adv Immunol 70:417–506

Rast JP, Anderson MK, Ota T, Litman RT, Margittai M, Shamblott MJ, Litman GW (1994) Immunoglobulin light chain class multiplicity and alternative organizational forms in early vertebrate phylogeny. Immunogenetics 40:83–99

Rast JP, Haire RN, Litman RT, Pross S, Litman GW (1995) Identification and characterization of T-cell antigen receptor related genes in phylogenetically diverse vertebrate species. Immunogenetics 42: 204–212

Rast JP, Amemiya CT, Litman RT, Strong SJ, Litman GW (1998) Distinct patterns of *IgH* structure and organization in a divergent lineage of chondrichthyan fishes. Immunogenetics 47:234–245

Rast JP, Litman GW (1998) Towards understanding the evolutionary origins and early diversification of rearranging antigen receptors. Immunol Rev 166:79–86

Reynaud C-A, Dahan A, Anquez V, Weill J-C (1989) Somatic hyperconversion diversifies the single V_H gene of the chicken with a high incidence in the D region. Cell 59:171–183

Roux KH, Greenberg AS, Greene L, Strelets L, Avila D, McKinney EC, Flajnik MF (1998) Molecular convergence of the nurse shark (new) antigen receptor (NAR) and unusual mammalian immunoglobulins. Proc Natl Acad Sci USA 95:11804–11809

Sakano H, Huppi K, Heinrich G, Tonegawa S (1979) Sequences at the somatic recombination sites of immunoglobulin light-chain genes. Nature 280:288–294

Shamblott MJ, Litman GW (1989) Genomic organization and sequences of immunoglobulin light chain genes in a primitive vertebrate suggest coevolution of immunoglobulin gene organization. EMBO J 8:3733–3739

Strong SJ, Mueller MG, Litman RT, Hawke NA, Haire RN, Miracle AL, Rast JP et al. (1999) A novel family of diversified immune-type genes encoding V and J regions as well as ITIMs. Proc Natl Acad Sci USA, in press

Thompson CB (1995) New insights into V(D)J recombination and its role in the evolution of the immune system. Immunity 3:531–539

Zheng B, Han S, Spanopoulou E, Kelsoe G (1998) Immunoglobulin gene hypermutation in germinal centers is independent of the RAG-1 V(D)J recombinase. Immunol Rev 162:133–141

Somatic Diversification

Evolution and Somatic Diversification of Immunoglobulin Light Chains

S.S. LEE[1], A. GREENBERG[2], and E. HSU[1]

1 Introduction

Immunoglobulin light (L) chains and L chain genes may seem the poor man's version of the heavy (H) chains: the specificity of some antibody combining sites may appear to reside largely in the H chain portion (ZHU et al. 1984), most of the secreted camel antibodies even function without L chains (HAMERS-CASTERMAN et al. 1993), and terminal deoxynucleotidyl transferase may (in humans) or may not (in mouse) add N regions to L chain (VICTOR and CAPRA 1994; MILSTEIN et al. 1992) – it is as if L chain junctional diversity were secondary in importance to that of the H chain. Nonetheless, it is in the L chain system that complementarity-determining regions (CDR) were initially recognized (WU and KABAT 1970), that immunoglobulin (Ig) gene rearrangement was demonstrated (HOZUMI and TONEGAWA 1976); it is also where somatic hypermutation in antibody sequences was first deduced (WEIGERT et al. 1970) and where the phenomenon of gene conversion in

[1] Department of Physiology and Pharmacology, State University of New York, Health Science Center at Brooklyn, 450 Clarkson Avenue, Box 31, Brooklyn, NY 11203-2098, USA
[2] Department of Microbiology and Immunology, University of Miami School of Medicine, Miami, FL 33136, USA

vertebrates was discovered (Reynaud et al. 1987). L chain genes may thus be viewed as genetically stripped-down versions of H chain genes that can better serve to elucidate genetic and evolutionary trends in antibody structure and diversity.

This review examines the nature of L chain somatic diversification in evolution by appraising poikilothermic model systems where multiple L chain types have been found: the NS3, NS4, and NS5 genes in the nurse shark and their homologs in the horned shark and the σ, ρ, and λ L chain genes in the clawed toad, *Xenopus*. For information on other cold-blooded vertebrates, there is a recent article describing Ig and T cell receptor (TCR) sequences so far isolated (Litman et al. 1999).

Mammalian and bird L chain types have long been defined as homologs of the human κ and λ chains. However, in phylogenetically more primitive animals the L chain classifications are not clear. First, the shark and *Xenopus* possess more than two L chain types. One of the shark types resembles mammalian κ (Greenberg et al. 1993) and the other two are arguably λ-like (Shamblott and Litman 1989a,b). In *Xenopus* the ρ sequences cluster with mammalian κ sequences (Zezza et al. 1991), and the most recently isolated L chain gene appears to have λ characteristics (Haire et al. 1996) whereas related sequences have not been identified for σ. Thus, κ-like genes have been identified in the most primitive vertebrates, but any relationship to λ sequences is more difficult to conclude due to their greater divergence in evolution (Hsu and Steiner 1992).

The humoral immune response in cold-blooded vertebrates is different from that in mammals in several respects, the consistent findings being that the antibodies are much less heterogeneous and binding affinity does not increase greatly with time or repeated boosts (for a review, see Hsu 1998). Limited heterogeneity and poor maturation have given the impression that the poikilothermic antibody repertoire is not as great as in mammals, which has been strengthened by the discovery of lack of combinatorial diversity in cartilaginous fishes (Hinds and Litman 1986; Shamblott and Litman 1989b). The Ig gene organization is such that rearrangement does not take place between clusters (Fig. 1), and in addition, many of the clusters consist of gene segments that are already joined in the germline, decreasing the potential junctional diversity in the Ig pool (Kokubu et al. 1988). It would seem that these organizational restrictions, apparent at the germline level, should have a great impact on expressed diversity. Ig diversity is difficult to evaluate in H chains of cold-blooded vertebrates because so many genes are involved. This review examines Ig diversity in a fish and an amphibian, showing the influence of somatic mechanisms, the extent of which has been not reported for any mammal.

2 L Chain Genes and Gene Organization in Cartilaginous Fishes

L chain sequences have been isolated from representatives of the two subclasses of Chondrichthyes, Elasmobranchii (the horned shark *Heterodontous francisci*, the

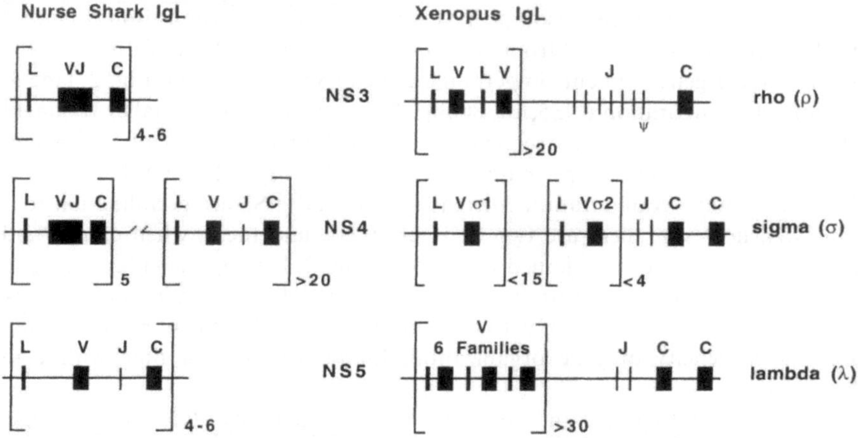

Fig. 1. Germline gene organization of the nurse shark and *Xenopus* IgL loci. The nurse shark L chain gene types, NS3, NS4, and NS5 are arranged as multiple clusters (GREENBERG et al. 1993; GREENBERG 1994). All of the NS3 genes appear to be germline-joined, whereas the NS5 genes are not. In the NS4 some of the clusters are germline-joined while the majority are not (see text; S. Lee, unpublished data). In *Xenopus* the gene organization of the three L chain types resembles that of mammals (ZEZZA et al. 1991, 1992; STEWART et al. 1993; SCHWAGER et al. 1991; HAIRE et al. 1996). The estimated numbers of genes are taken from these sources; that for ρ genes is in part from JI et al., 1999 who have deduced a sixth functional J gene from cDNA sequence and place it 5′ of the known JH, a pseudogene sequence (as indicated) and the C exon (STEWART et al. 1993). Rearrangements to the pseudogene have also been isolated (JI et al., 1999)

nurse shark *Ginglymostoma cirratum*, the sandbar shark *Carcharhinus plumbeus*, the little skate *Raja erinacea*), and Holocephali (spotted ratfish, *Hydrolagus colliei*) (SHAMBLOTT and LITMAN 1989a,b; SCHLUTER et al. 1989; GREENBERG et al. 1993; RAST et al. 1994). The L chain loci are all in the cluster organization. LITMAN and coworkers have defined three L chain isotypes in the horned shark – types I, II, and III; the homologs in nurse shark are NS5, NS3, and NS4, respectively (Table 1).

There are two skate L chain isotypes, homologs of the horned shark type I and type II genes, and both are arranged as joined VJ in the germline. Although the horned shark type II genes are also germline-joined, their type I genes are not, which suggests that the *Raja* VJ fusion occurred after radiation of elasmobranchs in the Triassic (RAST et al. 1994). These fusion events are rare; the horned shark and nurse shark diverged up to 180 million years ago (CAPETTA et al. 1993), and in

Table 1. Comparison of nurse shark and horned shark L chain nucleotide sequences (percentage identity)

	Type I	Type II	Type III	NS3	NS4	NS5
Type I		60	57	62	60	**77**
Type II			57	**80**	57	61
Type III				59	**80**	54
NS3					59	56
NS4						54

general, the homologs of each isotype have similar genomic organization. Figure 2 shows PCR products amplified from erythrocyte genomic DNA, conforming to the sizes for V and J gene segments in germline-joined (NS3, ca. 315bp; Fig. 2, lane 3) or nonjoined configuration (NS5, ca. 800bp, Fig. 2, lane 9). The smaller fragment size was predicted from cDNA sequences, and the larger fragment of 800bp suggests an intron of about 500bp separating V and J. Thus type I and NS5 are in the traditional germline configuration whereas type II and NS3 are germline-joined.

In NS4, however, there are two PCR products, and these, when cloned and sequenced, were found to be joined (ca. 315bp, Fig. 2, lane 6) and nonjoined (ca. 850bp, Fig. 2, lane 6, arrow) germline genes. The type III homolog in horned shark has been reported by Lee et al. (1998) to be nonjoined, so that the nurse shark NS4 exhibits heterogeneity of gene organization, found for the first time within an L chain type.

Detection of both joined and nonjoined germline genes by PCR is not a simple matter, as illustrated by the NS4 experiment – the smaller PCR product is preferentially amplified. Figure 2 shows the results after 20 cycles; after 40 cycles the NS4 nonjoined band can barely be distinguished. To establish that all NS3 genes are germline-joined we eluted PCR products after 20 cycles in fractions from 300bp to 1.2kb and reamplified with combinations of five different 5′ oligonucleotide primers in leader/FR1/CDR1 and four 3′ primers in the J. In NS5 two bands appeared, but only the larger (nonjoined) DNA fragment was digested with restriction endonucleases *Kpn*I and *Xho*I (Fig. 2, lanes 11, 12), which detect conserved sites in FR2 and FR3, respectively. The amplified NS5 germline genes detected by this set of primers are therefore all nonjoined.

In both horned shark and nurse shark most L chain transcripts appear to originate from those with nonrearranged germline gene segments, type I and NS4, respectively. These two types are not homologs. In the case of NS4, the higher expression may be so because it is the largest gene family in the nurse shark.

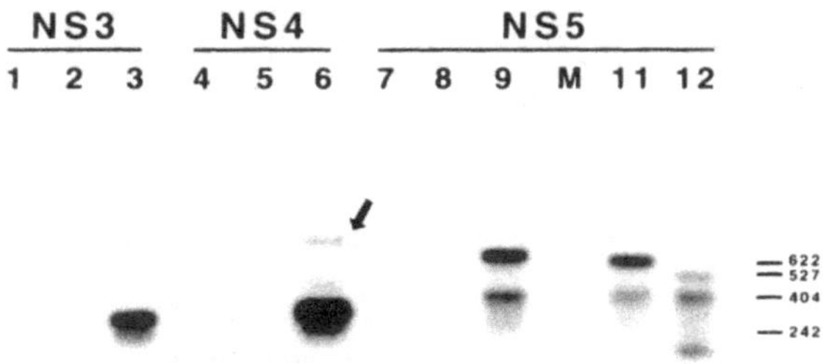

Fig. 2. Amplification of germline sequences of nurse shark L chain. PCR amplification was performed on erythrocyte DNA from an individual shark. Three sets of primers were used, the 5′ primer in the FR1 regions of NS3, NS4, or NS5, and the 3′ primer in the J sequences. *Lane 1*, 5′ NS3 alone; *lane 2*, 3′ NS3 alone; *lane 3*, 5′ and 3′ NS3 primers; *lane 4*, 5′ NS4 alone; *lane 5*, 3′ NS4 alone; *lane 6*, 5′ and 3′ NS4 primers; *lane 7*, 5′ NS5 alone; *lane 8*, 3′ NS5 alone; *lane 9*, 5′ and 3′ NS5 primers. The PCR products from amplification with the 5′ and 3′ NS5 primers (*lane 9*) were digested with *Kpn*I (*lane 11*) and *Xho*I (*lane 12*). The PCR products were transferred onto a nylon filter and each section hybridized with the appropriate probe. *Arrow*, the NS4 genes that are not germline-joined

There are estimated to be > 20 NS4 genes, 4–6 NS3 genes, and 4–6 NS5 genes, as found by genomic Southern blotting (GREENBERG et al. 1993; GREENBERG 1994).

The existence of germline-joined genes, in addition to the lack of combinatorial diversity as a result of the cluster organization, implies a lower molecular diversity in cartilaginous fish antibodies. This notion in turn would seem to explain the lack of affinity maturation in shark antibody responses (MÄKELÄ and LITMAN 1980). However, in examining nurse shark L chain gene transcripts we have found that, in the absence of combinatorial diversity there is unusually high junctional diversity, and although some L chain genes do not rearrange at all, they do diversify radically by somatic hypermutation.

2.1 Junctional Diversity in Shark L Chains

In the Ig light-chain literature, the CDR3 are not notable for junctional diversity. Unlike H chain junctions, little coding end processing occurs in L chains, and the site of recombination in L chain takes place within a limited area in the proximity of the recombination signal sequences (RSS; NADEL and FEENEY 1995). Few P regions are present, and N regions, if they exist at all, as they do in human L chains, are not extensive (MILSTEIN et al. 1992; VICTOR and CAPRA 1994). Thus the CDR3 junctional diversity is almost entirely germline-derived, except the codon generated by the somatic joining of the two gene segments. CDR3 lengths tend to be uniform, because these are determined by the number of nucleotides 3′ of the conserved $TG^{T}/_{C}$ in V and 5′ of the conserved $TT^{T}/_{C}$ in J sequences; and in multigene families the duplicated V genes do not usually carry deletions or insertions at the 3′ end, which forms the CDR3, while multiple L chain J germline gene segments do not differ in their 5′ ends by more than one or two nucleotides (KABAT et al. 1991). In fact, the L chain CDR3 sequences appeared to be so uniform, especially in length, to permit antibody combining site models to be constructed by reducing the somatically generated loops to only three canonical structures, one fewer, even, than the four given the L chain germline-encoded CDR1 (CHOTHIA et al. 1989). In contrast, of the six CDR H chain CDR3 was so diverse that modelling was were possible.

Most vertebrate L chain CDR3 vary over two or three codon lengths. We have found that, unlike previously described L chain pools from other vertebrates, the CDR3 variability of NS4 is extraordinarily high. In Fig. 3, PCR products were obtained from erythrocyte DNA and from peripheral blood leukocyte (PBL) cDNA. An end-labeling assay to determine CDR3 lengths was employed (DESRAVINES and HSU 1994). As shown in Fig. 3, PCR primers in the leader sequence and in J produced DNA fragments of about 376 bp from PBL cDNA. There is a conserved *Ava*II site in the NS4 framework 3 (FR3) that yields fragments of 262 bp and approx. 114 bp, the latter being the 3′ end of the L chain sequence containing the CDR3. The fragments were end-labeled, electrophoresed, and compared to a λ phage sequence for size. Fragments of 99, 102, 105, 108, 111, 114, 117, and 120

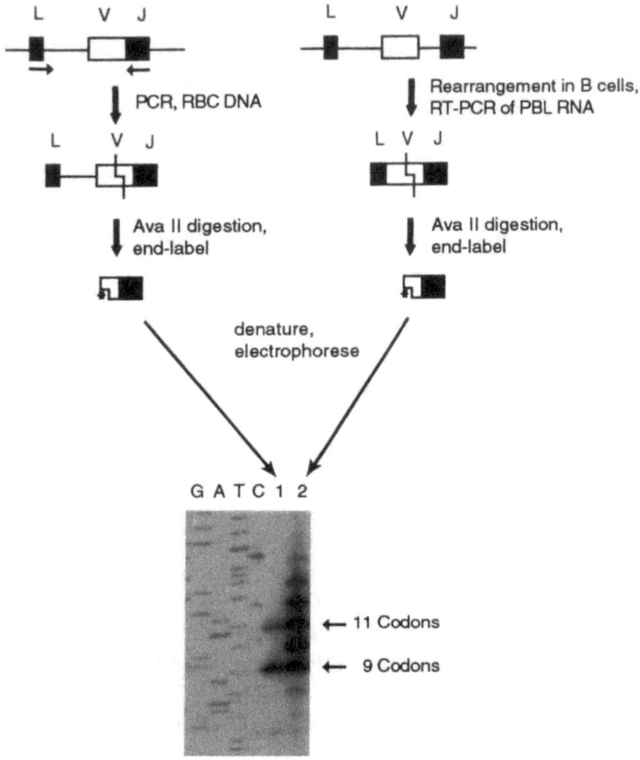

Fig. 3. End-labeling of PCR products from germline and expressed NS4 sequences. *Left*, PCR primers in the leader sequence and the J gene segment amplify fragments of 600 bp from nurse shark genomic DNA. These are the germline-joined genes. The PCR products were isolated and digested with *Ava*II, end-labeled with ^{32}P-dGTP and Klenow, denatured, and loaded onto a 5% acrylamide-urea gel. *Right*, Reverse transcriptase PCR was performed on PBL RNA from the same individual, using the same primers. Fragments of about 376 bp were isolated, digested with *Ava*II, end-labeled, denatured, and electrophoresed. *Photo*, M13mp18 phage sequence, reaction performed with −40 primer, read as shown, G, A, T, C. The sizes of the phage DNA were compared to the nurse shark fragments. The cDNA fragments ranged from 99 to 120 bp, and the CDR3 codon lengths are indicated (Desravines and Hsu 1994)

bases were distinguished (lane 2), and these correspond to fragments containing eight CDR3 lengths of 7–14 codons (see figure legend).

Since such a large spread of the shark NS4 CDR3 length variation has not been observed in other species, the DNA fragments were cloned and analyzed. In a small sample of 12 cDNA sequences six CDR3 of 9–14 codons were found, supporting the observations from the end-labeling technique. In comparison, PCR products obtained from erythrocyte DNA are restricted to two fragment sizes; the CDR3 sizes of 8 and 11 codons correspond to the germline-joined NS4 genes isolated from the same animal that carry *Ava II* sites (Lee et al. 1998).

Human λ appears to be the most diverse with CDR3 sizes of 9–12 codons (20 sequences from Kabat et al. 1991). The variety of human λ CDR3 lengths is due to the variable lengths of the 3′ coding end of the V gene segments. Most expressed mouse λ sequences are encoded by two germline genes (Vλ1, Vλ2) with similar 3′ coding ends; of 32 cDNA sequences listed (Kabat et al. 1991) all murine sequences carry CDR3 of 9 codons.

In contrast, in the chicken, with one single functional V and J germline gene segments, the rearranged CDR3 range from 9–12 codons (Reynaud et al. 1987). Since sequences cloned from 3-week-old chicken are somewhat more variable than those from 18-day embryos, the CDR3 length diversity in this case is probably a result of the gene conversion process.

At this time not many germline NS4 genes have been analyzed, and therefore the CDR3 length diversity of the shark cDNA sequences could be due either to gene conversion events or to variable coding end sequences of the multiple NS4 V and J gene segments. It is also possible that coding end processing occurs to a greater extent than found in tetrapod L chains. The nurse shark expressed sequences show an unusually large range of L chain CDR3 loop variability, exceeding that so far found in other vertebrates, and belie the impression created by the presence of germline-joined genes in chondrichthyan fish. In *Raja* the two L chain isotypes are reported to be entirely germlined-joined (Anderson et al. 1995), but perhaps other, unjoined *Raja* L chain genes have yet to be isolated.

The presence of three isotypes and high NS4 junctional diversity potentially give the nurse shark the most extensively diverse of L chain primary repertoires.

2.2 Somatic Hypermutation in Germline-Joined Shark L Chain Genes

Somatic hypermutation in cold-blooded vertebrates was first studied in *Xenopus* H chains (Wilson et al. 1992). The mutation rate was examined and estimated to be within an order of magnitude of that of mouse, and thus the availability of mutants was not the restricting factor for the poor affinity maturation observed during the *Xenopus* antibody response. With a tepid R/S ratio in the CDR compared to the FR, it was suggested that, although mutants were generated after immunization, they were poorly selected. One unusual feature in the base substitutions is a bias for alterations at GC rather than AT; changes from GC constituted 93% of the 56 mutations. Similar results can be seen in the horned shark H chain (Du Pasquier et al. 1998).

A third rearranging gene system for secreted and cell surface receptors was discovered in the nurse shark (Greenberg et al., 1995). The new antigen receptor (NAR) expressed sequences hypermutates, but the mechanism is apparently different from horned shark H chain, as there is no GC bias (Du Pasquier et al. 1998).

The studies on NAR establish once and for all that hypermutation has been an important mechanism of diversification in the most primitive vertebrate immune systems. It was suggested that in the absence of combinatorial and junctional diversity, the germline-joined Ig genes may also diversify by somatic hypermutation (ANDERSON et al. 1995). In the nurse shark we have found a L chain type consisting of a few genes where this can be convincingly demonstrated.

There are four to six NS3 genes in a nurse shark individual (S. Lee, not shown), and one combination of PCR primers amplifies only two sequences, as found in 19 clones obtained from the germline of this animal (Fig. 4, gNS3–1, gNS3–9). Of 15 cDNA sequences obtained from the same individual, five are identical to the germline sequences; there are 1–10 nucleotide differences found in the other ten. In 4770 bp of coding sequence 50 substitutions were found, and this frequency (1.0%) greatly exceeds that of Taq-induced errors (0.03%) using the same primers for other experiments. Thus, two of the substitutions may have originated in vitro, and this does not seriously affect the conclusions reached.

The NS3 substitutions are different in nature from those found in Ig from other vertebrates. There is no clear preference for transitions (22 substitutions) over transversions (28 substitutions). There appears to be a bias for alterations at G or A as opposed to T or C (Table 2); in this, NS3 is unlike *Xenopus* or horned shark H chain, both of which exhibit a GC preference. The GA bias in NS3 (74%) may be interpreted to resemble that found in sheep (69%), where somatic hypermutation is suggested to be an antigen-independent process occurring during B cell development (REYNAUD et al. 1995). However, a more conservative interpretation is that the bias in NS3 may result partly from its GA-rich CDR3, in which 18/27 germline positions are purines, and which contains more mutations (23/50) than CDR1 or CDR2. Of the 23 substitutions in CDR3, 18 are alterations from GA. Discounting the CDR3 mutations, the remaining changes in FR1-FR3 do not show a significant bias, and it is more similar in this way to shark NAR (DU PASQUIER et al. 1998).

The changes observed in the NS3 cDNA were probably generated by hypermutation rather than gene conversion since no consistent pattern of changes appear and so few genomic NS3 templates are available. One unusual pattern in the substitutions is the high incidence of tandem mutations; of 28 sites containing the 50 mutations, 14 consist of two to four nucleotide tandem mutations. In contrast, none of the 45 *Xenopus* H chain mutations occurred in tandem (WILSON et al. 1992), and of the 55 horned shark H chain mutations there were only 4 di- and trinucleotide changes (HINDS-FREY et al. 1993). In sheep L chain genes tandem mutations do occur (18 tandem mutations/128 sites in 21 sequences), but not to as remarkable an extent as the nurse shark NS3 L chain. Such an observation has only been made in nurse shark NAR sequences (M. Flajnik, personal communication) and in antioxazolone L chain sequences from mice deficient for a gene for mismatch repair, postmeiotic segregation 2 (*Pms2*; WINTER et al. 1998).

Many of the mutations (19/50) occur within the motif RGYW, where R is G or A, Y is T or C, and W is T or A, a putative hotspot for hypermutation in Ig (ROGOZIN and KOLCHANOV 1992). All the CDR1 changes occur at these positions. In contrast, in CDR3, where half of the NS3 mutations are found, most of the changes (20/23) occur outside the RGYW motif. Many (14/23) are clustered primarily at the fourth and fifth codon of CDR3, in the middle of the loop, and all are replacement changes. Although most of the residues of the L chain CDR3 loop

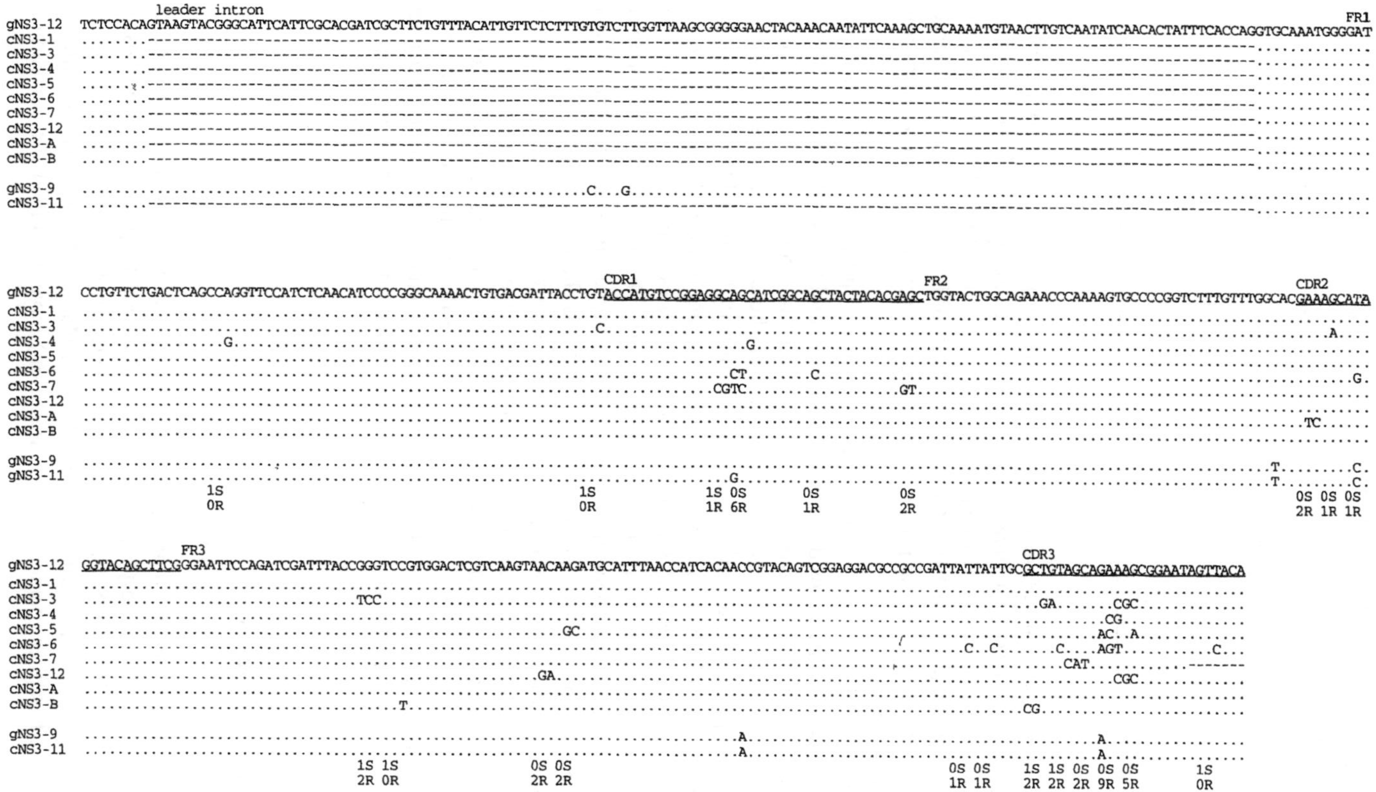

Fig. 4. Germline and expressed NS3 L chain sequences. Two kinds of germline NS3 sequences were obtained from 21 cloned PCR fragments; these are gNS3–1 and gNS3–9. PCR amplifications with erythrocyte DNA and with PBL RNA were carried out with primers in the leader and in the J. All cDNA sequences are aligned to gNS3–1; *dashes*, gaps in the leader intron and elsewhere; *dots*, identities. FR1, CDR1, FR2, CDR2, FR3 are *underlined and labeled*. Nucleotide substitutions that result in synonymous (*S*) or replacement changes (*R*) for the codon affected are enumerated at the bottom

Table 2. Nucleotide substitutions in NS3 L chain V sequences

To	From				
	G	A	T	C	Total
G		9	2	3	14
A	6		0	1	7
T	3	3		2	8
C	9	7	5		21
Total	18	19	7	6	50

have been observed in contact with ligand, from crystallographic studies of antigen-antibody complexes (PADLAN 1994), these positions are among the favored ones.

The overall replacement to silent substitution ratio is 4 (40R:10S), but if separated by distribution over FR (6R:6S = 1) and CDR (34R:4S = 8.5), there is a large R/S ratio for the CDR. These preliminary results suggest that the mutations are selected for in NS3.

In summary, the pattern of mutation in *Xenopus* and horned shark H chain sequences appear to differ considerably from those found in the nurse shark NS3 L chain sequences. The presence of tandem mutations and clear lack of GC bias is shared with the nurse shark NAR system. From these preliminary results it is highly surprising that B cell somatic hypermutation in H chain from a representative of one elasmobranch order should differ so significantly from L chain of another. At this time, the NS3 mutation characteristics appear to resemble the NAR most closely, and both are from the nurse shark. One possibility may be that the repair enzyme system theoretically linked to the hypermutation process differs among animals. However, the expression of the NS3 L chain has not been studied, and the function of NAR is not known, and it is possible that both are under a developmental selection pressure that may go undetected in the H chains.

From this small sample of sequences it is clear that, although the NS3 L chain genes are germline-joined, they nonetheless diversify extensively by somatic hypermutation. It has been suggested that the NAR changes are antigen-driven (DIAZ et al. 1998), but this is currently difficult to demonstrate for NS3.

3 *Xenopus* L Chains

There are three *Xenopus* L chain types, initially defined by monoclonal antibodies (HSU et al. 1991). These polypeptides differ in their electrophoretic properties in sodium dodecyl sulfate (SDS) polyacrylamide gel electrophoresis (PAGE) as well as by preferential association with various H chain classes. Recent experiments by Haire and Du Pasquier (personal communication) have examined the correlation between the L chain genes and the proteins thus defined, and these are referred to as σ, ρ, and λ (Table 3).

Unlike the cartilaginous fishes, all the *Xenopus* IgH (Schwager et al. 1988) and IgL loci are organized as that of mouse and man (Fig. 1). Based on data derived from genomic and cDNA sequences, the ρ locus consists of more than 20 Vρ and at least 6 Jρ gene segments and one Cρ exon (Zezza et al. 1991, 1992; Stewart et al. 1993; Ji et al., 1999). The σ locus consists of two V families, multiple Vσ1 and a few Vσ2, and two linked Cσ genes (Schwager et al. 1991). The λ locus is much more complex, including 6 Vλ families with a total of more than 30 gene segments, 2 Jλ gene segments and 2 Cλ exons (Haire et al. 1996).

This potential diversity, with so many different V and J elements, appears to be on par with mammalian L chain systems. However, in antibody responses where the L chain heterogeneity has been analyzed, the *Xenopus* L chains appear extremely restricted (Hsu et al. 1991).

3.1 *Xenopus* L Chain Heterogeneity

The sizes of the L chain polypeptides, as deduced from the cDNA sequences, are not readily correlated with their apparent molecular masses, as measured by electrophoretic mobility in SDS-PAGE (Table 3). The discrepancy may arise from sequence idiosyncracies having to do with local charge density or with hydrophobicity (Helenius and Simons 1972). The amino acid sequences of σ and λ predict more hydrophobicity, especially in CDR3, than for ρ (Kyte and Doolittle 1982), and this may cause the differential electrophoretic mobility. The few overlapping bands in both regions could be accounted for by individual sequence compositional differences of this nature.

Xenopus Ig immunoprecipitated with rabbit antiserum after lipopolysaccharide stimulation of splenocytes included all three subpopulations (Hsu et al. 1991). Analysis by two-dimensional gel electrophoresis revealed a complex pattern of 45 distinct bands, a number similar to that of mouse κ chains secreted after lipopolysaccharide stimulation (Weiss et al. 1984); Ig secreted after mitogen stimulation are not mutated, and therefore these experiments are comparable. The *Xenopus* components were identified by aligning and comparing them with bands obtained

Table 3. Properties of *Xenopus* L chains

L Chain Type	Apparent mass in SDS-PAGE[a] (kDa)	Predicted from cDNA[b] (Da)	H chain association	Mab[a]
ρ	29	28,595	μ, υ, χ	13 C6
σ	27	28,319	μ, χ	13 B2
λ	25	27,896	μ, υ, χ	1 E9

[a] Analyses from monoclonal antibodies reported in Hsu et al. 1992.
[b] Representative sequences were from Stewart et al. (1993), Schwager et al. (1991), Haire et al. (1996). Sizes were computed using individual amino acid values.

by the monoclonal antibodies. Nineteen unique bands of 25 belonged to ρ, 8 of 16 to σ, and 10 of 16 to λ; the other bands were overlapping between two or all groups (Fig. 5).

One would expect λ to be the most diverse and σ the least. Instead, the number and diversity of proteins secreted from stimulated splenocytes appear to be comparable between these two L chain isotypes. There is an additional interesting discrepancy between the λ proteins secreted by the splenocyte population and the circulating λ proteins. Two-dimensional gel electrophoresis revealed, in animal after animal, one predominant λ band in the serum; most of the λ bands found in the culture supernatant were absent in serum. Barring an experimental artifact created by the labeling technique, it appears that one λ gene or family is selectively used in vivo by many *Xenopus*.

Most of the IgY antibodies contain ρ. Anti-IgY monoclonal antibodies immunoprecipitated L chains mostly comigrating in the distinct ρ region. Conversely, anti-ρ or anti-λ immunoprecipitated all three H chains, whereas almost no υ H chain were found with anti-σ. Thus, μ and χ are associated with all three L chain types while υ is associated mostly with ρ and some λ. It is not clear why this preferential association exists – perhaps the receptors with specificity formed with σ L chain tend to bind ligands that are T-independent, and this isotype switch therefore does not occur in these cells.

The L chains of anti-hapten antibodies immunoprecipitated with antigen (dinitrophenyl, DNP) were analyzed by SDS-PAGE in isogenic *Xenopus* (Hsu et al. 1991). All four animals produced antibodies carrying L chains that were deduced to be only of the ρ type, as anti-σ and anti-λ antibodies did not bind the dominant 28–29kDa anti-DNP bands (E. Hsu, unpublished results). By two-dimensional gel electrophoresis there are at most one to three different L chains produced by one individual. In mammals, anti-hapten L chains carry leucine at position 96, which forms a binding pocket within the combining site (Rudikoff et al. 1981; S. Weiss, personal communication: The anti-DNP specificity of the Sp7 hybridoma is abrogated with a substitution in the λ1 L chain from leucine to tryptophan at position 96.); alternative amino acids, especially those with aromatic side chains, create too a shallow site. Although the six germline Jρ do not have leucine in that

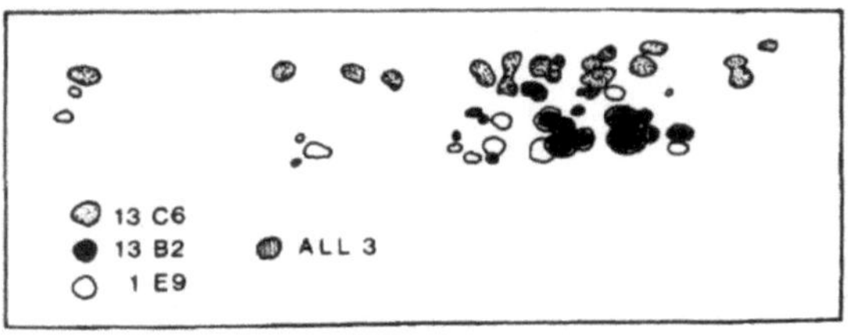

Fig. 5. Two-dimensional gel electrophoresis of *Xenopus* L chains. Metabolically labeled proteins from spleen culture were immunoprecipitated using rabbit antiserum or monoclonal antibodies to the three L chain types and electrophoresed. First dimension (horizontal) isoelectric focusing, pH 3–10, second dimension SDS-PAGE. The spots obtained from the monoclonal antibodies were compared with those from rabbit antiserum. The graphic interpretation (Hsu et al. 1991) is presented

position, half of all the ρ cDNAs carry a leucine at position 96 generated by rearrangement (Ji et al., 1999). The λ and σ junctions seldom have leucine at that position. Among these animals the anti-DNP L chain band in SDS-PAGE varies by a few hundred daltons; no two have the same electrophoretic mobility. This is interesting, as it suggests that the sequence contents differ. Thus, there is sufficient diversity in the *Xenopus* primary repertoire such that, per individual, different L chains were involved in the response. Nonetheless, for any one individual's response, the IgY antibodies were oligoclonal, as measured by L chain components.

Xenopus IgY responses to a number of antigens (DNP, sheep red blood cells, phosphorylcholine) are limited in heterogeneity (Wabl and Du Pasquier 1976; Du Pasquier and Wabl 1978), and it was suggested that somatic diversification in cold-blooded vertebrates is influenced by factors outside the immune system (Du Pasquier 1982). The current notion is that a poor selection of emerging mutants prevents the *Xenopus* response from acquiring heterogeneous, high-affinity antibodies (Wilson et al. 1992).

4 Concluding Remarks

The ability of cartilaginous fishes to diversify their Ig is perceived as inferior to tetrapods by virtue of what appears to be restrictions inherent in the gene organization. The results that we present here demonstrate that there is considerable molecular diversity in expressed shark Ig sequences, and that these result from somatic diversification mechanisms whose activities exceed what has been previously described in warm-blooded animals. This was possible only by inspecting the L chain cDNA sequences.

Whereas the cluster organization of nurse shark IgL loci may reduce combinatorial diversity, the extent of the junctional diversity found in cDNA is considerably greater than any other vertebrates. Some nurse shark L chain clusters contain germline-rearranged VJ, and while this arrangement precludes junctional diversity, these particular genes are extensively modified by somatic hypermutation. In sum, the potential repertoire could not be readily and accurately predicted from such information as gene organization, the numbers of genes, and even the apparent limitations of some somatic diversification mechanisms.

The shark Ig diversification may even result in a different spectrum of antibody-combining sites. The NS4 L chains may, by virtue of the extensive CDR3 length variability, have a different relationship to H chain than do the other, conventional L chain types, including those of tetrapods.

Xenopus L chain genes, on the other hand, carry a large number and diversity of genetic elements comparable to mammals. The protein heterogeneity is not correlated with what is anticipated from the genetic information, but from two-dimensional gel electrophoresis, the overall diversity appears to be similar to mouse. Despite a *richesse* in genes and diversification pathways in *Xenopus*, the antibody response is nonetheless markedly different from mammals. In neither the shark nor the amphibian *Xenopus* can the poor heterogeneity and affinity maturation in antibody responses be attributed to an impoverished primary repertoire; this is demonstrated by the extensive molecular diversity of their L chains.

Acknowledgements. We thank Martin Flajnik, Eduardo Padlan, and Barbara Stitt for discussions and Siegfried Weiss, Robert Haire, and Louis Du Pasquier for generously sharing results with us. We also thank Louis Du Pasquier for critical reading of the manuscript. This work was supported by grant MCB 9996049 from the National Science Foundation.

References

Anderson MK, Shamblott MJ, Litman R, Litman GW (1995) Generation of immunoglobulin light chain gene diversity in *Raja erinacea* is not associated with somatic rearrangement, an exception to a central paradigm of B cell immunity. J Exp Med 182:109–119

Cappetta H, Duffin C, Zidek J (1993) Chrondichthyes. In: The Fossil Record 2 (Benton MJ, ed), Chapman and Hall, London, pp 593–609

Chothia C, Lesk AM, Tramontano A, Levitt M, Smith-Gill S, Air G, Sheriff S, Padlan EA, Davies D, Tulip WR, Colman PM, Spinelli S, Alzari P, Poljak RJ (1989) Conformations of immunoglobulin hypervariable regions. Nature 342:21–26

Desravines S, Hsu E (1993) Measuring CDR3 length variability in individuals during ontogeny. J Immunol 168:219–225

Diaz M, Greenberg AS, Flajnik MF (1998) Somatic hypermutation of the new antigen receptor gene (NAR) in the nurse shark does not generate the repertoire: possible role in antigen-driven reactions in the absence of germinal centers. Proc Natl Acad Sci USA 95:14343–14348

Du Pasquier L (1982) Antibody diversity in lower vertebrates – why is it so restricted? Nature 296: 311–313

Du Pasquier L, Wabl MR (1978) Antibody diversity in amphibians: inheritance of isoelectric focusing antibody patterns in isogenic frogs. Eur J Immunol 8:428–433

Du Pasquier L, Wilson M, Greenberg AS, Flajnik MF (1998) Somatic mutation in ectothermic vertebrates: musings on selection and origins. Curr Top Microbiol Immunol 229:199–216

Greenberg AS (1994) Evolution of the antigen receptor family. Ph.D Thesis University of Miami. Miami FL

Greenberg AS, Avila D, Hughes M, Hughes A, McKinney EC, Flajnik MF (1995) A new antigen receptor gene family that undergoes rearrangement and extensive somatic diversification in sharks. Nature 374:168–173

Greenberg AS, Steiner L, Kasahara M, Flajnik MF (1993) Isolation of a shark immunoglobulin light chain cDNA clone encoding a protein resembling mammalian κ light chains: implications for the evolution of light chains. Proc Natl Acad Sci USA 90:10603–10607

Haire RN, Ota T, Rast JP, Litman R, Chan FY, Zon L, Litman G (1996) A third Ig chain gene isotype in *Xenopus laevis* consists of six distinct V_L families and is related to mammalian λ genes. J Immunol 157:1544–1550

Hamers-Casterman C, Atarhouch T, Muyldermans S, Robinson G, Hamers C, Bajyana, Songa E, Bendahman N, Hamers R (1993) Naturally occurring antibodies devoid of light chains. Nature 363:446–448

Helenius A, Simons K (1972) The binding of detergents to lipophilic and hydrophilic proteins. J Bio Chem 247:3656–3661

Hinds KR, Litman GW (1986) Major reorganization of immunoglobulin V_H segmental elements during vertebrate evolution. Nature 320:546–549

Hinds-Frey KR, Hikaru N, Litman RT, Litman GW (1993) Somatic variation precedes extensive diversification of germline sequences and combinatorial joining in the evolution of immunoglobulin heavy chain diversity. J Exp Med 178:815–824

Hozumi N, Tonegawa S (1976) Evidence for somatic rearrangement of immunoglobulin genes coding for variable and constant regions. Proc Natl Acad Sci USA 73:3628–3632

Hsu E (1998) Mutation, selection, and memory in B lymphocytes of exothermic vertebrates. Immunol Rev 162:25–36

Hsu E, Lefkovits I, Flajnik M, Du Pasquier L (1991) Light chain heterogeneity in the amphibian *Xenopus*. Mol Immunol 28:985–994

Hsu E, Steiner LA (1992) Primary structure of immunoglobulins through evolution. Curr Opin Struct Biol 2:422–431

Ji Y, Desravines S, Hsu E (1999) Junctional diversity in *Xenopus* immunoglobulin light chains. Mol Immunol, in press

Kabat EA, Wu TT, Perry HM, Gottesman KS, Foeller C (1991) Sequences of proteins of immunological interest. NIH Publication, 5th ed

Kokubu F, Litman R, Shamblott, Hinds K, Litman GW (1988) Diverse organization of immunoglobulin V_H gene loci in a primitive vertebrate. EMBO J 7:3413–3422

Kyte J, Doolittle RF (1982) A simple method for displaying the hydropathic character of a protein. J Mol Bio 157:105–132

Litman GW, Anderson MK, Rast JP (1999) Evolution of antigen binding receptors. Ann Rev Immunol (in press)

Lee A, Singh J, Flajnik MF, Hsu E (1998) A unique germline organization of Ig genes in shark. FASEB Lett 12(5) LB59 (addendum)

Mäkelä O, Litman GW (1980) Lack of heterogeneity in anti-hapten antibodies of a phylogenetically primitive shark. Nature 287:639–640

Milstein C, Even J, Jarvis JM, Gonzalez-Fernandez A, Gheradi E (1992) Non-random features of the repertoire expressed by the members of one V_{ox} gene family and of the V-J recombination. Eur J Immunol 22:1627–1634

Nadel B, Feeney AJ (1995) Influence of coding-end processing in V(D)J recombination. J Immunol 155:4322–4329

Padlan EA (1990) On the nature of antibody combining sites: unusual structural features that may confer on these sites an enhanced capacity for binding ligands. Proteins 7:112–124

Rast JP, Anderson MK, Ota T, Litman RT, Margittai M, Shamblott MJ, Litman GW (1994) Immunoglobulin light chain class multiplicity and alternative organizational forms in early vertebrate phylogeny. Immunogenetics 40:83–99

Reynaud CA, Anquez V, Grimal H, Weill JC (1987) A hyperconversion mechanism generates the chicken light chain preimmune repertoire. Cell 48:379–388

Reynaud CA, Garcia C, Hein WR, Weill JC (1995) Hypermutation generating the sheep immunoglobulin repertoire is an antigen-independent process. Cell 80:115–125

Rogozin IB, Kolchanov NA (1992) Somatic hypermutagenesis in immunoglobulin genes. II Influence of neighbouring base sequences on mutagenesis. Biochim et Biophys Acta 1171:11–18

Rudikoff S, Satow Y, Padlan E, Davies D, Potter M (1981) Kappa chain structure from a crystallized murine Fab: role of joining segment in hapten binding. Mol Immunol 18:705–711

Schluter SF, Hohman VS, Edmundson AB, Marchalonis JJ (1989) Evolution of immunoglobulin light chains: cDNA clones specifying sandbar shark constant regions. Proc Natl Acad Sci USA 86: 9961–9965

Schwager J, Bürckert N, Schwager M, Wilson M (1991) Evolution of immunoglobulin light chain genes: analysis of *Xenopus* IgL isotypes and their contribution to antibody diversity. EMBO J 10: 505–511

Shamblott MJ, Litman GW (1989a) Complete nucleotide sequence of primitive vertebrate immunoglobulin light chain genes. Proc Natl Acad Sci USA 86:4684–4688

Shamblott MJ, Litman GW (1989b) Genomic organization and sequences of immunoglobulin light chain genes in a primitive vertebrate suggest coevolution of immunoglobulin gene organization. EMBO J 8:3733–3739

Victor K, Capra JD (1994) An apparently common mechanism of generating antibody diversity: length variation of the V_L-J_L junction. Mol Immunol 0161:5890–5894

Wabl MR, Du Pasquier L (1976) Antibody patterns in genetically identical frogs. Nature 264:642–644

Weigert MG, Cesari IM, Yonkovich SJ, Cohn M (1970) Variability in the lambda light chain sequences of mouse antibody. Nature 228:1045–1047

Weiss S, Lehman K, Raschke WC, Cohn M (1984) Mice completely suppressed for the expression of immunoglobulin κ light chain. Proc Natl Acad Sci USA 81:211–215

Wilson M, Hsu E, Marcuz A, Courtet M, Du Pasquier L, Steinberg C (1992) What limits affinity maturation of antibodies in *Xenopus*, the rate of somatic mutations or the ability to select mutants? EMBO J 11:4337–4347

Winter DB, Phung QH, Umar A, Baker SM, Tarone RE, Tanaka K, Liskay RM, Kunkel TA, Bohr VA, Gearhart PJ (1998) Altered spectra of hypermutation in antibodies from mice deficient for the DNA mismatch repair protein PMS2. Proc Natl Acad Sci USA 95:6953–6958

Wu TT, Kabat EA (1970) An analysis of the sequences of the variable regions of Bence-Jones proteins and myeloma light chains and their implications for antibody complementarity. J Exp Med 132: 211–250

Zezza DJ, Mikoryak CA, Schwager J, Steiner LA (1991) Sequence of C region of L chains from *Xenopus laevis* Ig. J Immunol 146:4041–4047

Zezza DJ, Stewart SE, Steiner LA (1992) Genes encoding *Xenopus laevis* Ig L chains implications for the evolution of κ and λ chains. J Immunol 149:3968–3977

Zhu D, Lefkovits I, Köhler G (1984) Frequency of expressed immunoglobulin light chain genes in lipopolysaccharide-stimulated Balb/c spleen cells. J Exp Med 160:971–986

TCR/CD3 Complex

Evolution of the T Cell Receptor Signal Transduction Units

T.W.F. Göbel and L. Bolliger

1 Introduction

The mammalian T cell receptor (TCR)/CD3 complex has been the focus of intense studies for many years. It consists of six different chains, which form hetero- and homodimers to constitute the complete TCR/CD3 complex (see Bentley and Mariuzza 1996; Lanzavecchia et al. 1999; Terhorst et al. 1995; Wange and Samelson 1996 for recent reviews on different aspects). The clonotypic TCR genes have now been isolated from a number of vertebrate species, including cartilaginous, bony fish, amphibia, and chickens (see Charlemagne et al. 1998; Chen et al. 1995 for review). Paradoxically, information regarding the nonmammalian CD3 genes is limited to three cloned CD3 homologues (Bernot and Auffray 1991; Dzialo and Cooper 1997; Göbel and Fluri 1997) and the chicken ζ chain (Göbel and Bolliger 1998).

All TCR chains cloned to date contain short cytoplasmic domains lacking intrinsic enzymatic activity and positively charged transmembrane residues (Charlemagne et al. 1998; Rast et al. 1997). It is therefore likely that the TCR heterodimer of lower vertebrates also associates with signal transducing elements.

Basel Institute for Immunology, Basel, Switzerland

The analysis of these associated chains not only complements existing mammalian data but allows the phylogeny of the TCR/CD3 complex to be understood and sets a basis for the evolution of protein domains.

This review focuses principally on the analysis of the chicken TCR associated signal transduction units since it is the only nonmammalian species in which all components of an T cell antigen receptor have been characterized. The descriptions of the ζζ-homodimer and the CD3 chains are the foundation for speculations regarding the evolution, structure, and function of the TCR/CD3 complex.

2 Nonmammalian TCR Signaling Homologues

2.1 CD3ε

Chicken CD3ε is the only nonmammalian homologue cloned to date (Göbel and Fluri 1997). The overall homology of the extracellular domain is lower than that in mammals, but the area defined by the first cysteine and the membrane proximal region show higher conservation (Table 1, Fig. 1). The membrane proximal area harboring the CXXC motif, (X represents any amino acid) is highly conserved (see below). The size of the chicken extracellular domain is about ten amino acids shorter than in mammals, explaining the lower molecular weight (Table 2). In particular, the two cysteines which are thought to form a disulfide bridge creating an immunoglobulin (Ig)-like fold are separated by only 34 amino acids. This is a short distance for an Ig-like domain (45–60 amino acids in general), suggesting that CD3ε does not form a classical Ig-like structure. This hypothesis is strengthened by the fact that a chimeric protein containing the chicken CD3ε extracellular domain fused to human IgG is highly susceptible to papain digestion, which would not be predicted for a protein with a tight Ig-fold domain (Göbel, unpublished).

All CD3ε proteins lack N-linked glycosylation, suggesting that the addition of N-linked carbohydrates prevents efficient TCR assembly by increasing its "bulkiness". In fact, a CD3ε construct tagged with an N-terminal Flag epitope was efficiently expressed in T cells but did not assemble with other TCR/CD3 com-

Table 1. Amino acid identity of nonmammalian and human signal transduction units

Protein	Human	Total	Ex	Tm	Cy	ITAM 1	ITAM 2	ITAM 3
chCD3γ/δ	CD3γ	44	38	44	57	67		
chCD3γ/δ	CD3δ	44	37	52	52	67		
xCD3γ/δ	CD3γ	43	36	44	59	67		
xCD3γ/δ	CD3δ	39	35	35	51	60		
chCD3ε	CD3ε	42	31	52	58	87		
chζ	ζ	59	44	81	56	67	81	87

Sequences were aligned using the Clustal program and the PAM 250 matrix. ch, chicken; x, *Xenopus laevis*; Ex, extracellular; Tm, transmembrane; Cy, cytoplasmic.

```
GGQEE--------FAVEISGTTVTITCP-SSGDDIKWKP----------DPALGDNNKYIIQ Chicken
DGNEEMGGITQTPYKVSISGTTVILTCPQYPGSEILWQHNDKNIGGDEDDKNIGSDEDHLSL Human
DDAENIE------YKVSISGTSVELTCPLDSDENLKWEKNGQELPQKHD--------KHLVL Mouse

NH---DSSPLTVSCTAGDQEHT-----MYLNAKVCANCEELDTFTVVGIIAADLLITLGVLI Chicken
KEFSELEQSGYYVCYPRGSKPEDANFYLYLRARVCENCMEMDVMSVATIVIVDICITGGLLL Human
QDFSEVEDSGYYVCYTPASNKNT---YLYLKARVCEYCVELDLTAVAIIIIVDICITLGLLM Mouse

LVYYFSKNKKGQSR-----AAAGSRPRAQKMQRPPPVPNPDYEPIRKGQRDVYAGLEHRGF Chicken
LVYYWSKNRKAKAKPVTRGAGAGGRQRGQNKERPPPVPNPDYEPIRKGQRDLYSGLNQRRI Human
VIYYWSKNRKAKAKPVTRGTGAGSRPRGQNKERPPPVPNPDYEPIRKGQRDLYSGLNQRAV Mouse
```

Fig. 1. Alignment of chicken, human, and mouse CD3ε chains. The alignment was performed using the Clustal program with the PAM 250 matrix. *Shading*, conserved residues. *Bar above sequence*, position of the transmembrane region

Table 2. Biochemical properties of nonmammalian CD3 and ζ-chains

Protein	N-Gly	pI	Pred.	M_r	
				NR	R
ch CD3γ/δ	1	6.3	16.7	20, 19	19
x CD3γ/δ	1	5.1	16.4	18	18
ch CD3ε	0	5.9	16.7	17	19
x CD3ε	?	?	?	17	18
ch ζ	0	9.4	16.2	32	16

N-Gly, number of N-linked glycosylation sites; pI, predicted isoelectric point; Pred., predicted M_r based on protein sequence; NR, R, apparent M_r under nonreducing and reducing conditions, respectively.

ponents, supporting the idea that the size and charge of CD3ε is critical for its interactions with other TCR/CD3 components.

The transmembrane region containing a negatively charged residue is followed by a cytoplasmic area of low conservation and a highly conserved C-terminus of the protein (Fig. 1, Table 1). This C-terminus harbors the immunoreceptor-based tyrosine activation motif (ITAM; Reth 1989), a proline rich region known to be an SH3 binding site (XPPXP; Pawson 1995) and an endoplasmic reticulum retention motif (YXXLXXR; Mallabiabarrena et al. 1995).

It is interesting to note that CD3ε and CD3γ/δ are not only expressed in T cells, but also cytoplasmically in natural killer (NK) cells. This feature has also been conserved from chicken and possibly *Xenopus laevis* to man, where embryonic NK cells express cytoplasmic CD3 proteins, and adult NK cells can upregulate CD3 components following activation (Göbel et al. 1994; Lanier et al. 1992). The functional relevance of these intracytoplasmic CD3 proteins is not known; however, it is clear that NK cell development and function are normal in CD3 deficient mice (Renard et al. 1995; Wang et al. 1999). This shared expression of CD3ε by T cells and NK cells underlines the close developmental and functional relationship of these cells.

2.2 CD3γ/δ

Calculations based on the sequence divergence between human CD3γ and CD3δ predict that these genes arouse by a gene duplication about 200 million years ago, implying that most nonmammalian vertebrates reflect a situation before this gene duplication and thus harbor two instead of three CD3 genes (Krissansen et al. 1986). The isolation of a CD3 homologue from chicken and *Xenopus laevis* with equal homology to both mammalian CD3γ and CD3δ has substantiated this hypothesis (Fig. 2, Table 1). They were therefore designated CD3γ/δ chain in order to emphasize their ancestral nature (Bernot and Auffray 1991; Dzialo and Cooper 1997).

CD3γ/δ is thus predicted to combine characteristics which have later been distributed to CD3γ and CD3δ following the gene duplication. In fact, the extra-

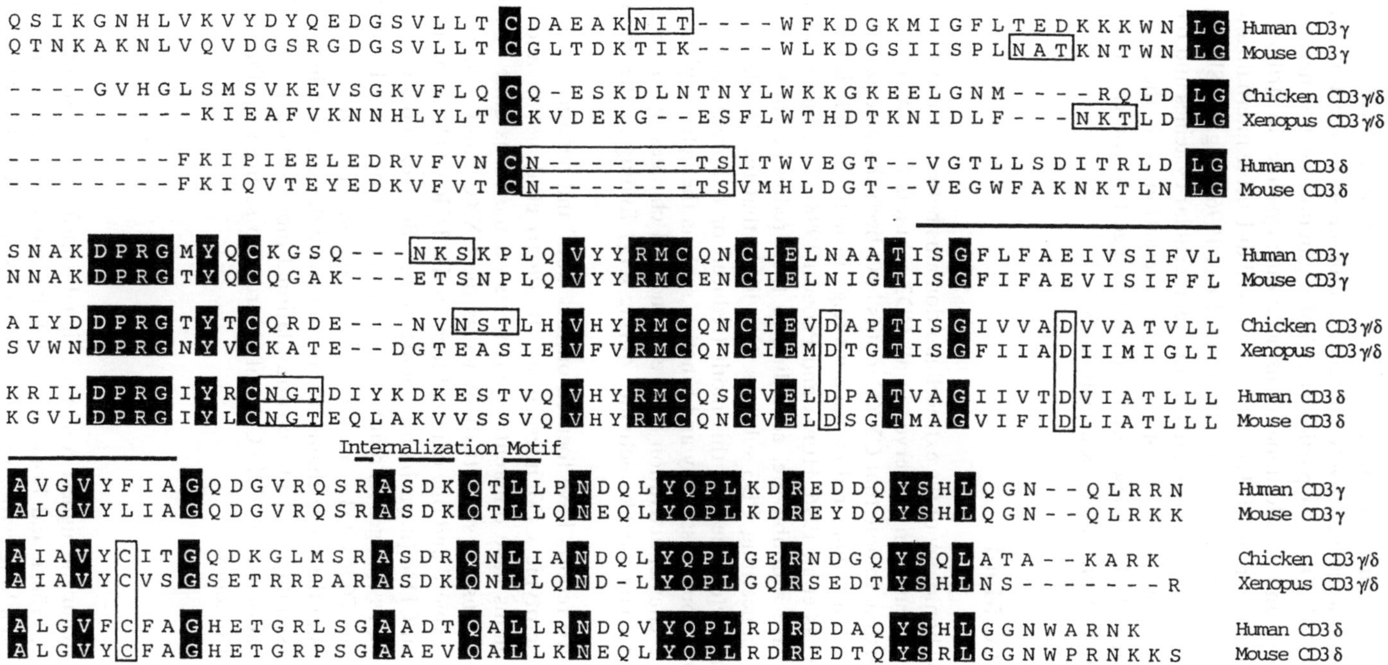

Fig. 2. Alignment of human and mouse CD3γ and CD3δ with nonmammalian CD3γ/δ. *Horizontal boxes*, putative N-linked glycosylation sites; *vertical boxes*, CD3δ-like features. The CD3γ internalization motif is indicated. See Fig. 1 for further details

cellular and transmembrane domains of CD3γ/δ contain CD3δ-like features including an additional negatively charged residue close to the transmembrane domain (CXXCXEXD), an aspartic instead of a glutamic acid as negative transmembrane charge and an additional transmembrane cysteine. In striking contrast the cytoplasmic domain of CD3γ/δ has a conserved di-leucine internalization motif which is found only in CD3γ (DIETRICH et al. 1994, 1996a; Fig. 2). Shared feature of all CD3γ, CD3δ, and CD3γ/δ are the four extracellular cysteines and the DPRG motif which is know to be involved in the interaction with the CD3ε chain (DIETRICH et al. 1996b). These results indeed suggest a conserved TCR/CD3 interaction pattern from chickens to mammals.

2.3 The CXXC Motif

Despite the low homology of the extracellular domains between different CD3 units the CXXC motif has been conserved in all CD3 subunits from *Xenopus laevis* to man. It is interesting to note that CXXC motifs have been characterized in proteins that are involved in cellular redox reactions (e.g., protein disulfide isomerase, thioredoxin; CHIVERS et al. 1996; HUPPA and PLOEGH 1998). This link to a family of proteins that have a double-life, according to their redox potential, allows new interpretations concerning CD3 subunits. It is likely that the double-character of the CD3 chains determined by this motif (reduced or oxidized) controls several aspects of CD3 subunit assembly and function such as the heterotypic interaction of various CD3 chains, activation induced structural changes by disulfide bridge reshuffling, and chaperonelike activities during CD3 folding in the endoplasmic reticulum.

Recently two reports have focused on this motif and demonstrated that it is involved in heterotypic CD3 interactions (BORROTO et al. 1998), and that the various CD3 chains have different requirements on the redox state of the motif (Bolliger and Johansson, submitted). Whereas the CXXC motif in murine CD3ε appears in one reduced and one oxidized form in the TCR/CD3, the motif requires to be oxidized in CD3δ (intrachain disulfide bridge; Bolliger and Johansson, submitted). Thus CD3ε changes its redox state while interacting with CD3δ. Since the biogenesis of the TCR/CD3 complex relies on the synchronized production of subcomplexes from the single chains that do not share the same intrinsic metabolic characteristics (stability, half life, posttranslational modifications) the TCR complex may have adopted the CXXC motif in order to secure its sufficient surface expression, and function.

2.4 ζ-Chain

The ζ-chain is by far the highest conserved signal transduction element of the TCR (Table 1, Fig. 3; GÖBEL and BOLLIGER 1998). Its structure, with a short extracellular domain and a long cytoplasmic domain containing three ITAM motifs and a

```
TVLGLTNPRLCYLLDGFLFIYAVIITALFVKAKLSQASEPQLLL-GQDD CHICKEN
QSFGLLDPKLCYLLDGILFIYGVILTALFLRVKFSRSAEPPAYQQGQNQ HUMAN
QSFGLLDPKLCYLLDGILFIYGVIITALYLRAKFSRSAETAANLQDPNQ MOUSE

VYNKLSRGHRDEYDVLGTRRGADLEKGGRHEQRRKNPHDTVYSSLQKDK CHICKEN
LYNELNLGRREEYDVLDKRRGRDPEMGGKP--RRKNPQEGLYNELQKDK HUMAN
LYNELNLGRREEYDVLEKKRARDPEMGGKQ-QRRRNPQEGVYNALQKDK MOUSE

MGEAYSEIGKKGEQRRRGKGNDAVYQGLSAATRDTYDALHMQPLPPR CHICKEN
MAEAYSEIGMKGE-RRRGKGHDGLYQGLSTATKDTYDALHMQALPPR HUMAN
MAEAYSEIGTKGE-RRRGKGHDGLYQGLSTATKDTYDALHMQTLAPR MOUSE
```

Fig. 3. Alignment of chicken, human, and mouse ζ-chains. See Fig. 1 for details

GTP-binding consensus motif, has been conserved (PETER et al. 1992; SANCHO et al. 1993).

Interestingly the transmembrane domain of the ζ-chain shows the highest identity of the protein (Table 1). This points to an essential role of this domain during TCR assembly and/or function. The transmembrane domain contains a cysteine, important for the formation of the interchain disulfide bridge and a dimerization motif, previously characterized for glycophorin A (BOLLIGER et al. 1997; LEMMON et al. 1992, 1994; Fig. 4). Mutagenesis of this dimerization motif together with molecular modeling allowed the dissection of regions of the ζ-chain transmembrane domain that interact with the different TCR subunits. In this model the transmembrane domains of the ζζ-homodimer interact with the mammalian CD3δ or CD3γ/δ transmembrane domain (Fig. 5). A mutational study of a phylogenetically conserved motif present in mammalian CD3γ, CD3δ, and CD3γ/δ transmembrane domains further supports this model. The opposite side of the ζζ-homodimer pointing away from the dimerization motif is likely to interact with the TCRαβ heterodimer (Fig. 5). Further support for this model comes from the recent demonstration that the plasma membrane proximal cytoplasmic region of the ζζ-homodimer masks the di-leucine degradation signal in the cytoplasmic tail of CD3γ

Chicken	LCYL **LD** GF LF IY **AV** II **T** ALFV
Mouse	LCYL **LD** GI LF IY **GV** II **T** ALYL
Human	LCYL **LD** GI LF IY **GV** IL **T** ALFL
Glycophorin A	.... **LI** .. **GV** .. **GV** .. **T** ....

Fig. 4. The ζ-chain dimerization motif. The chicken, mouse, and human ζ-chain transmembrane domains are compared to the dimerization consensus motif of human glycophorin A. *Boldface*, homologous amino acids

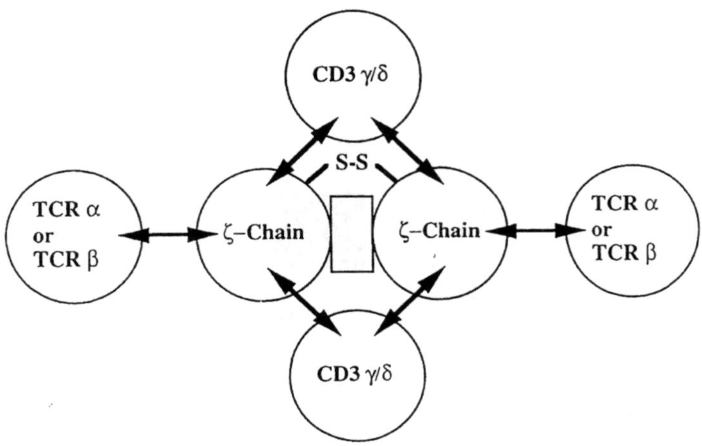

Fig. 5. Transmembrane organization of the TCR/CD3 complex. *Circles* (*in a top view*), α-helical transmembrane domains of the TCR/CD3 components; *rectangle*, dimerization motif interaction phase; *arrows*, the various interactions within the complex

and thus prevents its degradation (Dietrich and Geisler 1998). Taken together it seems likely that the ζζ-homodimer is not only critical for signaling but also orients the TCR/CD3 complex.

Since the chicken ζζ-homodimer was found to contain all the essential motifs for signaling and/or structure, it was tested whether it could replace the mouse ζ-chain utilizing the ζ-deficient mouse hybridoma MA5.8. The chicken ζζ-homodimer was able to assemble with mouse TCR and to fully restore the function of the mouse TCR when assayed for antibody, superantigen, and antigen reactivity (Göbel and Bolliger 1998). This successful reconstitution clearly shows the extraordinarily high conservation of intracytoplasmic signal transducing modules.

3 Biochemical Properties of the Nonmammalian CD3 Proteins

The CT3 monoclonal antibody has been extensively used to identify and characterize chicken T cells and to immunoprecipitate chicken TCR/CD3. It reacts with an epitope on the extracellular domain of CD3ε (Chen et al. 1986; Göbel and Fluri 1997). The CD3 chains can only be coimmunoprecipitated with the TCR chains under mild lysis conditions, indicating a weak association with the TCR. Upon surface iodination the 80-kDa TCR heterodimer is associated with three proteins of 20, 19, and 17kDa (Table 2).

The chicken CD3γ/δ chain represents the 20-kDa and 19-kDa proteins, since it appears in two forms at the cell surface (Table 2). These different forms represent glycosylation microhetereogeneity on a single N-linked glycosylation site of unknown relevance (Chen et al. 1986; Göbel, unpublished). Some of the protein is also covalently associated with CD3ε to form a 36-kDa heterodimer.

The CD3ε chain represents the 17-kDa protein without any N-linked glycosylation. Interestingly, CD3ε migrates slower under reducing than nonreducing conditions, which may be explained by the presence of intracellular disulfide bonds. It is important to note that this occurs only for CD3ε, although all CD3 chains contain the same number of extracellular cysteine residues, again implying a unique role for the cysteines in CD3ε. When transfected into COS cells or translated in a cell free system, CD3ε forms monomers, dimers, and multimers, which are not observed when CD3γ/δ is cotransfected (Huppa and Ploegh 1997; Göbel et al., in preparation). Likewise during the assembly in T cells CD3ε-ε homodimers and CD3γ/δ-ε heterodimers are observed; however, both CD3 chains appear as noncovalently linked monomers associated with the TCR on the cell surface. These homo- and heterodimers may thus represent important intermediates during the TCR/CD3 assembly. A covalently linked CD3ε-ε homodimer is also present on the cell surface as observed in mammals; however, it is not associated with the TCR (Blumberg et al. 1990; Jin et al. 1990; Göbel et al., in preparation).

The human C-terminal CD3ε sequence has been used to generate antipeptide-specific polyclonal antisera and a monoclonal antibody, which both crossreact with chicken and *Xenopus laevis* CD3ε, due to the high conservation of this epitope (Göbel et al., in preparation). As demonstrated for the chicken, the TCR chains in *Xenopus laevis* are associated with only two CD3 proteins, which share many features to their chicken homologues (Göbel et al., in preparation). It is thus possible for the first time to visualize the TCR/CD3 complex of an ectothermic vertebrate.

4 Genomic Organization of the CD3 Genes

The chicken CD3 genes are clustered on a 9.5-kb chromosomal region as opposed to one of about 50kb in man (Göbel et al., submitted). Whereas three CD3 genes are found in the mammalian CD3 cluster in the orientation CD3γ-CD3δ-CD3ε, only two genes are found in the chicken CD3 cluster (Tunnacliffe et al. 1987, 1988). Moreover, the chicken CD3 cluster is flanked by repetitive elements (chicken repeat element 1) and unrelated genes were identified on both sides of the CD3 cluster (Göbel et al., submitted).

The genomic organization of the CD3γ/δ encoding gene closely resembles that of mammalian CD3δ with a five exon structure (Saito et al. 1987; van den Elsen et al. 1986). This five exon structure most likely represents the primordial gene structure of a CD3 gene since this core is present in every CD3 gene (Table 3). Based on this highly conserved gene structure common to all CD3 genes, a new numbering of the CD3 exons is suggested in Table 3. All introns of CD3γ/δ and CD3ε represent type 1 introns, except the type 0 intron which is consistently found to split the ITAM motif (Malissen and Schmitt-Verhulst 1993).

An alternatively spliced form of CD3γ/δ can easily be detected in T cells. Most of the gene encoding the extracellular domain is not transcribed in this form, due to a cryptic splice signal in the second exon, whereas the rest of the molecule is identical to the full-length CD3γ/δ. The physiological relevance of this form is presently unknown (Göbel et al., unpublished).

A region of 2kb separates CD3γ/δ and CD3ε, which are transcribed in opposite orientations (Göbel et al., submitted). Chicken CD3ε spans a total of 5.5kb and consists of seven exons, including one miniexon which only codes for seven amino acids, whereas two and three miniexons are present in CD3ε of mouse and man, respectively (Clevers et al. 1988a,b).

The chicken CD3 genes lack classical promoter elements and their promoter is tissue nonspecific, and as in mammals the T cell specificity is most likely ensured by the enhancer region located between CD3γ/δ and CD3ε (Clevers et al. 1988b; van den Elsen et al. 1986; Göbel et al., unpublished). This 2-kb enhancer region contains multiple repeats of transcription factor binding sites implied in T cell function such as TCF-1, GATA-3, and IKAROS (Clevers et al. 1993). It will be interesting to analyze whether the CD3γ/δ and CD3ε utilize the same enhancer or two different enhancer elements located in the same region.

Table 3. Genomic organization of the CD3 genes

	CD3γ			CD3δ				CD3ε			
	Exon	Human	Mouse	Exon	Human	Mouse	Chicken	Exon	Human	Mouse	Chicken
5UT + SP + Ex	γ1a	132 (18.3)	> 159 (18.3)	δ1	151 (18.3)	153 (18.3)	98 (24.3)	ε1a	65 (0)	64 (0)	45 (0)
	γ1b	24 (8)	24 (8)					ε1b	109 (15.3)	102 (16.3)	115 (15.3)
								ε1c	21 (7)	18 (6)	21 (7)
								ε1d	15 (5)	15 (5)	
								ε1e	18 (6)		
Ex	γ2	228 (76)	228 (76)	δ2	219 (73)	219 (73)	216 (72)	ε2	249 (83)	216 (72)	201 (67)
Ex + TM + Cy	γ3	132 (44)	132 (44)	δ3	132 (44)	132 (44)	132 (44)	ε3	168 (56)	168 (56)	153 (51)
CY	γ4	44 (14.7)	44 (14.7)	δ4	44 (14.7)	44 (14.7)	44 (14.7)	ε4	47 (15.7)	47 (15.7)	47 (15.7)
						Type 0 intron					
CY + 3 UT	γ5a	83 (21)	83 (21)	δ5	156 (21)	151 (21)	300 (20)	ε5	694 (18)	844 (18)	1280 (18)
	γ5b	2170 (0)	> 228 (0)								

Numbers represent the nucleotide length (bp) and the encoded amino acids (paranthesis). All introns represent type 1 introns unless shown, 5UT, 3UT, 5′ and 3′ untranslated region; sp, signal peptide; ex, extracellular domain; tm, transmembrane domain; cy, cytoplasmic domain.

5 Evolution of the CD3 Genes

The three CD3 genes found in mammals are most likely the result of two successive gene duplications. We predict that an ancestral form of a CD3 gene would already have several features present in the mammalian CD3 genes: a five exon genomic organization; four extracellular cysteine residues; the conserved transmembrane motif to interact with the ζ-chain and a negatively charged transmembrane residue; an ITAM motif, a di-leucine internalization motif, and possibly a proline rich region in the cytoplasmic domain. All nonmammalian TCR heterodimers contain three positively charged transmembrane residues, thus suggesting that this primordial CD3 protein and the ζζ-homodimer would constitute such a primordial TCR/CD3 complex (Fig. 6). This CD3 gene has successively duplicated to form CD3γ/δ and CD3ε as found in *Xenopus laevis* and chicken, followed by a second duplication of CD3γ/δ to form mammalian CD3γ and CD3δ.

Interestingly, the pre-TCR and the γδ TCR in mammals associate with only two instead of three CD3 similarly to the nonmammalian TCR/CD3 complex. Although there is some debate about the constituents of the pre-TCR (Berger et al. 1997; Jacobs 1997), it is evident from knockout studies that CD3γ but not CD3δ is essential for the pre-TCR function (Dave et al. 1997; Haks et al. 1998). Moreover, evidence obtained from mutant mouse strains and biochemical analyses suggests that the TCR of γδ T cells only contains CD3γ and not CD3δ (Dave et al. 1997). It should be emphasized that this is not true for the majority of αβ T cells. Several experiments including transfection studies, mutant mouse strains, and analyses of immunodeficient patients have clearly established a unique role for all three CD3 proteins (Dave et al. 1997; Haks et al. 1998; Hall et al. 1991; PachecoCastro et al. 1998; Wang et al. 1999). A striking difference between the αβ TCR of mammals and other vertebrates is the lack of a solvent exposed loop in the TCRβ chain of nonmammalian vertebrates. This loop is also absent in the γδ TCR heterodimer. When this loop was mutated the TCR surface expression was markedly reduced (Fuller-Espie et al. 1998). Thus it is likely that this loop in the mammalian TCRβ chain somehow affects the interaction of the TCR heterodimer with the CD3 chains, creating an essential role for the third CD3 protein.

6 Models for TCR/CD3 Structure and Function

Enormous progress has been made during the past 10 years in the field of TCR biology: from its cloning to its three-dimensional structure. Novel biochemical techniques led to the discovery of compartmentalization of TCR components into detergent insoluble membrane domains upon engagement (membrane rafts), and to the demonstration that the orientation of cytoplasmic signaling molecules is critical for the activation of the signaling cascade (Caplan and Baniyash 1996; Graef

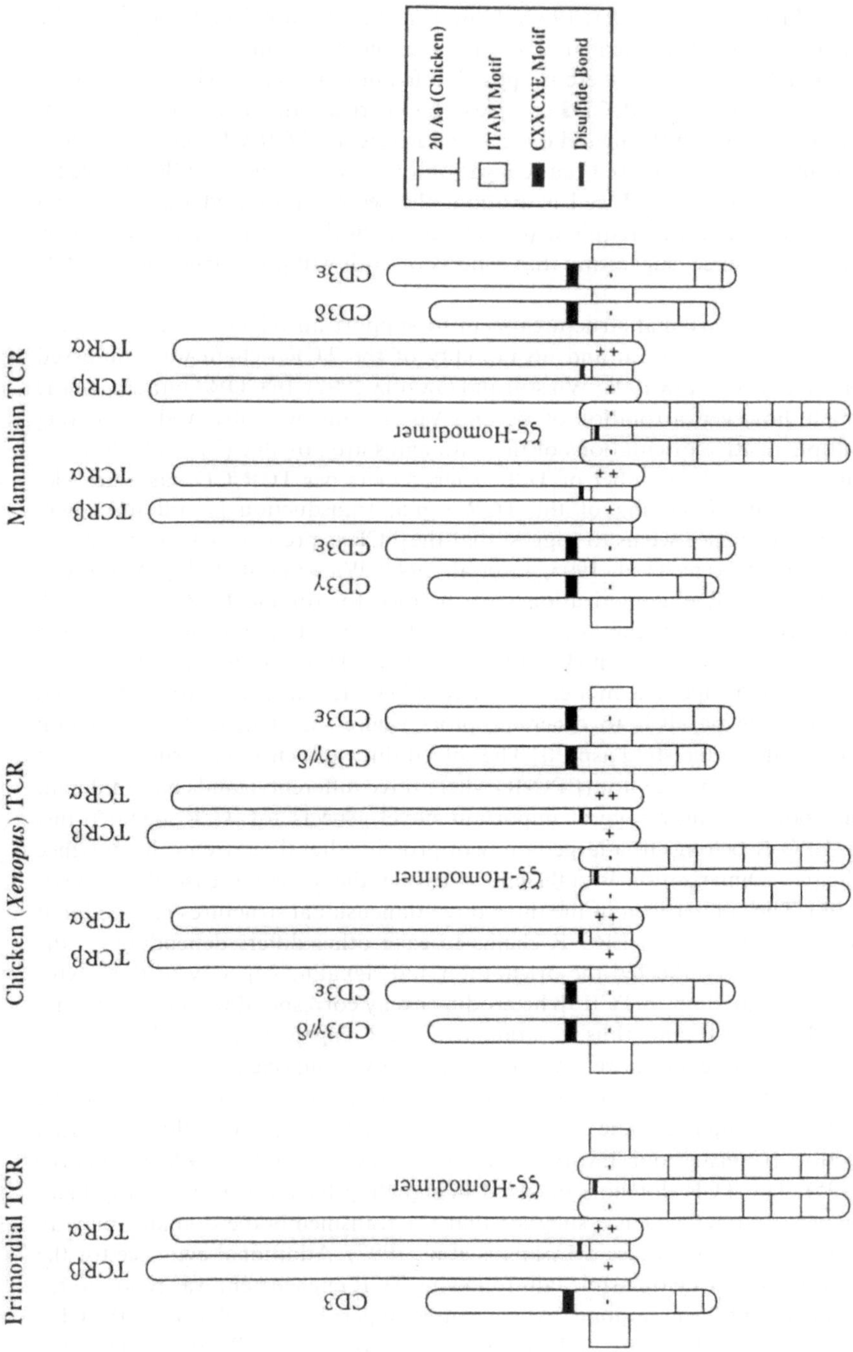

Fig. 6. Hypothetical Scheme of the TCR/CD3 complex evolution. The size of the chicken TCR and CD3 chains and the positions of the motifs in the chicken chains are drawn to scale. The scheme represents a model concerning the possible stoichometry and constituents of various TCR complexes, for individual chain interactions refer to Fig. 5

et al. 1997; SIMONS and IKONEN 1997). However, it is still not understood how the engagement signal is translocated across the plasma membrane.

Historically two models were proposed: a ligation model, in which the MHC/peptide complex forces TCR/CD3 complex oligomerization, and a conformational change model, in which the ligand directly influences the TCR/CD3 conformation. While the first model proposed that coligation or cross-linking of TCR complexes was the driving force for T cell activation, the second proposed that an initial conformation change was required for activation. Both models are supported by experimental evidence, suggesting that aspects of both will prove true for the TCR function.

The three-dimensional structures seem to support an oligomerization model, since no domain dislocation and no mobility of the TCRβ chain were observed (BENTLEY and MARIUZZA 1996; WILSON and GARCIA 1997). In CDR3 mutants of the Vα domain, however, a rotation of Vα and Vβ domains was observed, indicating that multiple relative orientations of these domains are possible (LI et al. 1997).

Unfortunately the number of TCR heterodimers per TCR/CD3 as a key element for the understanding of the TCR signal transduction is still unknown. Experimental evidence seems to suggest that the TCR is present as a superdimer at the cell surface (EXLEY et al. 1995; LI et al. 1997; WANG et al. 1998). Moreover, ligand induced TCR oligomerization was observed for soluble TCR (REICH et al. 1997). Experiments with the *vav* oncogene established that the oligomerization could also be inside-out driven (FISCHER et al. 1998; HOLSINGER et al. 1998).

In order to propose a model for TCR signal transduction, we discuss and incorporate some parallels to other receptors where the structural transduction pathway revealed interesting aspects. The solved three-dimensional structure of the human erythropoietin receptor (EPO-R) where three different ligands have different triggering potential may suggest important novel aspects for TCR signal transduction. EPO-R belongs to a large family of proteins that dimerize in the presence of the ligand. Dimerization is sufficient to trigger the signaling cascade (KLEMM et al. 1998). The comparison of the three three-dimensional structures revealed that the orientation of the two EPO-R chains to each other differs depending on the ligand, and uniquely affected the orientation and signaling capacity of their cytoplasmic tails. In analogy, a TCRαβ heterodimer may correspond to one monomeric chain of the EPO-R and thus a productive TCR engagement could lead to a reorientation of one or more TCR/CD3 complex components. In fact, recent reports showed that such a reorganization can occur and that the nature of the ligand (cross-linking antibodies, superantigen and MHC-peptide) influences signal transduction (CAPLAN and BANIYASH 1996; JOHANSSON et al. 1999; KISHIMOTO et al. 1995). The TCR domains involved in signaling have not been clearly identified; however, some evidence suggests that the transmembrane domains play an important role in this process (MANOLIOS et al. 1997). Additional evidence for the reorganization of the cytoplasmic tails following TCR engagement was reported for an ITAM deficient ζζ-homodimer, which caused hyperphosphorylation of the CD3 ITAMs, suggesting that the "bulkiness" of the cytoplasmic tails affect the accessibility to adapters required for signaling (VAN OERS et al. 1998).

Conformational change is also essential for G protein coupled seven transmembrane receptors (e.g., chemokine receptors, adrenergic receptors). Some aspects of the TCR/CD3 complex resemble these receptors. A heterotrimeric G protein is associated with CD3ε, and TCR signaling can be modulated by adrenergic receptors (De Blasi et al. 1995; Stanners et al. 1995). Even if the TCR/CD3 complex contains eight instead of seven transmembrane domains, it may function similarly to seven transmembrane receptors and transduce signals by conformational change. A rate-limiting step for the initiation of the signaling could be the GTP hydrolysis associated with the GTP-binding site of the ζζ-homodimer.

In conclusion, the consideration of non-TCR classical alternatives is as helpful as the phylogenetic analyses of the TCR/CD3 complex. Both strategies lead to novel insights of TCR/CD3 structure and function, which may provoke new experiments, challenge current views of the TCR and provide the basis for new targets of T-cell mediated diseases.

Acknowledgements. We thank Dr. L. Du Pasquier for critical reading of the manuscript. The Basel Institute for Immunology was founded and is supported by F. Hoffmann-La Roche & Co. Ltd., CH 4005 Basel, Switzerland.

References

Bentley GA, Mariuzza RA (1996) The structure of the T cell antigen receptor. Annu Rev Immunol 14:563–590

Berger MA, Dave V, Rhodes MR, Bosma GC, Bosma MJ, Kappes DJ, Wiest DL (1997) Subunit composition of pre-T cell receptor complexes expressed by primary thymocytes: CD3δ is physically associated but not functionally required. J Exp Med 186:1461–1467

Bernot A, Auffray C (1991) Primary structure and ontogeny of an avian CD3 transcript. Proc Natl Acad Sci USA 88:2550–2554

Blumberg RS, Ley S, Sancho J, Lonberg N, Lacy E, McDermott F, Schad V, Greenstein JL, Terhorst C (1990) Structure of the T-cell antigen receptor: Evidence for two CD3 ε subunits in the T-cell receptor-CD3 complex. Proc Natl Acad Sci USA 87:7220–7224

Bolliger L, Johansson B, Palmer E (1997) The short extracellular domain of the T cell receptor ζ chain is involved in assembly and signal transduction. Mol Immunol 34:819–827

Borroto A, Mallabiabarrena A, Albar JP, Martinez AC, Alarcon B (1998) Characterization of the region involved in CD3 pairwise interactions within the T cell receptor complex. J Biol Chem 273: 12807–12816

Caplan S, Baniyash M (1996) Normal T cells express two T cell antigen receptor populations, one of which is linked to the cytoskeleton via ζ chain and displays a unique activation-dependent phosphorylation pattern. J Biol Chem 271:20705–20712

Charlemagne J, Fellah JS, DeGuerra A, Kerfourn F, Partula S (1998) T-cell receptors in ectothermic vertebrates. Immunol Rev 166:87–102

Chen C-H, Ager LL, Gartland GL, Cooper MD (1986) Identification of a T3/T cell receptor complex in chickens. J Exp Med 164:375–380

Chen CH, Six A, Kubota T, Tsuji S, Kong F-K, Göbel T, Cooper MD (1995) T cell receptors and T cell development. Curr Top Microbiol Immunol 212:37–54

Chivers PT, Laboissiere MCA, Raines RT (1996) The CxxC motif; imperatives for the formation of native disulfide bonds in the cell. EMBO J 15:2659–2667

Clevers H, Dunlap S, Saito H, Georgopoulos K, Wileman T, Terhorst C (1988a) Characterization and expression of the murine CD3-epsilon gene. Proc Natl Acad Sci USA 85:8623–8627

Clevers HC, Dunlap S, Wileman TE, Terhorst C (1988b) Human CD3-epsilon gene contains three miniexons and is transcribed from a non-TATA promoter. Proc Natl Acad Sci USA 85:8156–8160

Clevers HC, Oosterwegel MA, Georgopoulos K (1993) Transcription factors in early T-cell development. Immunol Today 14:591–596

Dave VP, Cao Z, Browne C, Alarcon B, Fernandez-Miguel G, Lafaille J, de la Hera A, Tonegawa S, Kappes DJ (1997) CD3δ deficiency arrests development of the αβ but not the γδ T cell lineage. EMBO J 16:1360–1370

De Blasi A, Parruti G, Sallese M (1995) Regulation of G protein-coupled receptor kinase subtypes in activated T lymphocytes. Selective increase of beta-adrenergic receptor kinase 1 and 2. J Clin Invest 95:203–210

Dietrich J, Geisler C (1998) T cell receptor ζ allows stable expression of receptors containing the CD3γ leucine-based receptor sorting motif. J Biol Chem 273:26281–26284

Dietrich J, Hou X, Wegener AMK, Geisler C (1994) CD3γ contains a phosphoserine-dependent di-leucine motif involved in down-regulation of the T cell receptor. EMBO J 13:2156–2166

Dietrich J, Hou XH, Wegener AMK, Pedersen LO, Odum N, Geisler C (1996a) Molecular characterization of the di-leucine-based internalization motif of the T cell receptor. J Biol Chem 271: 11441–11448

Dietrich J, Neisig A, Hou X, Wegener A-MK, Gajhede M, Geisler C (1996b) Role of CD3γ in T cell receptor assembly. J Cell Biol 132:299–310

Dzialo R, Cooper MD (1997) An amphibian CD3 homologue of the mammalian CD3γ and δ genes. Eur J Immunol 27:1640–1647

Exley M, Wileman T, Müller B, Terhorst C (1995) Evidence for multivalent structure of T cell antigen receptor complex. Mol Immunol 11:829–839

Fischer KD, Kong YY, Nishina H, Tedford K, Marengere LE, Kozieradzki I, Sasaki T, Starr M, Chan G, Gardener S, Nghiem MP, Bouchard D, Barbacid M, Bernstein A, Penninger JM (1998) Vav is a regulator of cytoskeletal reorganization mediated by the T-cell receptor. Curr Biol 8:554–562

Fuller-Espie S, Hoffman Towler P, Wiest DL, Tietjen I, Spain LM (1998) Transmembrane polar residues of TCRβ chain are required for signal transduction. Int Immunol 10:923–933

Göbel TWF, Bolliger L (1998) The chicken TCR ζ chain restores the function of a mouse T cell hybridoma. J Immunol 160:1552–1554

Göbel TWF, Chen C-LH, Shrimpf J, Grossi CE, Bernot A, Bucy RP, Auffray C, Cooper MD (1994) Characterization of avian natural killer cells and their intracellular CD3 protein complex. Eur J Immunol 24:1685–1691

Göbel TWF, Fluri M (1997) Identification and analysis of the chicken CD3ε gene. Eur J Immunol 27:194–198

Graef IA, Holsinger LJ, Diver S, Schreiber SL, Crabtree GR (1997) Proximity and orientation underline signaling by the non-receptor tyrosine kinase ZAP-70. EMBO J 16:5618–5628

Haks MC, Krimpenfort P, Borst J, Kruisbeek AM (1998) The CD3γ chain is essential for development of both the TCRαβ and TCRγδ lineages. EMBO J 17:1871–1882

Hall C, Berkhout B, Alarcon B, Sancho J, Wileman T, Terhorst C (1991) Requirements for cell surface expression of the human TCR/CD3 complex in non-T cells. Int Immunol 3:359–368

Holsinger LJ, Graef IA, Swat W, Chi T, Bautista DM, Davidson L, Lewis RS, Alt FW, Crabtree GR (1998) Defects in actin-cap formation in Vav-deficient mice implicate an actin requirement for lymphocyte signal transduction. Curr Biol 8:563–572

Huppa JB, Ploegh HL (1997) In vitro translation and assembly of a complete T cell receptor-CD3 complex. J Exp Med 186:393–403

Huppa JB, Ploegh HL (1998) The eS-Sence of -SH in the ER. Cell 92:145–148

Jacobs H (1997) Pre-TCR/CD3 and TCR/CD3 complexes: decamers with differential signalling properties? Immunol Today 18:565–569

Jin Y-J, Koyasu S, Moingeon P, Steinbrich R, Tarr GE, Reinherz EL (1990) A fraction of CD3ε subunits exists as disulfide-linked dimers in both human and murine T lymphocytes. J Biol Chem 265:15850–15853

Johansson B, Palmer E, Bolliger L (1999) The extracellular domain of the ζ-chain is essential for TCR function. J Immunol 162:878–885

Kishimoto H, Kubo RT, Yorifuji H, Nakayama T, Asano Y, Tada T (1995) Physical dissociation of the TCR-CD3 complex accompanies receptor ligation. J Exp Med 182:1997–2006

Klemm JD, Schreiber SL, Crabtree GR (1998) Dimerization as a regulatory mechanism in signal transduction. Annu Rev Immunol 16:569–592

Krissansen GW, Owen MJ, Verbi W, Crumpton MJ (1986) Primary structure of the T3 gamma subunit of the T3/T cell antigen receptor complex deduced from cDNA sequences: evolution of the T3 gamma and delta subunits. EMBO J 5:1799–1808

Lanier LL, Chang C, Spits H, Phillips JH (1992) Expression of cytoplasmic CD3ε proteins in activated human adult natural killer (NK) cells and CD3 γ, δ, ε complexes in fetal NK cells. J Immunol 149:1876–1880

Lanzavecchia A, Iezzi G, Viola A (1999) From TCR Engagement to T Cell Activation: A Kinetic View of T Cell Behavior. Cell 96:1–4

Lemmon MA, Flanagan JM, Treutlein HR, Zhang J, Engelmean DM (1992) Sequence specificity in the dimerization of transmembrane α-helices. Biochemistry 31:12719–112725

Lemmon MA, Treutlein HR, Adams PD, Brünger AT, Engelman DM (1994) A dimerization motif for transmembrane α-helices. Nature Struct Biol 1:157–163

Li H, Lebedeva MI, Ward ES, Mariuzza RA (1997) Dual conformations of a T cell receptor Vα homodimer: implications for variability in Vα Vβ domain association. J Mol Biol 269:385–394

Malissen B, Schmitt-Verhulst A-M (1993) Transmembrane signalling through the T-cell-receptor-CD3 complex. Curr Opin Immunol 5:324–333

Mallabiabarrena A, Jimenez MA, Rico M, Alarcon B (1995) A tyrosine-containing motif mediates ER retention of CD3-ε and adopts a helix-turn structure. EMBO J 14:2257–2268

Manolios N, Collier S, Taylor J, Pollard J, Harrison LC, Bender V (1997) T-cell antigen receptor transmembrane peptides modulate T-cell function and T cell-mediated disease. Nat Med 3:84–88

Pacheco-Castro A, Alvarez-Zapata D, Serrano-Torres P, Regueiro JR (1998) Signaling through a CD3 γ-deficient TCR/CD3 complex in immortalized mature CD4+ and CD8+ T lymphocytes. J Immunol 161:3152–3160

Pawson T (1995) Protein modules and signalling networks. Nature 373:573–580

Peter ME, Hall C, Ruhlmann A, Sancho J, Terhorst C (1992) The T-cell receptor ζ chain contains a GTP/GDP binding site. EMBO J 11:933–941

Rast JP, Anderson MK, Strong SJ, Luer C, Litman RT, Litman GW (1997) α, β, γ, and δ T cell antigen receptor genes arose early in vertebrate phylogeny. Immunity 6:1–11

Reich Z, Boniface J, Lyons D, Borochov N, Wachtel E, Davis M (1997) Ligand-specific oligomerization of T-cell receptor molecules. Nature:617–620

Renard V, Ardouin L, Malissen M, Milon G, Lebastard M, Gillet A, Malissen B, Vivier E (1995) Normal development and function of natural killer cells in CD3εδ5/δ5 mutant mice. Proc Natl Acad Sci USA 92:7545–7549

Reth M (1989) Antigen receptor tail clue. Nature 338:383–384

Saito H, Koyama T, Georgopoulos K, Clevers H, Haser WG, LeBien T, Tonegawa S, Terhorst C (1987) Close linkage of the mouse and human CD3 γ- and δ-chain genes suggests that their transcription is controlled by common regulatory elements. Proc Natl Acad Sci USA 84:9131–9134

Sancho J, Peter ME, Franco R, Danielian S, Kang JS, Fagard R, Woods J, Reed JC, Kamoun M, Terhorst C (1993) Coupling of GTP-binding to the T cell receptor (TCR) ζ-chain with TCR-mediated signal transduction. J Immunol 150:3230–3242

Simons K, Ikonen E (1997) Functional rafts in cell membranes. Nature 387:569–572

Stanners J, Kabouridis PS, McGuire KL, Tsoukas CD (1995) Interaction between G proteins and tyrosine kinases upon T cell receptor.CD3-mediated signaling. J Biol Chem 270:30635–30642

Terhorst C, Simpson S, Wang B, She J, Hall C, Huang M, Wileman T, Eichmann K, Holländer G, Levelt C, Exley M (1995) Plasticity of the TCR-CD3 complex. In: Bell JI, Owen MJ, Simpson E (eds) T cell receptors. Oxford University Press, Oxford

Tunnacliffe A, Buluwela L, Rabbitts TH (1987) Physical linkage of three CD3 genes on human chromosome 11. EMBO J 6:2953–2957

Tunnacliffe A, Olsson C, Buluwela L, Rabbitts TH (1988) Organization of the human CD3 locus on chromosome 11. Eur J Immunol 18:1639–1642

van den Elsen P, Georgopoulos K, Shepley BA, Orkin S, Terhorst C (1986) Exon/intron organization of the genes coding for the δ chains of the human and murine T-cell receptor/T3 complex. Proc Natl Acad Sci USA 83:2944–2948

van Oers NS, Love PE, Shores EW, Weiss A (1998) Regulation of TCR signal transduction in murine thymocytes by multiple TCR ζ-chain signaling motifs. J Immunol 160:163–170

Wang B, Wang N, Whitehurst CE, She J, Chen J, Terhorst C (1999) T lymphocyte development in the absence of CD3ε or CD3γδεζ. J Immunol 162:88–94

Wang J, Lim K, Smolyar A, Teng M, Liu J, Tse AG, Hussey RE, Chishti Y, Thomson CT, Sweet RM, Nathenson SG, Chang HC, Sacchettini JC, Reinherz EL (1998) Atomic structure of an αβ T cell

receptor (TCR) heterodimer in complex with an anti-TCR Fab fragment derived from a mitogenic antibody. EMBO J 17:10–26
Wange RL, Samelson LE (1996) Complex complexes: Signaling at the TCR. Immunity 5:197–205
Wilson IA, Garcia KC (1997) T-cell receptor structure and TCR complexes. Curr Opin Struct Biol 7: 839–848

Subject Index

Current Topics in Microbiology and Immunology

Volumes published since 1989 (and still available)

Vol. 207: **Prusiner, Stanley B. (Ed.):** Prions Prions Prions. 1995. 42 figs. VII, 163 pp. ISBN 3-540-59343-8

Vol. 208: **Farnham, Peggy J. (Ed.):** Transcriptional Control of Cell Growth. 1995. 17 figs. IX, 141 pp. ISBN 3-540-60113-9

Vol. 209: **Miller, Virginia L. (Ed.):** Bacterial Invasiveness. 1996. 16 figs. IX, 115 pp. ISBN 3-540-60065-5

Vol. 210: **Potter, Michael; Rose, Noel R. (Eds.):** Immunology of Silicones. 1996. 136 figs. XX, 430 pp. ISBN 3-540-60272-0

Vol. 211: **Wolff, Linda; Perkins, Archibald S. (Eds.):** Molecular Aspects of Myeloid Stem Cell Development. 1996. 98 figs. XIV, 298 pp. ISBN 3-540-60414-6

Vol. 212: **Vainio, Olli; Imhof, Beat A. (Eds.):** Immunology and Developmental Biology of the Chicken. 1996. 43 figs. IX, 281 pp. ISBN 3-540-60585-1

Vol. 213/I: **Günthert, Ursula; Birchmeier, Walter (Eds.):** Attempts to Understand Metastasis Formation I. 1996. 35 figs. XV, 293 pp. ISBN 3-540-60680-7

Vol. 213/II: **Günthert, Ursula; Birchmeier, Walter (Eds.):** Attempts to Understand Metastasis Formation II. 1996. 33 figs. XV, 288 pp. ISBN 3-540-60681-5

Vol. 213/III: **Günthert, Ursula; Schlag, Peter M.; Birchmeier, Walter (Eds.):** Attempts to Understand Metastasis Formation III. 1996. 14 figs. XV, 262 pp. ISBN 3-540-60682-3

Vol. 214: **Kräusslich, Hans-Georg (Ed.):** Morphogenesis and Maturation of Retroviruses. 1996. 34 figs. XI, 344 pp. ISBN 3-540-60928-8

Vol. 215: **Shinnick, Thomas M. (Ed.):** Tuberculosis. 1996. 46 figs. XI, 307 pp. ISBN 3-540-60985-7

Vol. 216: **Rietschel, Ernst Th.; Wagner, Hermann (Eds.):** Pathology of Septic Shock. 1996. 34 figs. X, 321 pp. ISBN 3-540-61026-X

Vol. 217: **Jessberger, Rolf; Lieber, Michael R. (Eds.):** Molecular Analysis of DNA Rearrangements in the Immune System. 1996. 43 figs. IX, 224 pp. ISBN 3-540-61037-5

Vol. 218: **Berns, Kenneth I.; Giraud, Catherine (Eds.):** Adeno-Associated Virus (AAV) Vectors in Gene Therapy. 1996. 38 figs. IX,173 pp. ISBN 3-540-61076-6

Vol. 219: **Gross, Uwe (Ed.):** Toxoplasma gondii. 1996. 31 figs. XI, 274 pp. ISBN 3-540-61300-5

Vol. 220: **Rauscher, Frank J. III; Vogt, Peter K. (Eds.):** Chromosomal Translocations and Oncogenic Transcription Factors. 1997. 28 figs. XI, 166 pp. ISBN 3-540-61402-8

Vol. 221: **Kastan, Michael B. (Ed.):** Genetic Instability and Tumorigenesis. 1997. 12 figs.VII, 180 pp. ISBN 3-540-61518-0

Vol. 222: **Olding, Lars B. (Ed.):** Reproductive Immunology. 1997. 17 figs. XII, 219 pp. ISBN 3-540-61888-0

Vol. 223: **Tracy, S.; Chapman, N. M.; Mahy, B. W. J. (Eds.):** The Coxsackie B Viruses. 1997. 37 figs. VIII, 336 pp. ISBN 3-540-62390-6

Vol. 224: **Potter, Michael; Melchers, Fritz (Eds.):** C-Myc in B-Cell Neoplasia. 1997. 94 figs. XII, 291 pp. ISBN 3-540-62892-4

Vol. 225: **Vogt, Peter K.; Mahan, Michael J. (Eds.):** Bacterial Infection: Close Encounters at the Host Pathogen Interface. 1998. 15 figs. IX, 169 pp. ISBN 3-540-63260-3

Vol. 226: **Koprowski, Hilary; Weiner, David B. (Eds.):** DNA Vaccination/Genetic Vaccination. 1998. 31 figs. XVIII, 198 pp. ISBN 3-540-63392-8

Vol. 227: **Vogt, Peter K.; Reed, Steven I. (Eds.):** Cyclin Dependent Kinase (CDK) Inhibitors. 1998. 15 figs. XII, 169 pp. ISBN 3-540-63429-0

Vol. 228: **Pawson, Anthony I. (Ed.):** Protein Modules in Signal Transduction. 1998. 42 figs. IX, 368 pp. ISBN 3-540-63396-0

Vol. 229: **Kelsoe, Garnett; Flajnik, Martin (Eds.):** Somatic Diversification of Immune Responses. 1998. 38 figs. IX, 221 pp. ISBN 3-540-63608-0

Vol. 230: **Kärre, Klas; Colonna, Marco (Eds.):** Specificity, Function, and Development of NK Cells. 1998. 22 figs. IX, 248 pp. ISBN 3-540-63941-1

Vol. 231: **Holzmann, Bernhard; Wagner, Hermann (Eds.):** Leukocyte Integrins in the Immune System and Malignant Disease. 1998. 40 figs. XIII, 189 pp. ISBN 3-540-63609-9

Vol. 232: **Whitton, J. Lindsay (Ed.):** Antigen Presentation. 1998. 11 figs. IX, 244 pp. ISBN 3-540-63813-X

Vol. 233/I: **Tyler, Kenneth L.; Oldstone, Michael B. A. (Eds.):** Reoviruses I. 1998. 29 figs. XVIII, 223 pp. ISBN 3-540-63946-2

Vol. 233/II: **Tyler, Kenneth L.; Oldstone, Michael B. A. (Eds.):** Reoviruses II. 1998. 45 figs. XVI, 187 pp. ISBN 3-540-63947-0

Vol. 234: **Frankel, Arthur E. (Ed.):** Clinical Applications of Immunotoxins. 1999. 16 figs. IX, 122 pp. ISBN 3-540-64097-5

Vol. 235: **Klenk, Hans-Dieter (Ed.):** Marburg and Ebola Viruses. 1999. 34 figs. XI, 225 pp. ISBN 3-540-64729-5

Vol. 236: **Kraehenbuhl, Jean-Pierre; Neutra, Marian R. (Eds.):** Defense of Mucosal Surfaces: Pathogenesis, Immunity and Vaccines. 1999. 30 figs. IX, 296 pp. ISBN 3-540-64730-9

Vol. 237: **Claesson-Welsh, Lena (Ed.):** Vascular Growth Factors and Angiogenesis. 1999. 36 figs. X, 189 pp. ISBN 3-540-64731-7

Vol. 238: **Coffman, Robert L.; Romagnani, Sergio (Eds.):** Redirection of Th1 and Th2 Responses. 1999. 6 figs. IX, 148 pp. ISBN 3-540-65048-2

Vol. 239: **Vogt, Peter K.; Jackson, Andrew O. (Eds.):** Satellites and Defective Viral RNAs. 1999. 39 figs. XVI, 179 pp. ISBN 3-540-65049-0

Vol. 240: **Hammond, John; McGarvey, Peter; Yusibov, Vidadi (Eds.):** Plant Biotechnology. 1999. 12 figs. XII, 196 pp. ISBN 3-540-65104-7

Vol. 241: **Westblom, Tore U.; Czinn, Steven J.; Nedrud, John G. (Eds.):** Gastroduodenal Disease and Helicobacter pylori. 1999. 35 figs. XI, 313 pp. ISBN 3-540-65084-9

Vol. 242: **Hagedorn, Curt H.; Rice, Charles M. (Eds.):** The Hepatitis C Viruses. 2000. 47 figs. IX, 379 pp. ISBN 3-540-65358-9

Vol. 243: **Famulok, Michael; Winnacker, Ernst-L.; Wong, Chi-Huey (Eds.):** Combinatorial Chemistry in Biology. 1999. 48 figs. IX, 189 pp. ISBN 3-540-65704-5

Vol. 244: **Daëron, Marc; Vivier, Eric (Eds.):** Immunoreceptor Tyrosine-Based Inhibition Motifs. 1999. 20 figs. VIII, 179 pp. ISBN 3-540-65789-4

Vol. 245/I: **Justement, Louis B.; Siminovitch, Katherine A. (Eds.):** Signal Transduction and the Coordination of B Lymphocyte Development and Function I. 2000. 22 figs. XVI, 274 pp. ISBN 3-540-66002-X

Vol. 245/II: **Justement, Louis B.; Siminovitch, Katherine A. (Eds.):** Signal Transduction on the Coordination of B Lymphocyte Development and Function II. 2000. 13 figs. XV, 172 pp. ISBN 3-540-66003-8

Vol. 246: **Melchers, Fritz; Potter, Michael (Eds.):** Mechanisms of B Cell Neoplasia 1998. 1999. 111 figs. XXIX, 415 pp. ISBN 3-540-65759-2

Vol. 247: **Wagner, Hermann (Ed.):** Immunobiology of Bacterial CpG-DNA. 2000. 34 figs. IX, 246 pp. ISBN 3-540-66400-9

Additional material from *Origin and Evolution of the Vertebrate Immune System*
ISBN 978-3-642-64078-0, is available at http://extras.springer.com

EXTRAS ONLINE

MIX
Papier aus verantwortungsvollen Quellen
Paper from responsible sources
FSC® C105338

If you have any concerns about our products,
you can contact us on
ProductSafety@springernature.com

In case Publisher is established outside the EU,
the EU authorized representative is:
Springer Nature Customer Service Center GmbH
Europaplatz 3, 69115 Heidelberg, Germany

Printed by Libri Plureos GmbH
in Hamburg, Germany